Systems Modeling and Complex Networks Analysis

Systems Modeling and Complex Networks Analysis

Edited by **Victor Nason**

Willford Press

New York

Published by Willford Press,
118-35 Queens Blvd., Suite 400,
Forest Hills, NY 11375, USA
www.willfordpress.com

Systems Modeling and Complex Networks Analysis
Edited by Victor Nason

International Standard Book Number: 978-1-68285-241-5 (Hardback)

Printed in China.

Contents

Preface

This book presents the complex subject of systems modeling and complex networks analysis in a comprehensible format. The chapters in this book are compiled in such a manner, that it will provide in-depth knowledge about the concepts of this field. Some of the significant topics covered in this extensive book are cloud data management, network designs, complex adaptive systems modeling, etc. It aims to serve as a resource guide for students and experts alike and contribute to the growth of the discipline.

This book is a result of research of several months to collate the most relevant data in the field.

When I was approached with the idea of this book and the proposal to edit it, I was overwhelmed. It gave me an opportunity to reach out to all those who share a common interest with me in this field. I had 3 main parameters for editing this text:

1. Accuracy - The data and information provided in this book should be up-to-date and valuable to the readers.
2. Structure - The data must be presented in a structured format for easy understanding and better grasping of the readers.
3. Universal Approach - This book not only targets students but also experts and innovators in the field, thus my aim was to present topics which are of use to all.

Thus, it took me a couple of months to finish the editing of this book.

I would like to make a special mention of my publisher who considered me worthy of this opportunity and also supported me throughout the editing process. I would also like to thank the editing team at the back-end who extended their help whenever required.

Editor

1

Complex Adaptive Systems Modeling: A multidisciplinary Roadmap

Muaz A Niazi

Correspondence:
muaz.niazi@gmail.com
Present address: Bahria University, Shangrila Road, E-8, Islamabad, Pakistan

Keywords: Complex networks, Agent-based modeling, Complex adaptive systems, Complex adaptive systems modeling, Social network analysis, Biological networks, Ecological networks, Citation networks, Individual-based modeling

PAC Codes: 07.05.Tp, 89.75.-k, 89.75.Fb

Mathematics Subject Classification (2010): 05C82, 68T42, 00A72, 92C42

Complex Adaptive Systems Modeling

I would like to start by extending a very warm welcome from the editorial board of Complex Adaptive Systems Modeling (CASM) journal, a new and exciting SpringerOpen/BioMed Central project. This article is the opening editorial and a call for papers for CASM. It also provides a gentle introduction to CASM as well as serves as a concise multidisciplinary roadmap and survey for CASM readership. The primary goal is to assist multidisciplinary researchers with an interest in modeling complex systems in finding more about relevant helpful resources and the numerous exciting ways in which other colleague researchers use modeling approaches in their respective disciplines. Finally I also provide advice for prospective authors regarding the preparation of manuscripts in a style appropriate to the highly multidisciplinary CASM readership. It is hoped that this will help seed fruitful cross-fertilization of ideas transgressing disciplinary boundaries.

Briefly, as a journal, CASM has several unique features :

1. A shorter peer-review time.
2. Open access publication with global visibility.
3. Online only journal with rapid publication.
4. A focus on well-written articles readable by multidisciplinary researchers.
5. Publication scope encompassing two key paradigms for modeling Complex Adaptive Systems: Agent-based Modeling and Complex Networks.
6. Highly multidisciplinary editorial board with researchers ranging from ecology to biology and from the social sciences to computer sciences.

As such, while there are several other complexity-related journals, CASM is unique in several ways; firstly, because CASM is primarily concerned with articles in two key modeling and simulation paradigms for Complex Adaptive Systems (CAS); systems which can be better understood as a notion rather than a strict definition (more on this later). Secondly, CASM is open access; open access implies visibility to a global audience allowing readers to freely access all articles without paying any article or subscription fees. In addition, CASM is an initiative under SpringerOpen/BioMed Central, a well-established open access platform publishing several high quality and impact factor journals such as BMC Bioinformatics (impact factor: 2.75) and BMC Biology (impact factor: 5.75).

Another unique feature of CASM is our goal to minimize review times as much as practically possible. We hope to achieve this by having a stricter scope and stringent cover letter requirements. CASM aims to serve as a high quality publication venue for a multidisciplinary readership of researchers interested in these two modeling and simulation paradigms for CAS; Agent-Based Modeling (ABM) (Holland 2006) and Complex Networks (CN) (Mitchell 2006). CASM is therefore open to papers from these two distinct communities of multidisciplinary researchers with an underlying common interest in modeling the same type of systems i.e. CAS. Thus, quite simply, any paper would be in the scope of CASM as long as it has been written with a clear focus on CAS using at least any one of these two modeling paradigms of interest. A discussion of the widespread nature and multidisciplinary interest in these paradigms will be provided in later sections of this article.

Outline

The article first outlines general guidelines for submissions to CASM in the section "Guidelines for submissions". This is followed by an overview of the two paradigms of interest and a multidisciplinary survey of literature. Finally, the paper is concluded.

Guidelines for submissions

Aims & scope

CASM is a highly multidisciplinary modeling and simulation journal that serves as a unique forum for original, high-quality peer-reviewed papers with a specific interest and scope limited to agent-based and complex network-based modeling paradigms for CAS. The scope of CASM essentially spans any and all domains of CAS. Possible areas of interest include the Life Sciences (E.g. Biological Networks and agent-based models), Ecology (E.g. Agent-based/Individual-based models), Social Sciences (Agent-based simulation, Social Network Analysis), Scientometrics (E.g. Citation Networks) to large-scale Complex Adaptive COmmunicatiOn Networks and environmentS (CACOONS) such as Wireless Sensor Networks (WSN), Body Sensor Networks, Peer-to-Peer (P2P) networks, pervasive mobile networks, service oriented architecture, smart grid and the Internet of Things. A paper in these areas is welcome as long as it uses either an agent-based (simulation) and/or complex-network based modeling approach.

Keywords

For reference, here I would like to give a sampling of alphabetically arranged keywords which may be of interest to CASM readership:

agent-based models, agent-based simulation, artificial life, biological networks, Boolean networks, citation networks, complex adaptive systems, complex network analysis, complex networks, computer networks, emergence, epidemiological networks, gene expression networks, gene regulatory networks, individual-based modeling, metabolic networks, (agent-based modeling of) multiagent systems, network modeling, nucleic acid networks, protein interaction networks, self-adaptation, self-assembly, self-healing, self-organization, signaling pathway networks, social network analysis, social networks, social simulation, systems biology.

General guidelines for article submissions

Any paper submitted to CASM should have the following elements:

1. A strict adherence to the aims and scope of CASM.
2. A solution to an interesting problem or a review of recent literature.
3. A comparison of results with existing work from the state-of-the-art.
4. Enough details for the readers to be able to replicate the results.
5. A well-written and thoroughly edited/proof-read coherent piece of writing.
6. Adherence to general ethical guidelines both in terms of data as well as in terms of authorship.
7. A clear focus on the following two key elements:
   (a) At least one specific area of CAS (E.g. ecology, social sciences, large scale communication networks, biological sciences etc.)
   (b) At least one of the two paradigms of interest:
      i. Agent-based Modeling
      ii. Complex Network modeling.
8. CASM has a strongly multidisciplinary editorial board and readership. Therefore authors need to ensure that their articles have been written in a style that is comprehensible by a broad and multidisciplinary audience. Authors are advised to avoid excessive usage of domain-specific jargon/terminologies unless all such terms have been formally introduced at the start of the article. In addition, any domain-specific concepts must be explained in the article and the authors should not assume any advanced knowledge on the part of the readers. Thus, as an example, an article on Gene Regulatory Networks should be written in a style accessible to a number of social science researchers using Social Network Analysis and vice versa. Likewise an article written by Computer Scientists for Wireless sensor networks must also take into consideration CASM readership of researchers from Social and Biological Sciences and so on. Interested readers are advised to consult an excellent article on this topic by Hoffman (Hoffman 2006).

Example topics in complex networks

- All types of complex networks such as biological, social, citation, ecological networks etc.
- Agent-based modeling using an ABM tool (such as NetLogo, RepastS etc.) of CACOONS such as modeling concepts related to the internet, intranet, P2P, ad-hoc and Wireless Sensor Networks as networks/graphs.
- Network generation, visualization, modeling and simulation.

Example topics in agent-based models

- Agent-based Modeling case studies in ecology, business, biology, social sciences, operations research, economics, computer sciences, communication networks, artificial intelligence, robotics etc.
- Software engineering of agent-based and complex network models.
- Verification and validation of agent-based and complex network models.
- Exploratory agent-based models for conducting feasibility and proof-of-concept studies.
- Agent-based models and their respective DescRiptivE Agent-based Models (DREAM) (Niazi 2011)[a] or text based descriptions of models (E.g. the ODD protocol (Grimm et al. 2006)) for inter-disciplinary model comparisons.
- Validated agent-based models for verified and validated models such as using empirical validation approaches or VOMAS (Niazi et al. 2009) or any other methodology.

Papers which may be rejected without formal review

Submissions which are not in line with the above criteria or those which are of a purely or largely theoretical/formal/mathematical nature or have been written in a very domain-specific style may be considered as out of scope of CASM and rejected without review. Here are some obvious examples of manuscripts which may be rejected without review:

1. Manuscripts which are not written in a style suitable for a multidisciplinary audience.
2. Papers with a focus on software multiagent systems without any relation to agent-based modeling from the CAS perspective. These would perhaps be out of scope for CASM and perhaps more suited to the scope of Springer's "Autonomous agents and multiagent systems" journal.
3. Communication network papers which use simulators (such as OPNET, OMNET++, Matlab, custom simulators in programming languages etc.) with little or no relation with agent-based modeling would perhaps be more suitable for communication networks journals such as the Springer "Peer-to-Peer Networking and Applications", Springer "Wireless Networks" or "EURASIP Journal on Wireless Communications and Networking" amongst others.
4. Papers written with only a formal/mathematical perspective without any focus on an ABM or a CN model.
5. Papers which are poorly organized or written.
6. Papers which include plagiarized material.

Editorial process

Having a highly multidisciplinary following and a clear focus on the two modeling techniques for CAS requires adherence to a somewhat stringent editorial policy. These include cover letter requirements as well as a clear adherence to the aims and scope. In addition, only such submitted papers will be reviewed which have a clear focus on modeling CAS and have been written with a writing style understandable by a broad audience. Each paper is handled according to an editorial flow which may be roughly depicted in Figure 1. It may also be noted that papers submitted to CASM may be checked using CrossRef/iThenticate for possible plagiarism before a formal review.

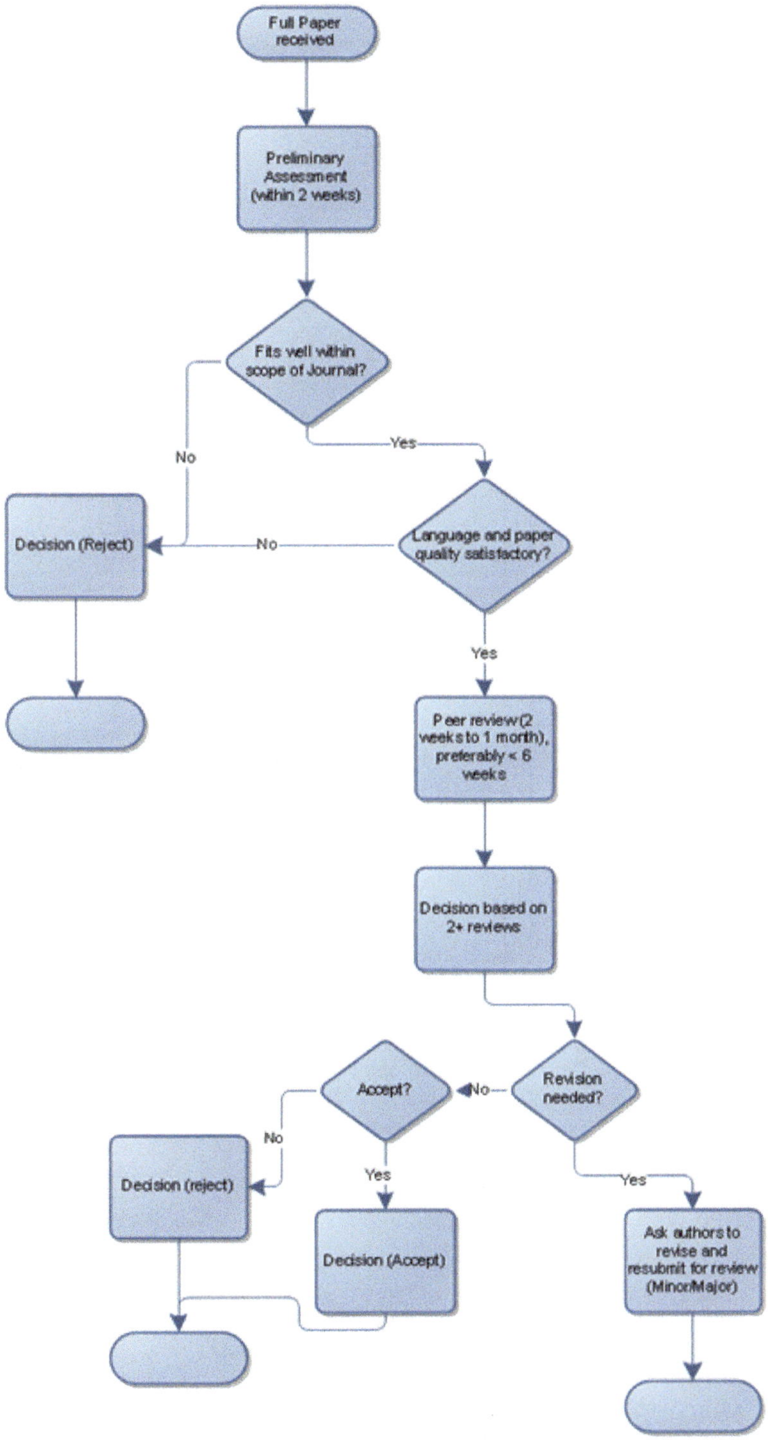

Figure 1 Editorial Process for CASM.

Cover letter requirements

In some journals, cover letters are considered a formality however, a well-written cover letter is one of the essential requirements for any submission to CASM. The cover letter may also help the editors in ensuring an adherence of the paper with publication ethics guidelines outlined by COPE[b]:

1. Names and contact details of all authors.
2. A single sentence summarizing the research contribution and significance of the paper in comparison with previous work (With 2-3 related references).
3. CASM strongly discourages redundant submissions. If there is any other related paper (by one of the authors) either currently under review in another venue (conference, workshop or journal etc.) or else has already been published/in press, then the cover letter must include a statement of changes which make the submitted paper different from that paper. The authors may briefly summarize how the submitted paper differs from the other paper(s). In addition, since the reviewers may not have access to the full text of the other paper(s), it is important to attach a soft copy of the paper during the manuscript submission process allowing the referees to evaluate the significance of the changes.
4. A brief list of individual contribution details of each author; essentially each author must have contributed significantly to the paper and all people who have contributed significantly must be listed as authors. Ghost, Gift, and other forms of unethical practices are strongly discouraged[c].
5. The cover letter may also note the corresponding author's justification of the particular order of authors. If the editorial office is not satisfied with this order, the authors may be asked to change the author order before or at any time during the formal manuscript handling process.
6. To ensure that the paper is not rejected as being out-of-scope, it is important to also briefly answer the following questions in the cover letter:
   (a) Which particular discipline (related to CAS) is being explored in the manuscript? (E.g. social sciences, economics, ecology, biology, CACOONS etc.)?
   (b) Which particular modeling technique is being used in the paper? (E.g. To be within scope, the paper must have either or both of Agent-based Simulation Model or Complex Networks).
   (c) Which particular modeling tool was used in the study? NetLogo (Wilensky 1999), Repast (North et al. 2007), Swarm (Minar et al. 1996), Network Workbench (Team 2006), Pajek (Batagelj and Mrvar 1998), Cytoscape (Shannon et al. 2003), Gephi (Bastian et al. 2009), Visone (Baur et al. 2002) or CiteSpace (Chen 2006) etc.
   (d) If the area of the paper is agent-based modeling then give details of how the verification and validation were conducted. This could include techniques such as empirical validation (Windrum et al. 2007) or validated agent-based modeling using a VOMAS (Niazi 2011) or any other technique suitable for ABM.

Modeling CAS

In this section, we briefly examine concepts related to CAS and the two modeling and simulation paradigms forming the scope of CASM.

CAS

The CAS concept is married to the interaction of a set of perhaps simple but numerous entities, components or agents which interact and adapt on the basis of non-linear interactions (Miller and Page 2007). This interaction is known to give rise to interesting and emergent phenomena. While it is rather difficult to contain all aspects of CAS in a single definition, our current understanding is that CAS are often found in nature and in nature-inspired artificial systems in close relation with or inspired by life in some way, a notion exemplified in one of the earliest definitions of CAS by Cowan and Feldman (1986):

> "Systems comprising large numbers of coupled elements the properties of which are modifiable as a result of environmental interactions"

CAS concepts are tied to an abstract concept of a society. As such, it can be noted that typical research articles with a focus in modeling CAS are ornate with the following concepts:

1. A large number of agents (E.g. Genes, societies, humans, animals, insects, software agents, data packets etc.).
2. Focus on somewhat simpler individual agents.
3. Focus on the nonlinear interaction between agents and global phenomena resulting from these interactions.

Agent-based modeling

In ABM, the basic idea is to develop a simulation model at some level of abstraction using concepts which represent interacting entities in the system to be modeled. ABM can range from Cellular Automata (CA) such as shown in Figure 2a to models with mobility such as shown in Figure 2b. Other examples include networked agents such as can be seen in Figure 2c.

Different activities which are associated with developing an ABM are similar to those required in a regular simulation model. However, there are some additional concepts required for modeling and simulation of CAS (Niazi and Hussain 2012). Banks and Chwif give general warnings about using simulation for manufacturing and some other areas in (Banks and Chwif 2010). Macal and North present successful approaches for teaching agent-based simulation in (Macal and North 2012). A tutorial on ABM has been presented in (Macal and North 2010). Railsback et al. gives an overview of Agent-based modeling platforms in (Railsback et al. 2006).

Complex networks

In terms of complex networks, the focus of studies is typically on data retrieved from the interactions of various CAS agents/components. So, while agent-based models are dynamic models, complex networks are typically developed from a large amount of data. The actual data sources can be as diverse as bio-chemical reactions, gene expressions, social networks, human cell phone mobility or even citation data etc. Activities in complex networks range from modeling, simulation, analysis and visualization of networks. Three different examples of using complex networks are given in Figure 3. Figure 3a shows the use of citation networks to analyze research areas using agent-based modeling. Figure 3b shows the structure of a retroviral RNA modeled as a complex network. Likewise, Figure 3c show the social network of top-cited authors in the consumer electronics domain.

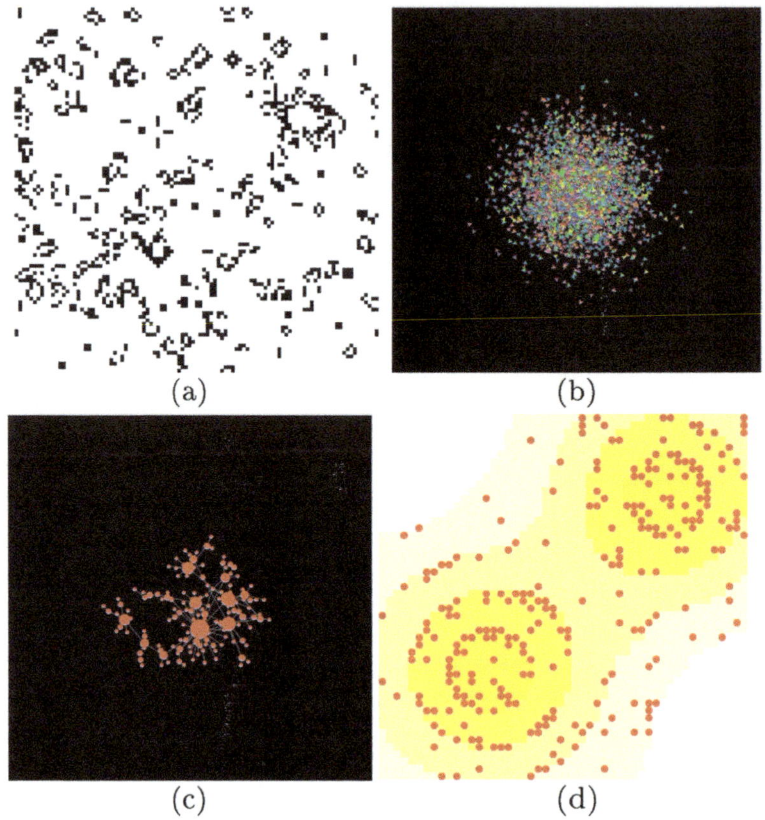

Figure 2 A sampling of different types of ABM using NetLogo. (**a**) Cellular Automata Model, (**b**) Mobile agents in an ABM, (**c**) Networked agents and (**d**) Sugarscape model in NetLogo.

A multidisciplinary road map

In this section, I shall give an overview of literature spread across various disciplines but having a focus on either Agent-based Modeling or Complex Network-based methods.

Social sciences

ABM

Agent-based modeling has found extensive usage in social sciences. Epstein and Axtell present a social sciences model for a society called Sugarscape (Epstein and Axtell 1996) as can be seen in Figure 2d. Bonabeau has noted ABM as an effective set of methods and techniques for simulating human systems (Bonabeau 2002). Kohler notes a clear focus on ABM in anthropology and sociology but does not note its significance in comparison with differential equation models (Kohler 2000). This work also notes skepticism towards simulation in general and ABM in particular. However, we may note that recent advances in validation and verification of models (Niazi et al. 2009) are focused on alleviating these problems such as discussed in (Miller and Page 2007).

Likewise Macy and Willer (Macy and Willer 2002) note that while sociologists often model processes in social sciences as interactions of mathematical variables, they have discovered ABM to be an alternate approach to modeling social life as interactions among influential and responsive agents. Gimblett demonstrates the integration of ABM with Geographical Information Systems (GIS) and its application on social and ecological data

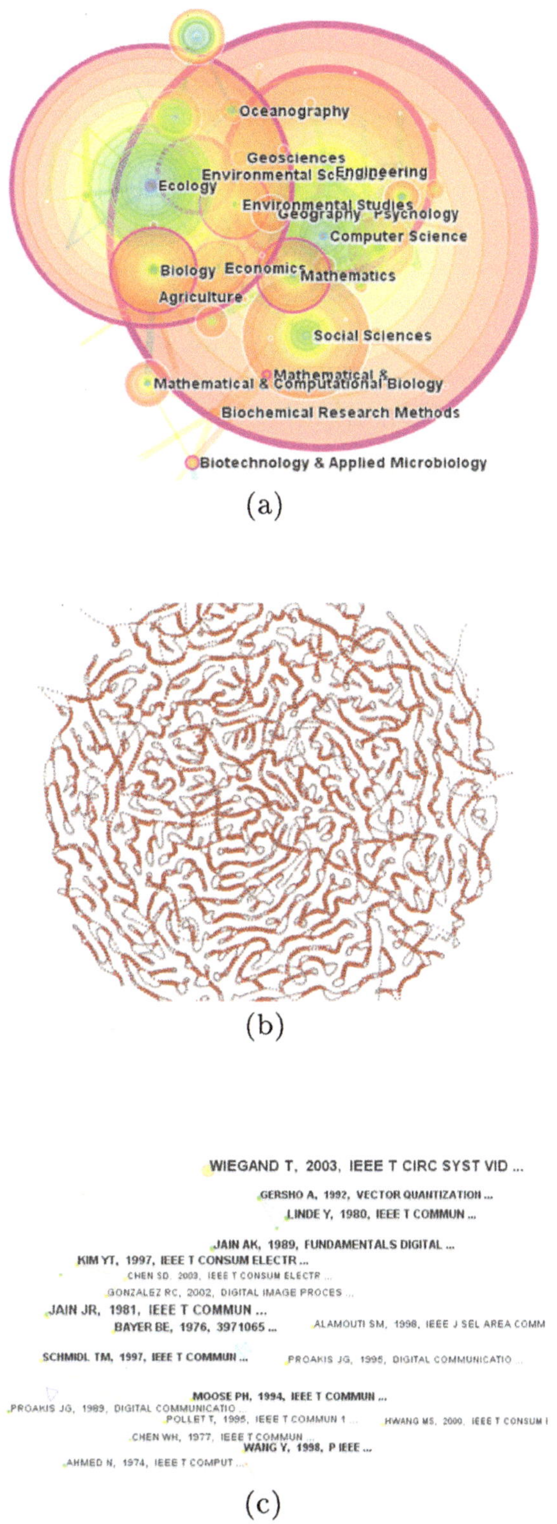

Figure 3 Example Complex Networks created using different tools from multidisciplinary data. (**a**) Complex network showing the prevalence of agent-based modeling in using scientometric data from ISI web of knowledge (Niazi and Hussain 2011b), (**b**) Complex Network of the entire genome of the RNA-based Feline Immunodeficiency Virus folded using RNAStructure and then plotted using NWB (Reuter and Mathews 2010) and (**c**) Social Network of the authors of top cited papers in the Consumer Electronics Domain (Niazi and Hussain 2011d).

(Gimblett 2002). Axelrod gives an overview of simulation in the social sciences (Axelrod and Tesfatsion 2006). More recent work includes (Gilbert and Troitzsch 2005). Davidson gives a Computer Sciences perspective to ABM (Davidsson 2002). Other recent work includes ABM for human socio-cultural behavior by Jiang et al. in (Jiang et al. 2012). A recent special issue on complex social systems was edited by Lopez et al. (López-Paredes et al. 2012). Another source of valuable information in the application of ABM is the journal JASSS. Tesfatsion has given an overview of developing agent-based models of computational economics in (Tesfatsion 2002,2003; Tesfatsion and Judd 2006). Whereas Luna and Stefansson give an overview of using Swarm as platform for economic simulations in (Luna and Stefansson 2000). Modeling of driver route behavior in transportation research is given by Dia in (Dia 2002). A tutorial on using ABM in social sciences has been presented by North and Macal in (North and Macal 2007).

CN

Publications with a focus on social networks are present in high-ranking journals such as the Proceedings of the National Academy of Sciences (Lewis et al. 2012). Earlier articles include one by Scott (1988) which argues the importance of the social network model. A relevant special issue in the 90's gave directions to the field in political sociology, interorganizational relations, social support, social influence, and epidemiology (Galaskiewicz and Wasserman 1993). A book and reference on the topic is by Yang and Knoke (2008). Another recent book on the topic is (De Nooy et al. 2011). There are other recent reviews such as the one by McGloin and Kirk in (McGloin and Kirk 2010). Gardy et al. use social network analysis in a tuberculosis network (Gardy et al. 2011). In case of supply chain networks, network analysis has been used by Kim et al. in (Kim et al. 2011). Suer et al. demonstrate how the use of networks can improve studies of primate behavior (Sueur et al. 2011).

Computer sciences

ABM

Agents have always been considered an integral part of artificial intelligence as exemplified by the classic artificial intelligence text by Russell and Norvig, often labelled as the "Agent book" (Russell and Norvig 2010). However, the exact notion of agency is often quite different in CAS as compared to multiagent systems (which are traditionally software agents) as compared with agents in an agent-based model which represent entities from the real world (Niazi and Hussain 2011b).

The particular sub-section of multiagent systems which has an overlap with agent-based modeling is the development of simulation models of multiagent systems using agent-based modeling tools such as NetLogo, RepastS, Swarm and Mason amongst others. In addition, agent-based models may be developed for other artificial CAS such as Wireless Sensor Networks, Swarm Robotic Networks, Self-assembling robots, Peer-to-peer networks and others. Due to a lack of any other term to classify these networks, we classify this set of communication networks as Complex *A*daptive *CO*mmunication *N*etworks and environment*S* (CACOONS). Examples include work by Vidal (2006). Agent-based tools have also been identified as effective in teaching multiagent concepts (Vidal et al. 2004) as well as developing experimental simulations of multiagent systems (Sklar 2007). Agent-based models have also been developed for wireless sensor networks (Niazi and

Hussain 2011a, 2011c) as well as for modeling self-organizing and large-scale communication networks (Niazi et al. 2009). A methodology for using simulation as a means of developing distributed agent systems has been present by Fortino and Russo in (Fortino and Russo 2012).

CN

Communication networks have a long history of being represented as graphs. Graphs are a precursor of complex networks since networks are essentially graphs "loaded" with additional information. Earlier work such as the famous algorithm by Dijkstra has also used graphs/networks (Dijkstra 1959). Use of social network analysis in sensors has been shown in (Tsugawa et al. 2012). Other examples include graph-based models for mobility by Tian et al. in (Tian et al. 2002) and models of internet topology by Calvert et al. in (Calvert et al. 1997). Another interesting aspect of using agents is as a means of programming Wireless Sensor Networks presented by Aiello et al. in (Aiello et al. 2011).

Life sciences and ecology

ABM

Grimm et al. present (Holcombe et al. 2012) pattern-oriented modeling by giving a unifying framework for deciphering the internal details of agent-based models of CAS (Grimm et al. 2005). Mansury et al. (2002) develop an ABM of "spatio-temporal search and agglomeration" to investigate "cell motility" dynamics as well as "aggregation" with the expectation of a CAS nature of tumors. Recently ABM has been explored in synthetic biology by Krishnamurty in (Krishnamurthy). Holcombe et al. note the strength of using ABM to model complex biological systems. Read et al. present an approach to calibration of an agent-based model of EAE, "a mouse proxy for multiple sclerosis" (Read et al. 2012). Qu et al. present a digital orange tree simulation using an ABM (Qu et al. 2012). Another work by Høye et al. demonstrates systematic modification of the digital version of a real landscape to produce artificial landscapes (Høye et al. 2012). Iantovics presents a large-scale hybrid medical diagnosis system (Iantovics 2012). An introduction to ABM in the domain of life sciences and ecology is by Railsback and Grimm in (Railsback and Grimm 2011).

CN

Biological networks are often studied in terms of their community structure similar to social networks (Girvan and Newman 2002). Efficient algorithms for identifying Boolean networks and related biological networks from DNA micro-array data have been proposed by Akutsu et al. in (Akutsu et al. 2000). Complex network tools such as Cytoscape enable conversion of biological data to complex networks (Cline et al. 2007). Cytoscape has a number of plug-ins allowing data retrieval from different databases as well as for performing other tasks (Maere et al. 2005). Among other methods, process diagrams have also been used to represent biological networks (Kitano et al. 2005). Different techniques have also been developed to co-cluster biological networks with gene expression data such as in (Hanisch et al. 2002).

VANTED is a tool for the visualization as well as structural analysis of biological networks usable for correlating with experimental data from biochemical experiments (Junker et al. 2006). Another option for the discovery of useful scientific information in biological networks is the utilization of topological motif structure discovery (Shen-Orr et al. 2002; Milo et al. 2002) in gene regulatory networks or protein interaction maps (Berg and Lassig 2004).

Conclusions

In this article, I have presented a multidisciplinary road map as well as an overview of Agent-based Modeling and Complex Networks for the modeling and simulation of Complex Adaptive Systems. I have also introduced CASM, a new multidisciplinary SpringerOpen/BioMed Central journal with aims and scope in these two areas. CASM reflects a high quality peer-reviewed venue for multidisciplinary researchers interested in ABM and CN. CASM also offers a faster publication time as well as a broad scope and an open access publication model. As such, quite simply, any submitted paper will be in the scope of CASM as long as it has a focus on the modeling and simulation of a CAS using either or both of ABM and CN methods.

Endnotes

[a]DREAM refers to non-textual formal descriptions of agent-based models such as those obtained by converting agent-based models to complex networks and subsequently applying quantitative centrality measures.

[b]http://publicationethics.org/resources/guidelines

[c]For reference, kindly review http://publicationethics.org/files/u2/2003pdf12.pdf

Competing interests

The authors declare that they have no competing interests.

Acknowledgements

I would like to thank the SpringerOpen/BioMed Central team in general and our Springer editor Matthew Amboy, in particular for their untiring efforts in helping develop CASM from an idea to a reality. I would also like to specifically thank (listed alphabetically) Prof. Lubomir Chitkusev, Dr. Carlos Gershenson, Prof. Michael Huhns, Prof. Amir Hussain, Prof. Tuncer Ören, Prof. José M. Vidal, Prof. Anatoly Temkin and Prof. Tanya Zlateva for their continual advice, critique and support.

References

Aiello, F, Fortino G, Gravina R, Guerrieri A: **A Java-based Agent Platform for Programming Wireless Sensor Networks.** *Comput J* 2011, **54**(3):439–454 doi:10.1093/comjnl/bxq019.

Akutsu, T, Miyano S, Kuhara S: **Algorithms for identifying Boolean networks and related biological networks based on matrix multiplication and fingerprint function.** *J Comput Biol* 2000, **7**(3-4):331–343.

Axelrod, R, Tesfatsion L: **Guide for Newcomers to Agent-Based Modeling in the Social Sciences.** In *Handbook of Computational Economics,* Vol. 2: Agent-Based Computational Economics. Edited by Tesfatsion, L, Judd K. New York: North-Holland; 2006:1647–1659.

Banks, J, Chwif L: **Warnings about simulation.** *J Simulation* 2010, **5**(4):279–291.

Bastian, M, Heymann S, Jacomy M: **Gephi: An Open Source Software for Explor- ing and Manipulating Networks.** In *Proceedings of the Third International ICWSM Conference (2009)*; 2009.

Batagelj, V, Mrvar A: **Pajek-program for large network analysis.** *Connections* 1998, **21**(2):47–57.

Baur, M, Benkert M, Brandes U, Cornelsen S, Gaertler M, Köpf B, Lerner J, Wagner D: **Visone Software for visual social network analysis.** In *Graph Drawing*. Vienna, Austria: Springer; 2002:554–557.

Berg, J, Lässig M: **Local graph alignment and motif search in biological networks.** *Proc Nat Acad Sci USA* 2004, **101**(41):14689.

Bonabeau, E: **Agent-based modeling: Methods and techniques for simulating human systems.** *Proc Nat Acad Sci USA* 2002, **99**(Suppl 3):7280.

Calvert, K, Doar M, Zegura E: **Modeling internet topology.** *Commun Mag, IEEE* 1997, **35**(6):160–163.

Chen, C: **CiteSpace II: Detecting and visualizing emerging trends and tran- sient patterns in scientific literature.** *J Am Soc for Inf Sci Technol* 2006, **57**(3):359–377.

Cline, M, Smoot M, Cerami E, Kuchinsky A, Landys N, Workman C, Christmas R, Avila-Campilo I, Creech M, Gross B, et al.: **Integration of biological networks and gene expression data using Cytoscape.** *Nat Protoc* 2007, **2**(10):2366–2382.

Cowan, G, Feldman M: **Preview of Workshop on Complex Adaptive Systems.** *Bull Santa Fe Institute* 1986, **1:**11.

Davidsson, P: **Agent based social simulation: A computer science view.** *J Artif Societies Social Simul* 2002, **5**(1).

De Nooy, W, Mrvar A, Batagelj V: *Exploratory social network analysis with Pajek*, Vol. 34. Cambridge, UK: Cambridge Univ Pr; 2011.

Dia, H: **An agent-based approach to modelling driver route choice behaviour under the influuence of real-time information.** *Transportation Res Part C: Emerging Technol* 2002, **10**(5-6):331–349.

Dijkstra, E: **A note on two problems in connexion with graphs.** *Numerische mathematik* 1959, **1:**269–271.

Epstein, J, Axtell R: *Growing Artificial, Societies: Social Science from the Bottom up.* Cambridge, MA, USA: The MIT Press; 1996.

Galaskiewicz, J, Wasserman S: **Social network analysis.** *Sociological Methods & Res* 1993, **22:**3–22.

Gardy, J, Johnston J, Sui S, Cook V, Shah L, Brodkin E, Rempel S, Moore R, Zhao Y, Holt R, et al.: **Whole-genome sequencing and social-network analysis of a tuberculosis outbreak.** *New England J Med* 2011, **364**(8):730–739.

Gilbert, G, Troitzsch K: *Simulation for the social scientist.* New York City, USA: Open Univ Pr; 2005.

Gimblett, H: *Integrating Geographic Information Systems and Agent-Based Modeling Techniques for Simulating Social and Ecological Processes.* USA: Oxford University Press; 2002.

Girvan, M, Newman M: **Community structure in social and biological networks.** *Proc Nat Acad Sci* 2002, **99**(12):7821.

Grimm, V, Berger U, Bastiansen F, Eliassen S, Ginot V, Giske J, Goss-Custard J, Grand T, Heinz S, Huse G, et al.: **A standard protocol for describing individual-based and agent-based models.** *Ecol Modell* 2006, **198:**115–126.

Grimm, V, Revilla E, Berger U, Jeltsch F, Mooij W, Railsback S, Thulke H, Weiner J, Wiegand T, DeAngelis D: **Pattern-oriented modeling of agent-based complex systems: lessons from ecology.** *Science* 2005, **310**(5750):987–991.

Fortino, G, Russo W: **ELDAMeth: A Methodology For Simulation-based Prototyping of Distributed Agent Systems.** *Inf Software Technol* 2012, **54**(6):608–624 doi:10.1016/j.infsof.2011.08.006.

Hanisch, D, Zien A, Zimmer R, Lengauer T: **Co-clustering of biological networks and gene expression data.** *Bioinformatics* 2002, **18**(suppl 1):S145–S154.

Hoffmann, R: **The metaphor, unchained.** *Am Scientist* 2006, **94**(5):406–407.

Holcombe, M, Adra S, Bicak M, Chin S, Coakley S, Graham A, Green J, Greenough C, Jackson D, Kiran M, et al.: **Modelling complex biological systems using an agent-based approach.** *Integr Biol* 2012, **4:**53–64.

Holland, J: **Studying complex adaptive systems.** *J Syst Sci Complexity* 2006, **19:**1–8.

Høye, T, Skov F, Topping C: **Interpreting outputs of agent-based models using abundance–occupancy relationships.** *Ecol Indicators* 2012, **20:**221–227.

Iantovics, B: **Agent-based medical diagnosis systems.** *Comput Inf* 2012, **27**(4):593–625.

Jiang, H, Karwowski W, Ahram T: **Applications of agent-based simulation for human socio-cultural behavior modeling.** *Work: A J Prev, Assess and Rehabilitation* 2012, **41:**2274–2278.

Junker, B, Klukas C, Schreiber F: **VANTED: A system for advanced data analysis and visualization in the context of biological networks.** *BMC Bioinformatics* 2006, **7:**109.

Kim, Y, Choi T, Yan T, Dooley K: **Structural investigation of supply networks: A social network analysis approach.** *J Operations Manage* 2011, **29**(3):194–211.

Kitano, H, Funahashi A, Matsuoka Y, Oda K: **Using process diagrams for the graphical representation of biological networks.** *Nat Biotechnol* 2005, **23**(8):961–966.

Kohler, T: *Dynamics in Human and Primate Societies: Agent-Based Modeling of Social and Spatial Processes.* USA: Oxford University Press; 2000.

Krishnamurthy, E: **Agent-based Models in Synthetic Biology: Tools for Simulation and Prospects.** *Int J Intell Syst.* 2012, **4**.

Lewis, K, Gonzalez M, Kaufman J: **Social selection and peer influence in an online social network.** *Proc Nat Acad Sci* 2012, **109:**68–72.

López-Paredes, A, Edmonds B, Klugl F: **Special Issue: Agent Based Simulation of Complex Social Systems.** *Simulation* 2012, **88:**4–6.

Luna, F, Stefansson B: *Economic Simulations in Swarm: Agent-based modelling and object oriented programming, Volume 14.* New York, USA: Springer; 2000.

Macal, C, North M: **Tutorial on agent-based modelling and simulation.** *J Simulation* 2010, **4**(3):151–162.

Macal, C, North M: **Successful approaches for teaching agent-based simulation.** *J Simulation* 2012 doi:10.1057/jos.2012.1.

Macy, M, Willer R: **From factors to actors: Computational sociology and agent- based modeling.** *Annu Rev Sociology* 2002, **28:**143–166.

Maere, S, Heymans K, Kuiper M: **BiNGO: a Cytoscape plugin to assess overrep- resentation of gene ontology categories in biological networks.** *Bioinformatics* 2005, **21**(16):3448–3449.

Mansury, Y, Kimura M, Lobo J, Deisboeck T: **Emerging patterns in tumor systems: simulating the dynamics of multicellular clusters with an agent-based spatial agglomeration model.** *J Theor Biol* 2002, **219**(3):343–370.

McGloin, JM, Kirk DS: **Social Network Analysis.** In *Handbook of Quantitative Criminology.* Edited by Piquero, AR, Weisburd D. New York: Springer; 2010:209–224.

Miller, J, Page S: *Complex Adaptive, Systems: An Introduction to Computational Models of Social Life.* Princeton, NJ, USA: Princeton Univ Pr; 2007.

Mitchell, M: **Complex systems: Network thinking.** *Artif Intelligence* 2006, **170**(18):1194–1212.

Milo, R, Shen-Orr S, Itzkovitz S, Kashtan N, Chklovskii D, Alon U: **Network motifs: simple building blocks of complex networks.** *Science's STKE* 2002, **298**(5594):824.

Minar, N, Burkhart R, Langton C, Askenazi M: **Santa Fe Institute, Santa Fe N: The swarm simulation system: A toolkit for building multi-agent simulations.** In *Advancing Social Simulation: The First World Congress.* Santa Fe, NM: Santa Fe Institute; 1996.

Niazi, MA: **Towards A Novel Unified Framework for Developing Formal, Net- work and Validated Agent-Based Simulation Models of Complex Adaptive Systems.** *PhD thesis* 2011.

Niazi, MA, Hussain A: **A novel agent-based simulation framework for sensing in complex adaptive environments.** *Sens J, IEEE* 2011a, **11**(2):404–412.

Niazi, MA, Hussain A: **Agent-based computing from multi-agent systems to agent-based models: a visual survey.** *Scientometrics* 2011b, **89**(2):479–499.

Niazi, MA, Hussain, A: **Sensing emergence in complex systems.** *IEEE Sens J* 2011c, **11**(10):2479–2480.
Niazi, M, Hussain A: **Social network analysis of trends in the consumer electronics domain.** In *Consumer Electronics (ICCE), 2011 IEEE International Conference on*: IEEE; 2011d:219–220.
Niazi, MA, Hussain A: *Cognitive Agent-based Computing-I: A Unified Framework for Modeling Complex Adaptive Systems Using Agent-based and Complex Network-based Methods*. Berlin, Germany: Springer Verlag; 2012.
Niazi, MA, Hussain A, Kolberg M: **Verification and validation of agent based simulation using, VOMAS approach.** In *Proceedings of the Third Workshop on Multi-Agent Systems and Simulation'09 (MASS '09), as part of MALLOW 09*. Torino, Italy; 2009.
North, MJ, Howe TR, Collier NT, Vos JR: **A Declarative Model Assembly Infrastructure for Verification and Validation.** In *Advancing Social Simulation: The First World Congress*. Edited by Takahashi, S, Sallach DL, Rouchier J. Heidelberg, FRG: Springer; 2007.
North, M, Macal C: *Managing business complexity: discovering strategic solutions with agent-based modeling and simulation*. USA: Oxford university press; 2007.
Qu, H, Norman D, Lu Z: **Digital orange tree: An intelligent agent based simulation model.** *Afr J Agric Res* 2012, **7**(5):690–710.
Railsback, S, Grimm V: *Agent-Based and Individual-Based Modeling: A Practical Introduction*. Princeton, New Jersey: Princeton Univ Pr; 2011.
Railsback, S, Lytinen S, Jackson S: **Agent-based simulation platforms: Review and development recommendations.** *Simulation* 2006, **82**(9):609–623.
Read, M, Andrews P, Timmis J, Kumar V: **Techniques for grounding agent-based simulations in the real domain: a case study in experimental autoimmune encephalomyelitis.** *Math Comput Modell Dynamical Syst* 2012, **18:**67–86.
Reuter, J, Mathews D: **RNAstructure: software for RNA secondary structure prediction and analysis.** *BMC Bioinformatics* 2010, **11:**129.
Russell, S, Norvig P: *Artificial Intelligence: A Modern Approach*: Prentice hall; 2010.
Scott, J: **Social network analysis.** *Sociology* 1988, **22:**109–127.
Shannon, P, Markiel A, Ozier O, Baliga N, Wang J, Ramage D, Amin N, Schwikowski B, Ideker T: **Cytoscape: a software environment for integrated models of biomolecular interaction networks.** *Genome Res* 2003, **13**(11):2498–2504.
Shen-Orr, S, Milo R, Mangan S, Alon U: **Network motifs in the transcriptional regulation network of Escherichia coli.** *Nature Genet* 2002, **31:**64–68.
Sklar, E: **NetLogo, a multi-agent simulation environment.** *Artif Life* 2007, **13**(3):303–311.
Sueur, C, Jacobs A, Amblard F, Petit O, King A: **How can social network analysis improve the study of primate behavior?** *Am J Primatology* 2011, **73**(8):703–719.
Team, N: **Network Workbench Tool.** 2006. [http://nwb.slis.indiana.edu].
Tesfatsion, L: **Agent-based computational economics: Growing economies from the bottom up.** *Artif life* 2002, **8:**55–82.
Tesfatsion, L: **Agent-based computational economics: modeling economies as complex adaptive systems.** *Inf Sci* 2003, **149**(4):262–268.
Tesfatsion, L, Judd K: *Handbook of Computational Economics: Agent-Based Computational Economics, Volume 2*. Amsterdam, Netherlands: North Holland; 2006.
Tian, J, Hahner J, Becker C, Stepanov I, Rothermel K: **Graph-based mobility model for mobile ad hoc network simulation.** In *Simulation Symposium, 2002. Proceedings. 35th Annual*. San Diego, California, USA: IEEE; 2002:337–344.
Tsugawa, S, Ohsaki H, Itoh Y, Ono N, Kagawa K, Takashima K, Imase M: **Toward large-scale and dynamic social network analysis with heterogeneous sensors in ambient environment.** In *Virtual Reality Workshops (VR), 2012*: IEEE; 2012:1–2.
Vidal, J: *Fundamentals of, Multiagent Systems: Using NetLogo Models*. South Carolina, Columbia, SC, USA: University of South Carolina; 2006.
Vidal, J, Buhler P, Goradia H: **The past and future of multiagent systems.** In *Proceedings of AAMAS Workshop on Teaching Multi-Agent Systems*. New York, NY: ACM; 2004.
Wilensky, U: **NetLogo.** 1999. Center for Connected Learning and Computer-Based Modeling, Northwestern University. Evanston, IL. [http://ccl.northwestern.edu/netlogo/].
Windrum, P, Fagiolo G, Moneta A: **Empirical validation of agent-based models: Alternatives and prospects.** *J Artifl Societies and Social Simul* 2007, **10**(2):8.
Yang, S, Knoke D: *Social Network Analysis*. Thousand Oaks, USA: Sage Publications, Inc; 2008.

2

GENESIS-CBA: an agent-based model of peer evaluation and selection in the internet interdomain network

Aemen Lodhi[1*], Amogh Dhamdhere[2] and Constantine Dovrolis[1]

*Correspondence: aemen.lodhi@gatech.edu
[1]School of Computer Science, Georgia Institute of Technology, Atlanta, USA
Full list of author information is available at the end of the article

Abstract

Purpose: We propose an agent-based model for peer selection in the Internet at the Autonomous System (AS) level. The proposed model, *GENESIS-CBA*, is based on realistic constraints and provider selection mechanism, with ASes acting in a myopic and decentralized manner to optimize a cost-related fitness function. We introduce a new peering scheme, *Cost-Benefit-Analysis*, which, unlike existing peering strategies, gives ASes the ability to analyze the impact of each peering link on their economic fitness. Using this analysis, ASes engage in only those peering relations that can have a positive impact on their fitness.

Methods: Our proposed model captures the key factors that affect network formation dynamics: highly skewed traffic matrix, policy-based routing, geographic co-location constraints, and the costs of transit/peering agreements. As opposed to analytical game-theoretic models, which focus on proving the existence of equilibria, GENESIS-CBA is a computational model that simulates the network formation process and allows us to actually compute distinct equilibria (i.e., networks) and to also examine the behavior of sample paths that do not converge.

Results: We find that such oscillatory sample paths occur in about 7% of the runs, and they always involve tier-1 ASes. GENESIS-CBA results in many distinct equilibria that are highly sensitive to initial conditions and the order in which ASes (agents) act.

Conclusions: Our results imply that we cannot predict the properties of an individual AS in the Internet. However, certain properties of the global network or of certain classes of ASes are predictable.

Keywords: Internet interdomain network; Peering strategies; Cost-benefit-analysis; Internet autonomous systems

Background

Tens of thousands of Autonomous Systems (ASes) interconnect in a complex and dynamic manner to form the Internet. Each AS is an independent entity under single administrative control, and is identified through a unique AS number (ASN), which is assigned by a central authority, the Internet Assigned Numbers Authority (IANA) (Internet Assigned Numbers Authority 2012). Each AS manages its own internal network topology and traffic routing (intradomain activity) and connects to the rest of the world by forming links with other geographically co-located ASes (interdomain activity). These ASes generate traffic destined for other ASes in the Internet, consume traffic destined for them, and some of them transit traffic, i.e., they carry traffic on behalf of other ASes to transport it to its destination.

ASes belong to different categories: enterprise networks (or stubs), e.g., university campuses, banks etc.; content sources, which generate much more traffic than they consume, e.g., Google, Facebook, YouTube etc.; access providers, which consume much more traffic than they generate, e.g., residential ISPs, Comcast, France Telecom; transit providers, which carry traffic on behalf of other ASes to its destination, e.g., Level 3 Communications, Cogent, etc. Various combinations of the aforementioned categories also exist, e.g., ASes that provide both transit and access services such as AT&T. Thus, the Internet is a network of ASes where traffic flows from one AS to another en route to its destination. A traffic flow may traverse multiple transit providers before reaching its destination AS. Additionally, ASes differ in geographic size (expanse) and economic parameters (e.g., transit prices).

ASes connect with one another mostly through two types of relations: customer-provider (or "transit") links and settlement-free peering links. Under the customer-provider relationship, the provider is responsible for connecting the customer with the rest of the Internet. The provider may itself have a transit provider and so on. Settlement-free peering (subsequently referred to as "peering" for brevity) is more subtle than a customer-provider relationship. In peering, two ASes (or peers) agree to exchange traffic between themselves and their customers over a shared link. Peers do not pay each other for this exchange; however, they share the cost of the link.[1] For example, ASes peering at an Internet Exchange Point (IXP), use the IXP's infrastructure for peering. Thus, the IXP's infrastructure forms the shared link between the peers. While they do not pay one another for the traffic that they exchange among themselves, they pay the IXP for usage of its infrastructure.

Peering is typically carried out with the objective of saving transit costs by diverting traffic from customer-provider links to peering links. ASes generally follow broad rules, referred to as "peering strategies" to determine their peering links. Examples of such rules-of-thumb include *Open* (peer with all co-located ASes except customers), *Selective* (peer with only those ASes that have similar traffic volume), *Restrictive* (do not peer with anyone unless the peering relationship is necessary to maintain global reachability). The highest level of the Internet consists of a clique of approximately 15 ASes fully meshed among themselves through peering links. This clique is known as the "Tier-1" clique and these ASes are known as "Tier-1 transit providers". Tier-1 providers do not require a transit provider as they can reach the entire Internet through their customers and peers. Figure 1 provides an illustration of these topological concepts. In Figures 1a and 1b x, y, z, w and T are ASes; T is a Tier-1 provider, x and y are transit providers, and w and z are stubs. Figure 1a shows the flow of traffic between transit providers and stubs through the Tier-1 provider. In Figure 1b, the introduction of a peering link between x and y causes the traffic to flow over a different path, avoiding the Tier-1 provider altogether. This results in reduced transit costs for x and y. Figure 1c illustrates the concept of the Tier-1 clique in the Internet. The ASes shown are Tier-1 providers, which form a complete mesh among themselves through peering links. Traffic between their customer trees is routed through these peering links.

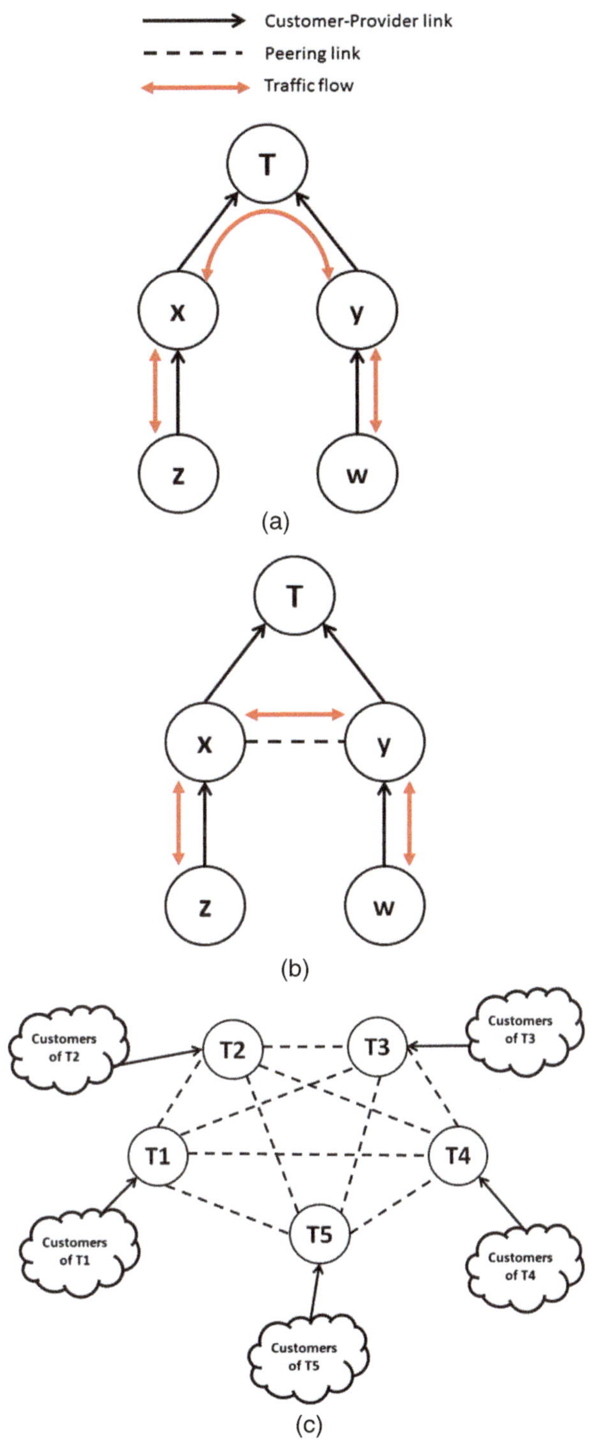

Figure 1 Illustration of Internet topological concepts.(**a**) Traffic flow before peering. (**b**) Traffic flow after peering. (**c**) Tier-1 clique.

Most interactions between ASes are local (unilateral in the case of provider selection and bilateral in the case of peering), decentralized and dynamic. These local connectivity decisions, however, can have a global impact on the economic viability of all ASes and the structure of the Internet. The Internet remains in a persistent state of flux, subject to changes in various exogenous factors, e.g., geographic expansion of ASes, new entrants in the connectivity market, removal of nodes from the network owing to bankruptcies, changes in technology, etc. How will the Internet change due to consolidation of content (Labovitz et al. 2010), large penetration of video streaming, falling transit prices (Internet Transit Prices: Historical and Projected 2010), expanding geographic footprint of content providers (Gill et al. 2008), cheap local availability of peering infrastructure at IXPs (Augustin et al. 2009)? We proposed a computational agent-based network formation model, called "GENESIS" (Lodhi et al. 2012b), as a tool to study such questions. GENESIS is modular and easily extensible, allowing researchers to experiment with different peering or provider selection strategies, traffic matrix characteristics, routing policies, cost parameters, etc.

In the context of the interdomain connectivity, ASes face an important question: which other ASes should they peer with? GENESIS does not empower agents to choose their peering strategy; instead, a peering strategy is assigned to each agent based on its hierarchical status: Tier-1 providers use *Restrictive*, all other transit providers use *Selective* and stubs use *Open* peering strategy. In this paper, we take the concept of choosing peers much further than just choosing a peering strategy and then peering with all those ASes that satisfy the chosen strategy. We modify the peering mechanism in GENESIS by introducing a *Cost-Benefit-Analysis* (CBA) scheme, wherein transit providers analyze the impact of each peering link on their economic fitness and then decide whether to peer or not. We refer to this modified version of GENESIS as GENESIS-CBA

CBA is not just another peering strategy like *Selective, Restrictive* or *Open*; it is a completely different way of deciding which ASes to peer with. Other peering strategies are rules-of-thumb and are used to approximate cost-benefit-analysis. They are used because they are easier to deploy. However, our work shows that such informal analysis can lead to situations where ASes end up choosing peering strategies detrimental to their long-term economic fitness (Lodhi et al. 2012c). With CBA, agents can engage in precise analysis of the impact of each peering link on their fitness.

In this paper, we first introduce the main features of GENESIS and then introduce our cost-benefit-analysis scheme, as proof-of-concept, to study the properties of the resulting networks. Starting from a population of ASes and an initial topology, the model executes an iterative process in which each AS acts asynchronously to optimize its set of provider and peering links. We focus on the dynamic behavior of the model, identify the cause of some oscillatory sample paths (about 10% of the runs), and analyze the variability of the resulting equilibria.

Related work

Most previous related work has focused on characterizing and modeling the AS-level Internet topology. Various graph theoretic models aim to reproduce observed Internet topological properties (e.g., power-law degree distribution (Faloutsos et al. 1999)) (Wang and Loguinov 2006; Yook et al. 2002; Wang X and Loguinov 2010). Another class of models takes a bottom-up approach (Li et al. 2005), modeling the optimization objectives and constraints of individual ASes, to create graphs that have the same topological properties (Li et al. 2005; Fabrikant et al. 2002; Chang et al. 2003; Corbo et al. 2009).

The GENESIS model was first introduced by Lodhi et al. (2012b). GENESIS differs from previously mentioned related work because it does not aim to be a topology generation model or tool; instead, it captures interdomain traffic flow, geographic constraints and economics to model the network formation process. The main difference between GENESIS (Lodhi et al. 2012b) and this work is the mechanism used by transit providers to determine their peers. In GENESIS, transit providers use rule-based peering strategies: Tier-1 transit providers are assigned *Restrictive* while all other transit providers are assigned *Selective* strategy. A transit provider using Restrictive strategy does not peer with any other agent unless the absence of a peering relationship results in a disconnected network; a transit provider using Selective strategy peers with agents which have similar traffic volume (it uses the ratio of total traffic volumes to determine the similarity of other agents to itself). Such rule-based peering strategies may lead to extraneous costs from redundant peering links and may prevent beneficial peering links from being formed as shown by Lodhi et al. in (2012a). GENESIS-CBA does away with such rule-based peering strategies and introduces a new *Cost-Benefit-Analysis* module to GENESIS. Additionally, GENESIS-CBA does not assign peering strategies to transit providers based on their hierarchical status; it empowers them to analyze the impact of each peering link on their fitness and then choose only those peers that can improve their fitness.

A large body of work on game-theoretic network formation models exists in the computer science and economics literature. We refer the interested reader to two recent books (Goyal 2007; Jackson 2008). Those models capture the objectives and potential strategies of each node as players in a non-cooperative game, and focus on proving the existence of (typically many) equilibria. The need for mathematical tractability, however, requires significant simplifications (such as lack of geographic constraints, simple cost functions, or uniform traffic flow between nodes. Consequently, the resulting networks are typically simple graphs, such as rings, trees or other regular structures. We avoid such simplifications in GENESIS-CBA. Additionally, few approaches in that line of work have focused on the dynamics of the network formation process and on the selection between multiple possible equilibria.

Our work is most related to agent-based computational models that simulate the dynamics of the network formation process, capturing the asynchronous and decentralized process through which nodes adjust their connectivity. Holme et al. propose an agent-based network growth model (Holme et al. 2008). However, they treat link formation as a random process, and do not capture link rewiring, settlement-free peering, and use hot-potato routing instead of realistic policy based shortest-path Internet routing. The model of Chang et al. (2006) uses hard-wired strategies for provider and peer selection, among other differences. The ITER model of Dhamdhere et al. (2010) is more similar to GENESIS-CBA, but it assigns a specific function to each agent (e.g., content provider, enterprise customer etc.) Additionally, peering strategies are pre-assigned to the agents. In our work both the hierarchical position of the agent and its peering strategies are an outcome of the model. Their goal is to study the evolutionary transition of the Internet from a *hierarchical* to a *flat* structure, while our focus is on the peer selection mechanism adopted by transit providers.

Methods

The GENESIS model

We consider a population of N agents (representing Autonomous Systems) which interconnect through two types of links: customer-provider and peering. Each agent has the following attributes: a set of geographic locations in which it is present (corresponding to IXPs), a certain traffic volume that it sends to and receives from every other agent, and certain economic parameters, such as the transit price it would charge to its potential customers at a given location. We do not assign a business function to agents *a priori*; an agent may act as a transit provider, content/access provider, enterprise organization, or any combination of these functions depending on its generated/consumed traffic, transit pricing, or geographical presence. An inter-network traffic matrix gives the average traffic volume sent from each agent to every other agent. The model only captures the interconnections between distinct agents and does not consider intradomain activity and connectivity. The geographic locations, transit prices and traffic flow of each agent remain constant during the timescales of interest in this model. We focus on the impact of the dynamics of provider and peer selection: how does the internetwork and the economic fitness of agents change with time as agents choose the best transit provider and peers for themselves? Table 1 provides a list of all notations used in the model.

Geographical presence

Each agent is placed in one or more geographic locations. These locations are roughly analogous to the IXPs in the real world. The *world* consists of G_M locations and an agent x is present at a set $G(x)$ of randomly assigned locations. Two agents are *co-located* if they share at least one location. For agent x, $O(x)$ denotes the set of agents that are co-located with x. A link between two agents can be formed only if they are co-located. An exception to this rule exists if x and y are non-colocated $Tier-1$ agents. This exception is introduced to ensure that the network is always connected.

Traffic matrix and transit traffic

The element T_{xy} of the N-by-N interdomain traffic matrix T denotes the average traffic volume generated by agent x and consumed by agent y. Overall, x generates traffic $V_G(x)$ and when aggregated across all consumers of traffic it is given by:

$$V_G(x) = \sum_{i=1}^{N} T_{xi} \tag{1}$$

Overall, x consumes traffic $V_C(x)$ and when aggregated across all producers of traffic it is given by:

$$V_C(x) = \sum_{i=1}^{N} T_{ix} \tag{2}$$

T_{xx}=0, i.e., we do not capture exchange of traffic locally within the agent. The *transit traffic* $V_T(x)$ of x is the traffic volume that is neither generated nor consumed by x; it only passes through x enroute to its destination.

The transit traffic of an agent depends on the underlying network topology and the routing algorithm. Even if the interdomain traffic matrix T is constant, the transit traffic

Table 1 List of abbreviations & notations

Notation/Abbreviation	Explanation
AS	Autonomous System
IXP	Internet Exchange Point
CBA	Cost Benefit Analysis
$O(x)$	Set of agents geographically co-located with x
$G(x)$	Set of geographic locations where x is present
$C(x)$	Set of agents which are customers of x
$S(x)$	Peering strategy of x
$R(x)$	Transit provider of x
T_{xi}	Traffic originating in x and destined for i
$V_G(x)$	Traffic generated by x
$V_C(x)$	Traffic consumed by x
$V_T(x)$	Traffic transiting x
$V_P(x)$	Traffic exchanged between x and its provider
$V(x)$	Total traffic managed by x
V_{xy}	Total traffic exchanged between x and y in then network
$TC(x)$	Transit cost of x
$TR(x)$	Transit revenue of x
$PC_{prv}(x)$	Private peering cost of x
$PC_{pub}(x)$	Public peering cost of x
$F(x)$	Fitness of x
$PP_{pub}(x)$	Public peers of x
$PP_{prv}(x)$	Private peers of x
$PP(x)$	All peers of x
$N(t)$	State of network at time t. The state of the network is defined by the topology of the network and the peering strategies of all agents
α	Peering cost per Mbps
β	Economies of scale exponent for peering traffic
τ	Economies of scale exponent for transit traffic
Ψ	Private peering threshold traffic volume (Mbps)

of an agent may change as the underlying topology changes. The total traffic volume of an agent $V(x)$ is given by the following expression:

$$V(x) = V_C(x) + V_G(x) + V_T(x) \tag{3}$$

Economic attributes

The economic attributes of an agent include its transit price, transit and peering costs, transit revenue and *fitness*.

Prices Each agent x maintains a vector of prices $P(x)$. Each element of the vector corresponds to the price the agent has in each location in which it is present. ASes generally maintain different prices at different geographic locations to reflect the underlying differences in competition. The price of agent x in each location is determined randomly, independent of its prices in other locations and the prices of co-located agents. The random prices are chosen from a specific range.

Transit cost Let x be a transit customer of y, and let $P_y(x)$ be the *lowest* transit price of y across all regions in which x and y are co-located. If $V_P(x)$ is the traffic exchanged between x and y, then the transit payment from x to y is:

$$TC(x) = P_y(x) \times V_P(x)^{\tau} \tag{4}$$

where τ is a transit traffic exponent that captures the *economies-of-scale* observed in practice. The transit revenue of y, $TR(y)$ is the aggregate of transit costs incurred by all transit customers of y. Let $C(y)$ denote the set of all transit customers of y. Then its transit revenue is given by:

$$TR(y) = \sum_{\forall x \in C(y)} TC(x) \tag{5}$$

Peering cost Agents engaging in settlement-free peering relations share the underlying peering costs. There are two primary mechanisms for peering — private and public — that have different cost structures. If the traffic exchanged between two peers is less than a threshold Ψ Mbps they peer publicly at an IXP, otherwise they peer privately. In public peering, agents exchange traffic over a common switching fabric at an IXP, aggregating traffic from different peering sessions through the same port, whereas in private peering they set up a direct link with each peer (Norton 2011). (DE-CIX) provides an example of Internet peering costs at a European IXP.

Private peering cost Let $V_{prv}(x, y)$ be the traffic exchanged between agent x and its peer y over a private peering link. The total cost of private peering for x is given by:

$$PC_{prv}(x) = \alpha \times \sum_{y} V_{prv}(x, y)^{\beta} \tag{6}$$

where α is the peering cost per Mbps and β is the peering traffic exponent that accounts for economies of scale.

Public peering cost Let $V_{pub}(x, z)$ be the traffic exchanged between agent x and its peer z over the public peering infrastructure. As x aggregates all its public peering traffic over the same port, the corresponding cost is:

$$PC_{pub}(x) = \alpha \left(\sum_{z} V_{pub}(x, z) \right)^{\beta} \tag{7}$$

Fitness The fitness of an agent represents its net profit,

$$F(x) = TR(x) - TC(x) - PC_{pub}(x) - PC_{prv}(x) \tag{8}$$

If an agent is a stub, i.e., an AS without any customers, the first term is zero and the agent's fitness will be negative.

Peering

Each agent x has a peering strategy $S(x)$. Two agents x and y are potential peers if *(a)* x and y are co-located (b) x and y do not have a customer-provider relationship (c) the traffic exchanged between x and y over the proposed peering link, $V_{PP}(x, y)$, will be greater than a certain threshold Ω Mbps. The above conditions define *universal* peering constraints for all pairs of agents. However, potential peers x and y can establish a peering relationship if and only if they satisfy the constraints of each other's peering strategies.

Unlike provider selection, where a customer unilaterally chooses its provider, peering is a bilateral decision process. Thus, two potential peers x and y can peer if and only if the constraints of both $S(x)$ and $S(y)$ are satisfied. Depeering, however, is a unilateral decision by one of the peers.

The set of all public and private peers of x is denoted by $PP_{pub}(x)$ and $PP_{prv}(x)$ respectively. $PP(x)$ denotes the set of all peers of x.

$$PP(x) = PP_{pub}(x) \cup PP_{prv}(x) \tag{9}$$

Let $y \in S(x)$ denote that y satisfies the constraints of peering strategy of x. Algorithm 1 shows the peering acquisition algorithm for node x. Algorithm 2 shows depeering decision by node x.

Algorithm 1 Peer acquisition

Construct $PP_{potential}(x) = \{y : y \in O(x),\ y \notin C(x),\ x \notin C(y),\ y \notin PP(x)\}$
For $\forall y \in PP_{potential}(x)$
  If $y \models S(x)$
    If $V_{xy} \geq \Psi$
      Do $PP_{prv}(x) = PP_{prv}(x) + y$
        $PP_{prv}(y) = PP_{prv}(y) + x$
    Else
      Do $PP_{pub}(x) = PP_{pub}(x) + y$
        $PP_{pub}(y) = PP_{pub}(y) + x$
End
Do $PP(x) = PP_{prv}(x) \cup PP_{pub}(x)$

Algorithm 2 Depeering

For $\forall y \in PP(x)$
  If $y \nvDash S(x)$
    If $y \in PP_{prv}(x)$
      Do $PP_{prv}(x) = PP_{prv}(x) - y$
Else
      Do $PP_{pub}(x) = PP_{pub}(x) - y$
End
Do $PP(x) = PP_{prv}(x) \cup PP_{pub}(x)$

In this paper, we consider two peering strategies: *Open* and *Cost-Benefit-Analysis* (described later). We emphasize that GENESIS is not limited to these particular strategies. In our previous work (Lodhi et al. 2012b) we have implemented and analyzed *Selective* and *Restrictive* strategies.

Provider selection

An agent must have a transit provider if it cannot reach all other agents in the network via its peers and customers. Agent x selects a provider y if: *(a)* x is co-located with y, *(b)* y is "larger" than x (explained next), *(c)* y is not a peer of x, *(d)* y is not a customer of an existing peer of x, and *(e)* y is the least expensive among all agents that satisfy the previous constraints. We say that an agent y is larger than an agent x if y is present in at least as many locations as x, and it carries more transit traffic than x. Formally, the provider of node x denoted by $R(x)$ is given by:

$$R(x) = \arg \min_{y \in O(x)} P(y) : V_T(y) > V_T(x), |G(y)| \geq |G(x)|, y \notin PP(y), y \notin C(PP(x)) \tag{10}$$

If an agent x cannot find a provider that fulfills the previous criteria, then x becomes a $tier - 1$ (T1) AS. In order to ensure a connected network, $T1$ agents form a clique using peering links, even if they do not overlap. A more detailed analysis and justification of this provider selection model has appeared in (Lodhi and Dovrolis 2010).

Routing

In GENESIS, interdomain traffic is routed as in the real-world Internet. Traffic follows the shortest path subject to two common policy constraints: "prefer customer over peer over provider links" and satisfy the "valley-free" routing property. This traffic routing scheme is illustrated in Figures 1a and Figures 1b.

Initial topology

The model is initialized using an initial network topology. To create the initial network topology, we select agents sequentially and at random. For a selected agent x, we determine its provider randomly from the set of agents that *(a)* overlap with x, *(b)* are not in the customer tree of x, and *(c)* have greater geographic expanse than x. If we cannot find a provider for an agent, then it joins the clique of tier-1 agents.

Network formation process

An execution, or sample path, of GENESIS proceeds in discrete time units called *iterations*. In each iteration, every agent *plays* asynchronously once. The order in which agents play during an iteration is determined at the start of the sample path, and it remains the same throughout that sample path. Every time an agent plays, it carries out the following actions in order: *(a)* Examine depeering with existing peers, *(b)* Examine peering with new peers, *(c)* Provider selection, and *(d)* Peering strategy update. At the end of each iteration, we recompute the fitness of each agent. If none of the agents has adjusted its connectivity and peering strategy in that iteration, then it is easy to show that there will be no changes in subsequent iterations, and we say that GENESIS has reached an *equilibrium*.

The *state* of the network at any point in time can be defined based on the connections and peering strategy of all agents. Two states **A** and **B** are distinct if they differ in terms of the underlying network topology or the peering strategy of one or more nodes. Even if we start with the same population of agents and the same initial topology, two sample paths can result in two distinct equilibria as a result of the different order in which nodes may play.

The network formation process in GENESIS is deterministic and so it can have one of two possible outcomes: either convergence to an equilibrium, or a limit cycle in which the network moves repeatedly through the same sequence of states. We describe the stability of the model in Section 'Stability of GENESIS-CBA'. The network formation process is formally described in Algorithm 3.

Algorithm 3 Network formation process

Input:
    $X \leftarrow$Ordered set of agents $\{x_0, x_1, x_2, ..., x_N\}$
Begin
$t \leftarrow 0$
Start Iteration $i \leftarrow 0$
Play agent x_i
        Examine depeering for agent x_i
        If x_i depeered any peer in examination above
            Do *Compute* $V_T(x_k)\ \forall x_k \in X$
        Examine peering for agent x_i
        If x_i acquired any new peer in examination above
            Do *Compute* $V_T(x_k)\ \forall x_k \in X$
        Examine provider selection for agent x_i
        If x_i changed its provider in examination above
            Do *Compute* $V_T(x_k)\ \forall x_k \in X$
If $i == N$
        Do *Compute* $F(x_k)\ \forall x_k \in X$
        Do *Record* $S(t)$
        If $t > 0$ && $N(t) == N(t-1)$
            End *Equilibrium*
        Else If $t > 1$ && $N(t) == N(t') : \exists t' < t - 1$
            End *Oscillation*
Else
            $t \leftarrow t + 1]$
            Go to Start Iteration
Else
        $i \leftarrow i + 1$
        Go to Play

Peering strategies in GENESIS-CBA

The economic objective of peering is to offload as much upstream transit traffic on provider links to settlement-free peering links thus saving transit costs. We consider the following two peering strategies in this paper:

Cost benefit analysis

When using Cost Benefit Analysis (CBA) peering strategy, an agent x separately evaluates the impact of each peering link (existing or potential) on its fitness, as opposed to broad rule-based peering strategies, e.g., Selective, Open, etc. For example, under Open strategy, x would peer with all potential peers regardless of the impact of each peering relationship on its fitness. Under CBA, x agrees to establish (or retain) only those peering links that can have a positive impact on its fitness. In GENESIS-CBA, only transit providers employ the CBA strategy.

In order to establish a new peering link using CBA, x evaluates each potential peer y by hypothetically establishing a peering link with it. Each new link in the network causes some traffic to be re-routed along different paths based on the interdomain routing policy. x computes the impact of the hypothetical peering link by updating its traffic flow, costs and revenue. x agrees to peer with y if and only if its fitness increases as a result of the hypothetical link. In the real world, such link evaluations are carried out through *peering trials*. Under a peering trial, two ASes establish a peering relationship for a short time period and evaluate the impact of the temporary link on their economic fitness. The link is retained if both peers find the relationship beneficial for their fitness, otherwise it is terminated. Similar evaluations are carried out for existing peers when using CBA. Each existing peer is hypothetically depeered and the impact of depeering on fitness is computed. The link is terminated if depeering improves fitness, otherwise it is retained.

Scalability of CBA We assume that an agent x can choose any number of peers from $PP(x)$ (from 0 to $n-1$). Traffic in the interdomain network is routed based on the shortest path prefer-customer-then-peer-then-provider routing policy. The addition (or removal) of a peering link may affect the flow of traffic on other peering links. Thus, for x, peering with agent y_i may impact the evaluation of agent y_j. We explain this by means of the following example. Let y_j be a customer of y_i. Let V_{xy_i} and V_{xy_j} denote the traffic exchanged between x and y_i and y_j respectively. If x peers with y_i, then traffic from both y_i and y_j flows over the x—y_i peering link as shown in Figure 2. The peering cost for x with the single peering link is given by Equation 11. If x peers with both y_i and y_j then traffic from both ASes traverses separate peering links as shown in Figure 3. The peering cost for x with both peering links is given by Equation 12 while Equation 13 shows that the cost under the two scenarios is different.

$$PC(x) = \alpha \times (V_{xy_i} + V_{xy_j})^{\beta} \tag{11}$$

$$PC(x) = \alpha \times (V_{xy_i}^{\beta} + V_{xy_j}^{\beta}) \tag{12}$$

$$\because (V_{xy_i} + V_{xy_j})^{\beta} < V_{xy_i}^{\beta} + V_{xy_j}^{\beta} \tag{13}$$

Hence the question is: how should x go about finding its peers from among potential peers? In order to find the optimal set of peers, x would have to evaluate each combination of peers. The number of such combinations is given by:

$$\sum_{k=1}^{n} \binom{n}{k} \tag{14}$$

Note that each combination would involve evaluation of k peering links by both agents involved. Given the large co-location density in many regions of the interdomain network, the number of evaluations required to determine the optimal set of peers may be infeasible.

CBA heuristic CBA involves computation of all traffic flows, costs and revenue by x. Hence, employing this strategy to evaluate all potential and existing peers is not scalable.

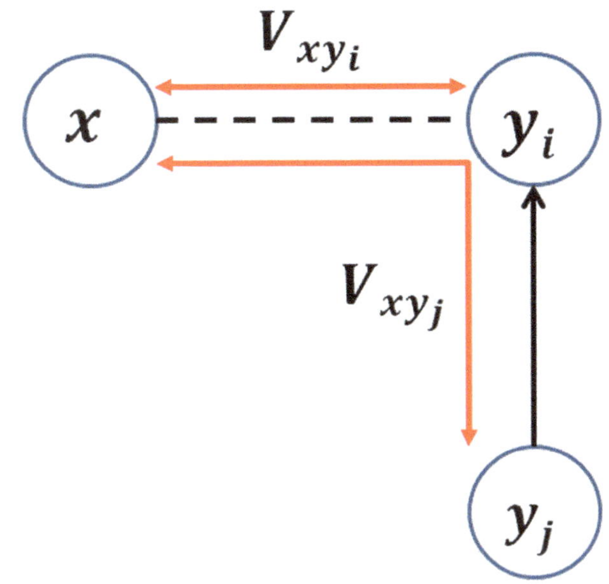

Figure 2 y_j in customer tree of peer of x.

We propose a simple heuristic to alleviate the scalability issues with CBA. Although our proposed heuristic does not compute the optimal set of peers, yet it allows agents to prioritize peers based on peering traffic aggregation and avoid redundant evaluations.

Our heuristic exploits a key feature of the *prefer-customer-then-peer-then-provider* routing policy employed in the Internet. It is based on the notion that if an AS x can reach an AS y through an existing peering link L_{xz} with another AS z, then the peering link L_{xy} would be redundant. Such a link would not save x any transit costs and would only add to its peering costs. Thus, if x finds that it has a peering path to y, it can refuse to peer with y

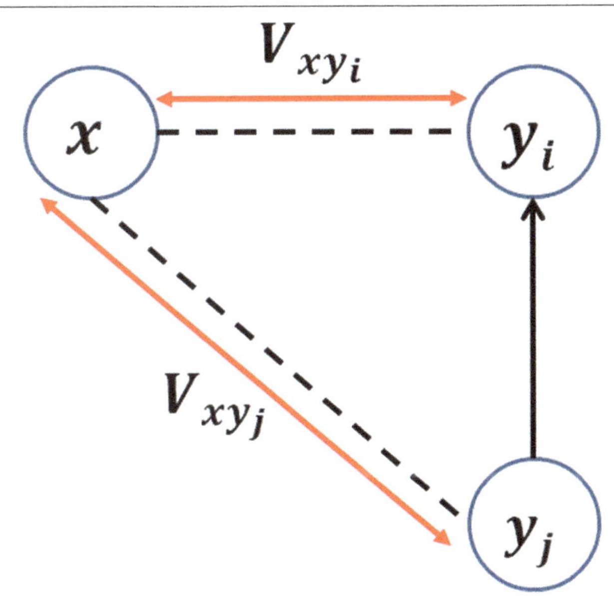

Figure 3 y_j peering with x.

without actually evaluating it. Note that if x is able to reach y through an existing peering link, then it implies that y is in the customer tree of an existing peer of x.

This heuristic is explained in Figure 4. The figure shows x and z as peers while y is in the customer tree of z. In Figure 4a, x is not a peer of y. In this network, traffic from x to y flows over the peering link L_{xz}. Thus, x is able to reach y over an existing peering link. If V_{xz} and V_{xy} is the total traffic volume exchanged between x and z, and x and y respectively, then the peering cost of x is given by:

$$PC(x) = \alpha \times (V_{xz} + V_{xy})^{\beta} \tag{15}$$

In Figure 4b, x peers with z. Thus, traffic between x and z flows over the direct peering link and now the peering cost of x is given by:

$$PC'(x) = \alpha \times (V_{xz})^{\beta} + \alpha \times (V_{xy})^{\beta} \tag{16}$$

Note that the peering cost of x is increased because of the additional peering link.

$$PC(x) < PC'(x) \because (V_{xz} + V_{xy})^{\beta} < (V_{xz})^{\beta} + (V_{xy})^{\beta} \tag{17}$$

Thus, if an agent x can reach another agent y through an existing peer higher up in the network hierarchy, it does not need to establish a redundant peering link with y. The redundant peering link does not reduce upstream transit costs; on the other hand it adds to the cost of peering by reducing the advantage of economies-of-scale. Therefore, x can decide not to peer with y without actually doing cost-benefit-analysis. Algorithms 4 and 5 give the procedures for depeering and peering respectively using CBA.

Algorithm 4 Depeering by agent *i* using CBA

Input $PP(i) = \{x_1, x_2, ..., x_n\}$

Sort $PP(i)$ hierarchically $PP(i) = \{x_1, x_2, ...x_t, x_{t+1}, ...x_z\} : x_t \notin C(x_{t+1})$

Do $PP(i) \leftarrow PP(i) - x_k : x_k \in C(x_j), j < k$ {Remove from $PP(i)$ all peers which are in customer trees of other existing peers}

Do CBA for $\forall x_j \in PP(i)$

  Do $F_{original} \leftarrow F(i)$

  Do Remove peering link between i and x_j

  Do Route traffic

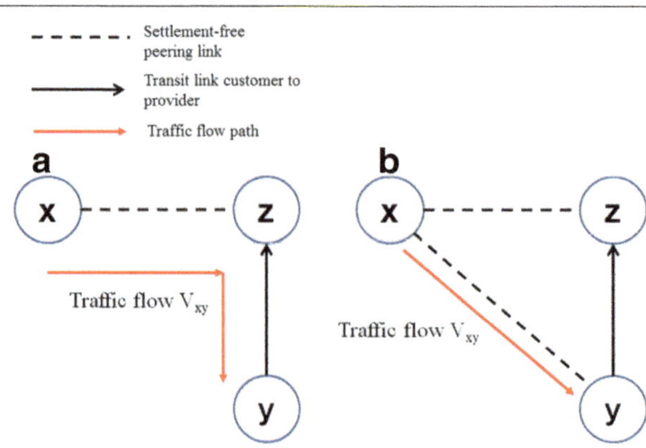

Figure 4 CBA heuristic illustrated. (**a**) *x* not peering with *y*. (**b**) *x* peering with *y*.

Do Compute $F(i)$

$F_{update} \leftarrow F(i)$

If $F_{update} > F_{original}$

Do $PP(i) \leftarrow PP(i) - x_j$

Algorithm 5 Peering by agent *i* using CBA

Do $PP'(i) \leftarrow j \ \forall j \notin C(k) \ \forall k \in PP(i)$

Do $PP'(i) = PP'_P(i) \bigcup PP'_S(i) : C(j) = \emptyset \ \forall j \in PP'_P(i),\ C(j) \neq \emptyset \ \forall j \in PP'_P(i),$ $PP_P(i) \bigcap PP'_S(i) = \emptyset$

Sort $PP'_P(i)$ hierarchically $PP'_P(i) = \{x_1, x_2, ...x_t, x_{t+1}, ...x_z\} : x_t \notin C(x_{t+1})$

Do CBA $\forall x_j \in PP'_P(i)$

Do $F_{original} \leftarrow F(i)$

Do Establish peering link between i and x_j

Do Route traffic

Do Compute $F(i)$

$F_{update} \leftarrow F(i)$

If $F_{update} > F_{original}$

Do $PP(i) \leftarrow PP(i) + x_j$

Do $PP_S(i) \leftarrow PP_S(i) - x_k \ \forall x_k \in C(x_j),\ x_k \in PP_P(i)$

Do CBA $\forall x_j \in PP'_S(i)$

Do $F_{original} \leftarrow F(i)$

Do Establish peering link between i and x_j

Do Route traffic

Do Compute $F(i)$

$F_{update} \leftarrow F(i)$

If $F_{update} > F_{original}$

Do $PP(i) \leftarrow PP(i) + x_j$

Open

An agent that uses this strategy agrees to peer with any other agent that is co-located with it, with the exception of its direct customers. In GENESIS-CBA, the Open peering strategy is followed by stubs. For stubs, Open peering does not always lead to the same network connectivity as CBA. However, since most stubs have limited geographic expanse, i.e., they are co-located with only few agents and thus have limited connectivity options, the difference in connectivity using either peering strategy is often small. Additionally, stubs aim to reduce their transit costs by peering with as many other agents as possible as they are not concerned with transit revenue. Thus, we improve the scalability of our model by limiting CBA adoption to transit providers only.

Parameterization of GENESIS-CBA

In the previous section we described the various components of GENESIS-CBA, without mentioning any parameter values. Next, we parameterize GENESIS-CBA to replicate the current state of the Internet as closely as possible. The parameter values of GENESIS-CBA are shown in Table 2, along with a brief explanation for each parameter. We refer to the *instance* of GENESIS-CBA with these parameter values as the *Default model.*

Geographic distribution

The *world* in the Default model consists of 50 locations. The maximum number of locations in which an agent can be present is 15. We parameterize the Default model such that the number of locations at which agents are present follows a heavy-tailed distribution. Only a few agents are present in a large number of locations; most agents are limited to one location only. This parameterization is based on data collected from PeeringDB (2012), an online database where ASes voluntarily contribute information about themselves.

Traffic matrix

The Default model employs heavy-tailed distributions for generated and consumed traffic as seen in the real world. We parameterize the traffic matrix based on real-world

Table 2 Input parameters for default model

Parameter, symbol, description	Value	Explanation
Number of ASes N	500	Limited by simulation time constraints
Number of geographic locations G_{Max}	50	Based on approximate ratio of IXPs to peering networks in the Internet. PeeringDB ratio 10.27. GENESIS ratio 10.0 (PeeringDB 2012)
Geographic expanse distribution	Zipf(1.6)	Based on data about number of participants at each IXP collected from PeeringDB (PeeringDB 2012). $G(x)$ assigned randomly to each node
Maximum expanse for an AS	15	
Generated traffic distribution	Zipf(1.2)	Produces a heavy-tailed distribution of outgoing traffic. With this distribution, 0.1% of the ASes generate nearly 28% of the total traffic. This is consistent with the behavior reported in (Chang et al. 2005; Feldmann et al. 2004) & Labovitz et al. (2010), which show that the traffic produced by high-ranking ISPs and content providers follows a Zipf distribution. $V_G(x)$ assigned randomly to each node
Consumed traffic distribution	Zipf(0.8)	Produces heavy-tailed distribution of incoming traffic, similar to measured traffic distribution at Georgia Tech.
Mean consumed traffic	500 Mbps	$V_C(x) \propto \|G(x)\|$, rationale being that a node with large expanse will also have a large number of access customers
Private peering threshold Ω	50 Mbps	Survey of peering strategies (Peering Strategy Survey 2011)
Transit cost multiplier range $P(x)$	$[35,45]/Mbps per iteration	Parameterized based on IP transit prices recorded by Telegeography (Telegeography 2012). $P(x)$ assigned randomly
Transit cost exponent τ	0.75	Parameterized based on data from Chang et al. (2006) and Peering Strategy Survey (2011)
Peering cost multiplier α	$20/Mbps per iteration	
Peering cost exponent β	0.40	
Selective peering ratio σ	2.0	

observations reported in (Chang et al. 2005; Feldmann et al. 2004; Labovitz et al. 2010). With this parameterization, 0.1% of all agents generate 28% of all the network traffic.

Economic components

We parameterize economic components of the model, e.g., transit cost per Mbps, peering cost per Mbps etc. based on real-world data. We use transit pricing data from Telegeography (2012) and peering cost per Mbps data from Amsterdam IXP (2012).

Results and discussion

We implemented GENESIS-CBA as a stand-alone single-threaded tool using C++. Our implementation did not use any existing network simulation or modeling tool. The results presented subsequently all correspond to this implementation. However, as stated earlier, GENESIS-CBA is modular and extensible and can easily incorporate new methods and policies.

Validation

The validation of models such as GENESIS-CBA is inherently difficult due to the lack of available data about the economics of ISPs, the interdomain traffic matrix, and the resulting traffic flow. Even the AS-level Internet topology is not accurately known because only a small fraction of peering links are visible through BGP route monitors. Consequently, we have carried out a best-effort approach to validate GENESIS-CBA, examining whether it produces certain well-known qualitative properties of the Internet. This effort should be viewed as a sanity check, rather than a comprehensive validation process, in view of the previous difficulties.

Average path length

The average path length of the AS-level Internet has remained almost constant at 4 AS-hops, at least during the last twelve years (Dhamdhere and Dovrolis 2011). GENESIS-CBA produces networks with approximately the same average path length (4.4 AS-hops) with 500 agents.

Degree distribution

Figure 5 shows the complementary CDF (C-CDF) of the degree distribution that results from the default GENESIS-CBA model. Though not a strict power law (the tail of the distribution is truncated due to the limited number of agents), it is clear that the degree distribution is highly skewed, with few agents having a much larger degree than most other agents.

Link Load Distribution

Figure 6 shows the distribution of link traffic loads at equilibrium. Most links in the network are peering links (among stubs) that carry small traffic volumes. There are, however, few links that carry several orders of magnitude more traffic; those are mostly transit and peering links between large transit providers, or between transit providers and major content producers. Akella et al. (2003) have observed a similarly skewed distribution of link loads in the Internet.

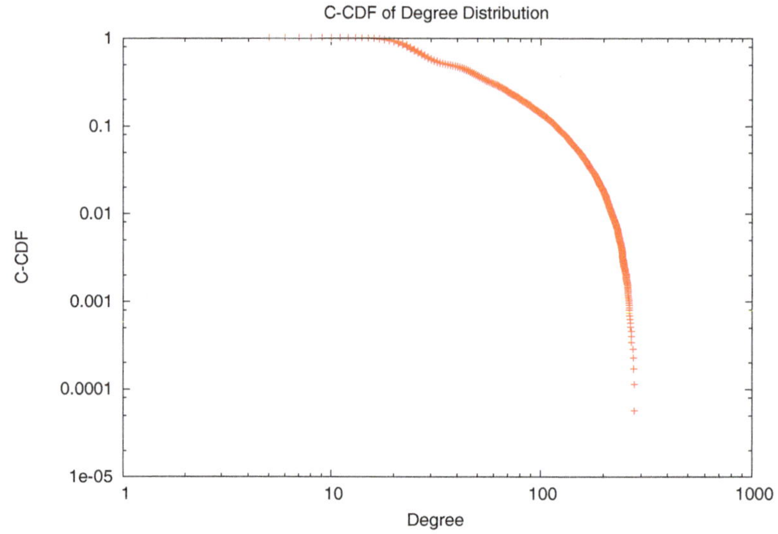

Figure 5 Degree distribution.

Fraction of transit providers

In the Default model, about 10% of the agents end up as transit providers at equilibrium, which is similar to the corresponding fraction reported in the measurement study by Dhamdhere et al. (2011).

Model characteristics

Convergence delay

We define the convergence delay as the number of iterations it takes for the sample path to converge to equilibrium.

We find that the convergence delay is less than 10 iterations for all sample paths that converge to equilibrium. We investigate why the network converges to equilibrium

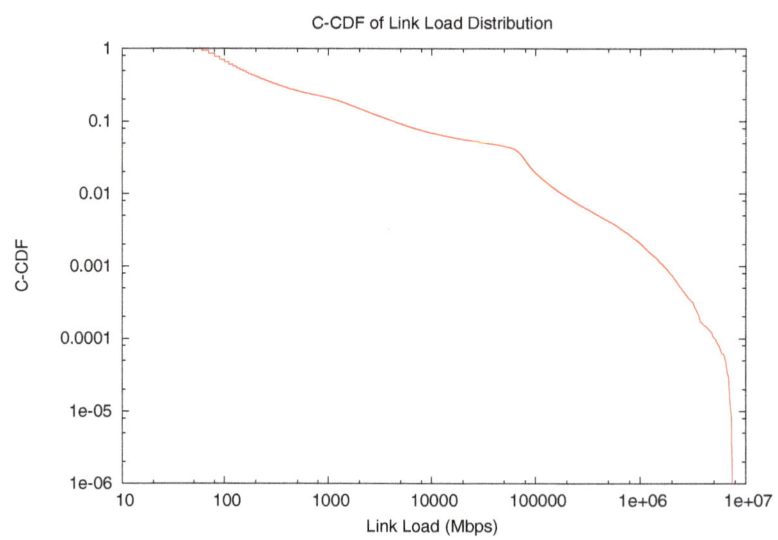

Figure 6 Link Load distribution.

so rapidly despite the fact that agents act asynchronously. Figure 7 shows the number of agents that acquire new peers, remove existing peers, change transit providers, or change peering strategies over time for a typical sample path of the model. The figure shows a rapid decline in the number of agents that switch transit providers and peering strategies (between CBA and Open). We find that the network converges to stability in a small number of iterations due to the rapid convergence of stubs to their preferred transit providers. As agents select the cheapest eligible transit provider, the number of transit providers in the network reduces significantly in the first few iterations. Most agents become "stubs" (agents without transit customers), and converge to the cheapest transit provider and peering strategy. The dynamics of transit provider selection in the model dictate that stubs never change their transit providers and cannot themselves become transit providers at a later time. As a result, their peering strategies do not change over time. The few remaining transit providers accumulate large transit volumes and compete for the remaining customers. As a result, some existing transit providers also become stubs. This effect reduces the number of transit providers, enhancing stability in the network.

Effect of population size

We find that the behavior of the model does not change significantly as the population size increases over a fixed number of geographic locations. Figure 8 shows this effect. Initially, with a small number of agents, the population density across all geographic locations is low. At this point, however, more transit providers are required to maintain connectivity between all agents across all locations. As the population size increases, geographic locations become saturated with agents providing transit at the same price, causing the number of transit providers to stabilize (agents always choose the cheapest eligible transit provider). The stability in the number of transit providers leads to stable network characteristics, e.g., the average path length. Similarly, for reasons described above, the

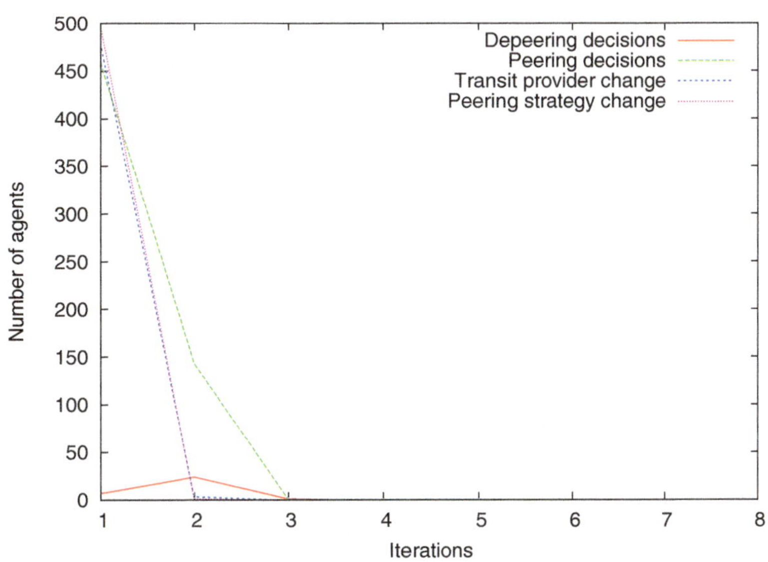

Figure 7 Convergence to equilibrium.

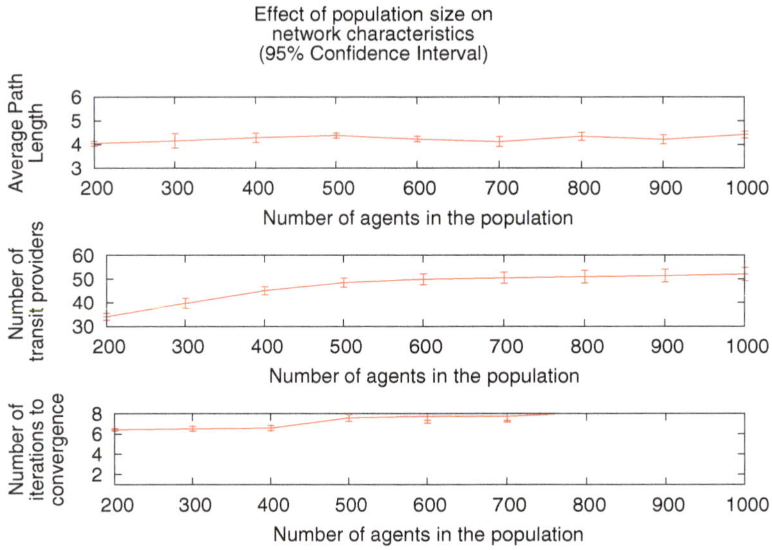

Figure 8 Scalability of the model.

convergence behavior of the model also does not change significantly with increasing population size.

Effect of geographic and traffic distribution

Our investigation reveals that co-location among agents has the most significant effect on the resulting networks. We vary the Zipf parameter of the geographic distribution to determine the effect on the resulting networks. An increase in the parameter value makes the geographic distribution more skewed. Figure 9 shows the effect of geographic distribution on network characteristics. We find that a less skewed distribution leads to increased co-location density (i.e., more agents are co-located with one another) resulting

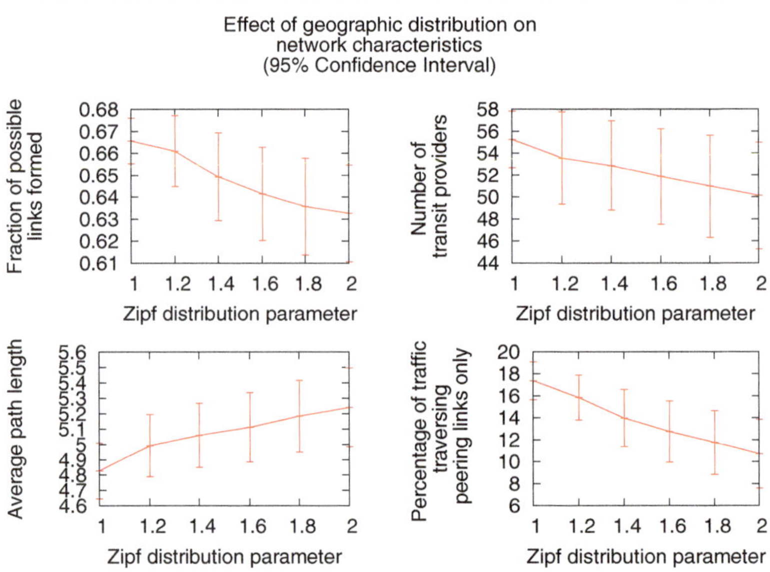

Figure 9 Sensitivity to geographic distribution parameters.

in a larger number of transit providers, greater connectivity, and reduced average path lengths. We also find that as a result of increased connectivity more traffic bypasses transit links and instead reaches its destination by traversing only peering links.

Stability of GENESIS-CBA

The network formation process in GENESIS-CBA is deterministic (the parameters of the population and the playing order does not change during a sample path), and so it can have one of two possible outcomes: convergence to an equilibrium, or a *limit cycle* in which the network moves repeatedly through the same sequence of states. An equilibrium results when none of the agents changes its connections or peering strategy during an iteration. On the other hand, a limit cycle (or oscillation) results when the state of the network $S(t_k)$ in the k'th iteration is identical to the state $S(t_{k-m})$ in the (k-m)'th iteration, with $m > 1$. The length of the limit cycle in that case is m. Note that there is no possibility for chaotic trajectories because the system has a finite number of states, and so if it does not converge to an equilibrium, it will eventually return to a previously visited state.

We ran many simulations with distinct node populations to evaluate how often the network formation process results in oscillations in practice. We observed that oscillations take place in about 7% of the sample paths. Even though this is not a large percentage, it is also not negligible and it has an important implication: if GENESIS-CBA is a good model for the Internet, we should not expect that the the interdomain connectivity in the latter will be stable. We return to this point at the end of this section.

Why study equilibria?

The Internet is in a constant state of flux, as new ASes and IXPs appear, and the traffic matrix, prices and costs vary with time. The reader may wonder, why study equilibria in a system that is constantly evolving?

The fact that GENESIS-CBA leads to an equilibrium, at least in 93% of the sample paths, means that as long as those exogenous parameters (set of nodes, locations, traffic matrix, costs, etc) remain constant, the network formation process converges to a stable point. If this convergence takes place quickly, compared to the time scales in which those exogenous parameters change, the evolution of the Internet can be thought of as a trajectory through a sequence of equilibria, with a new equilibrium resulting every time there is a change in one of the previous exogenous parameters.

The study of equilibria in the context of cost-benefit-analysis enables us to study the situation where each agent, by using cost-benefit analysis, has reached a situation where it has no further incentive to change its connectivity. Additionally, the study of equilibria gives us insight into the behavior of the model: what factors, e.g., initial topology, playing order of agents etc., influence the formation of the network for a given population setting; which factors have a greater impact on the network formation process; are the networks formed for different populations with the same statistical features qualitatively the same or different etc.

Oscillations

Causes of oscillations We analyzed many oscillatory sample paths, attempting to identify a common cause behind all of them. We find that oscillations occur primarily owing to the inability of agents using CBA to forecast with complete accuracy the network-wide effects of their actions. Agents using CBA can predict the immediate effect of formation (or termination) of a peering link. However, they cannot predict how other agents in the network will react to their actions. During the course of network formation a certain agent may be forced to revise its decisions multiple times, given the reactions of other agents. Thus, GENESIS-CBA produces a complex co-evolutionary network where limited foresight of agents may lead to complex oscillatory patterns.

We find that oscilltions are limited to the tier-1 clique owing to the CBA heuristic conditions of not peering with a peer's customers. We explain this pattern with a simple example shown in Figure 10.

In Figure 10a, x and y are two tier-1 transit providers peering with each other using CBA. x and y have customers a and b respectively. x and y do not peer with each other's customers as the CBA scheme prohibits such links. Consider the following sequence of actions:

1. x plays and evaluates its peering link with y using CBA. It does so by hypothetically depeering y and recomputing its fitness.
2. Assume that x satisfies the transit provider selection criteria for y. y chooses x as its transit provider after being depeered as shown in Figure 10b.
3. x recomputes its fitness after depeering y and finds that its fitness has increased. As y switched from peer of x to its customer, x's transit revenue increased and its peering cost decreased. x therefore decides to depeer y.
4. Since y is no longer a peer of x, it can peer with the customers of x. y evaluates peering with a. Since y can save transit costs by peering with a, its fitness increases. Thus, y decides to peer with a as shown in Figure 10c.
5. Note that the y—a peering link decreases the transit traffic of x, as traffic $a \longleftrightarrow b$ and $a \longleftrightarrow y$ does not transit through x any more. As a result of the decreased transit traffic, x no longer satisfies the provider selection criteria for y. Since GENESIS-CBA does not allow a partitioned network, a peering-by-necessity relationship is established between x and y. Once y peers with x, it no longer needs to peer with a, a customer of x. This brings us back to the initial network of Figure 10a.

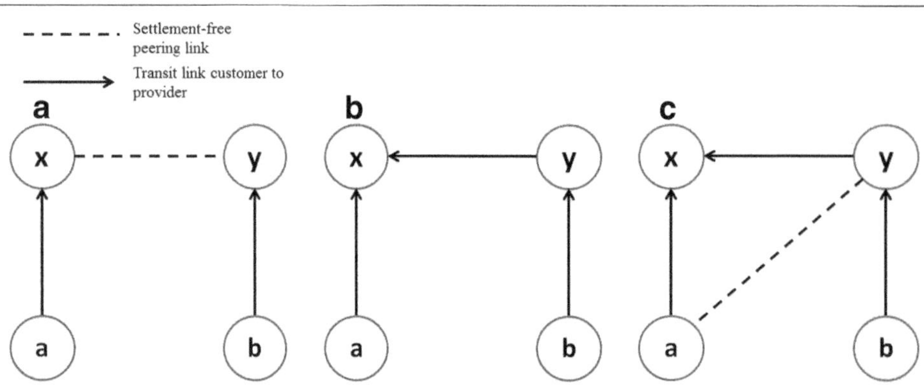

Figure 10 Illustration of oscillations in GENESIS-CBA. (**a**) Tier-1 peers. (**b**) Tier-1 depeered. (**c**) Reduction in transit traffic of one of the Tier-1 agents.

The previous example is only an abstraction of more complex oscillatory sample paths that occur in practice, involving more agents and more steps.

Network-wide effects of oscillations We next study the effect of oscillations on the entire network as well as the affected agents. We use the *Jaccard Similarity* metric to quantify the similarity between two network states. If L_A and L_B are the sets of links in states **A** and **B**, respectively, then the Jaccard similarity between the two states is defined as:

$$J_{AB} = \frac{|L_A \bigcap L_B|}{|L_A \bigcup L_B|}$$

Note that a link is defined by both its two end-points as well as its type (transit versus peering). The value of $J_{AB} = 0$ indicates that the two networks do not have any link in common, whereas $J_{AB} = 1$ indicates that the two states are exactly the same. In addition to the complete network, we also examine the sub-network composed only of transit providers, excluding stubs and their links. We refer to this sub-network as the *provider network*.

For each oscillatory sample path, we compute the minimum Jaccard similarity across all pairs of states during an oscillation. Figure 11 shows the distribution of these minimum similarity values, across all oscillatory sample paths. When we examine the complete network, the minimum Jaccard similarity is almost equal to one in all oscillations, i.e., the oscillation affects the connectivity of only few nodes. We observed that in 95% of oscillations the number of agents that undergo any change (connectivity and/or strategy) during the oscillation is less than 5% of the total network population. Note that all agents involved in oscillations are not transit providers; some agents are stubs that act as passive participants in the oscillation. When we only examine the provider network, the corresponding minimum Jaccard similarity values are lower but still higher than 0.90. In other words, oscillations cause more fluctuations in the connectivity of transit providers but those fluctuations are still mostly local in scope.

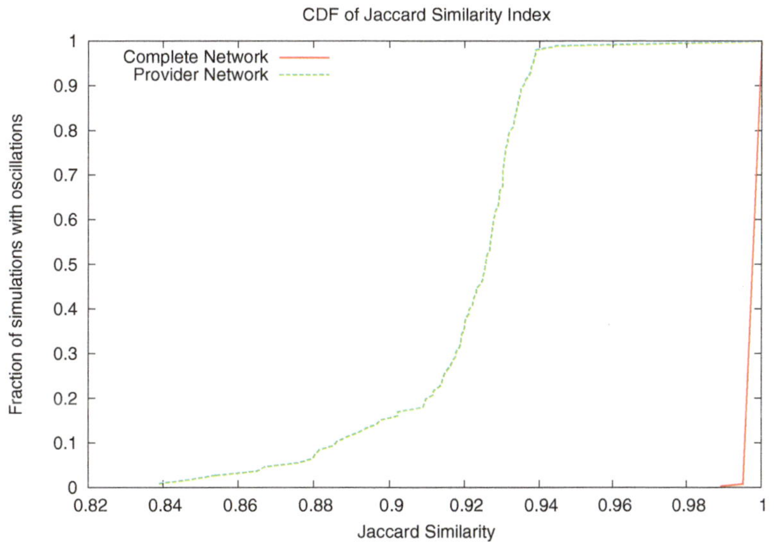

Figure 11 CDF of the Minimum Jaccard Similarity Index of network states during oscillations.

Length of oscillations We measured that 95% of the oscillatiory sample paths have a limit cycle length that is less than 15 iterations. This length is positively correlated with the number of affected ASes (Pearson correlation coefficient: 0.94). Thus, the more agents change connectivity or peering strategy during an oscillation, the longer it will take for the network to repeat its limit cycle.

How are oscillations relevant to the internet? Peering agreements and negotiations are typically shrouded in secrecy, yet peering disputes between high-profile nodes over traffic ratios, content types, and peering conditions often generate attention. For example, Cogent and Sprint-Nextel terminated their peering relationship in 2008 after they failed to resolve a peering dispute. However, the peering relationship between them was restored after some time. Similar disputes have been reported between Level-3 and Cogent (2005), Cogent and Telia (2008), and Level-3 and Comcast (2010). Peering disputes between large providers often have significant effects, as single-homed customers of those nodes are unable to reach one another.

The fact that GENESIS-CBA can produce oscillatory sample paths should not be considered an artifact of the model. As described earlier, oscillations in GENESIS-CBA occur because some providers switch between tier-1 and non-tier-1 status, which resembles the peering disputes between tier-1 providers seen in practice. In the real world, such disputes do not persist and some exogenous mechanisms (such as revised contracts, negotiations, regulation or legal actions) are applied to break the impasse. GENESIS-CBA does not capture those exogenous mechanisms, but it captures that the endogenous dynamics of the network formation process cannot always manage to produce a stable network, and so some additional mechanisms should be designed to resolve such oscillations when they appear.

Variability across equilibria

In this section we focus on those sample paths that converge to an equilibrium, and examine the differences between distinct equilibria. Specifically, given a population of agents and an initial topology, does GENESIS-CBA always result in the same equilibrium? If not, how different are the resulting equilibria? What are the causes of these differences? Can we predict the properties of an individual agent at equilibrium? Are there any global properties that are predictable with statistical significance? And who are the most predictable, or least predictable, agents in a network, in terms of fitness?

Distinct equilibria

The network that emerges from a GENESIS-CBA sample path highly depends on both the initial topology and the playing order. Specifically, given the same population of agents and the same initial topology, 90% of the sample paths that converge to an equilibrium produce a distinct equilibrium. If the initial topology also varies across sample paths, this percentage increases to 95%. So, if GENESIS-CBA is a good model of the Internet, we understand that it is very hard in practice to predict the evolution of the Internet, as even minor changes in the order in which ASes act can have global effects.

Differences between equilibria

To quantify the difference between distinct equilibria, we again use the Jaccard Similarity metric between the corresponding network states. Figure 12 shows the CDF of that metric

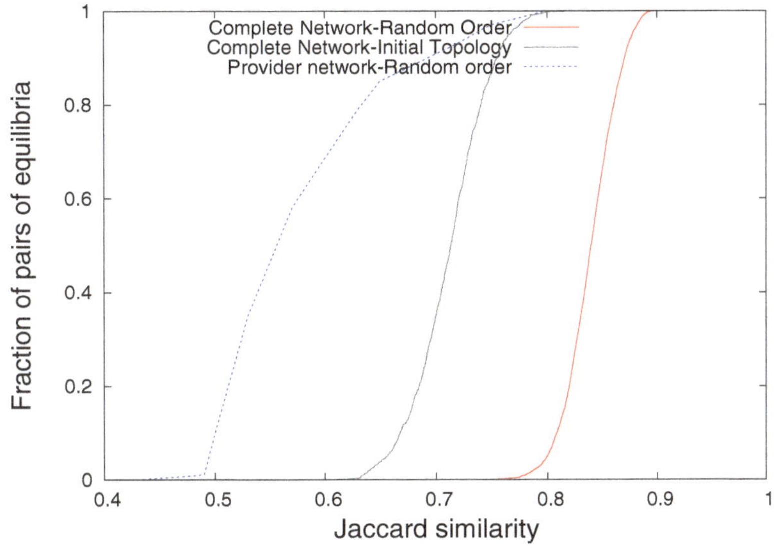

Figure 12 CDF of Jaccard similarity Index between equilibria.

for all pairs of distinct equilibria resulting from 100 sample paths. Note that the Jaccard Similarity is typically higher than 0.8 when we only vary the playing order. When we also vary the initial topology, the similarity metric is lower (between 0.65–0.8) but still high in absolute terms. So, even though there are many distinct equilibria, they are quite similar in terms of topology.

Figure 12 also shows the distribution of Jaccard similarity for the subset of the network that includes only transit providers (the "provider network"). In this case, the Jaccard similarity is significantly lower, with 45% of the equilibrium pairs being less than 55% similar. In other words, even though stubs often end up with the same transit provider across different equilibria, the hierarchy of transit providers and the peering links between them are much less predictable.

Variability of fitness distribution

We next focus on the distribution of fitness values in a network. To determine the similarity of fitness distributions across different equilibria, we applied the two-sample Kolmogorov-Smirnov (KS) test on fitness distributions across all pairs of equilibria. We did so both for the entire network, and for the provider network. We could *not* reject the null hypothesis that the distributions are identical for any pair of distributions at the 99% significance level. In other words, even though there are many distinct equilibria, the statistical distribution of the agents' fitness remains invariant. However, this does not mean that the fitness of an individual agent would remain the same across different equilibria; this is the subject of the next paragraph.

Variability of individual node fitness

Next, we look at the variability of the fitness of individual agents across different equilibria. We compute the Coefficient of Variation (CoV) of the fitness of each agent across 100 equilibria resulting from different playing orders. Figure 13 shows the CDF of CoV of fitness for different agent classes. The CoV is close to zero for 90% of agents

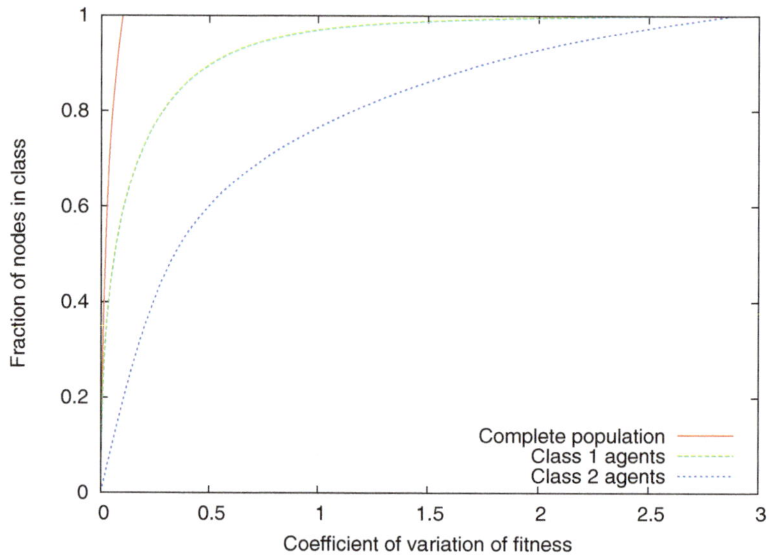

Figure 13 CDF of CoV of fitness for different agent classes.

indicating that most agents see almost identical fitness in all equilibria. However, 5% of the agents have CoV greater than 1.0 and 1% of the agents have CoV as high as 3.0. Who are these "most unpredictable" agents and what causes the large variability in their fitness?

We classify agents in two classes. *Class-1* agents are those that are either stubs at all equilibria, or that are transit providers at all equilibria. *Class-2* agents, on the other hand, are those that are transit providers in some equilibria and stubs in others. Figure 13 also shows the CoV distribution for these two classes of agents. Note that the CoV of Class-1 agents is close to zero for almost all agents, while the CoV of Class-2 agents is significant for most of them. In the following sections, we discuss the properties of the agents in each class in more detail.

Properties of Class-1 nodes The largest fraction of agents in this class are stubs (including agents that were transit providers initially but are stubs at equilibrium). It is not surprising that the fitness of stubs does not vary significantly across equilibria. These agents do not have revenues and their provider is always the cheapest transit provider at the locations they are present at. Additionally, they use the Open peering strategy, and so their set of peers consists mostly of other stubs that are co-located with them. Additionally, however, Class-1 includes some large, in terms of geographical expanse, transit providers (12~15 locations). These providers have a large set of customers and so significant transit traffic. Their large transit volume and expanse restricts their set of providers and peers, reducing the variability in their connectivity and fitness across equilibria. Note that tier-1 providers always belong to Class-1.

Properties of Class-2 nodes We observe that most agents in this class have mid-range expanse (6~10 locations) and relatively high prices. Thus, while they are mostly unattractive as transit providers for stubs, smaller transit providers may choose them as provider, depending on their expanse and transit traffic volume. The latter, however, changes dynamically as an agent gains or loses customers during a sample path. This aspect of the model creates a positive feedback effect where the larger the transit volume of a node is, the more it qualifies for being chosen as provider by other providers. However, the identity of the agents that will strongly benefit from this positive feedback depends on the order in which the agents play. In other words, the agents in Class-2 are mostly those that end up as transit providers in some equilibria simply because they were "lucky" in those sample paths to get some customers early in the network formation process, accumulating transit volume and then attracting even more customers. It is these agents that exhibit the largest variation in fitness and that act as the biggest source of variability in the resulting networks.

Conclusions

We proposed GENESIS-CBA, an agent-based network formation model that captures interdomain traffic flow, policy-based routing, geographic constraints, and the economics of transit and peering relations. GENESIS-CBA implements a new peering scheme that can replace existing peering strategies. The new scheme allows agents to carry out economic analysis for each peering link. Agents can use this analysis to form only those peering links that improve their economic fitness. We examined the convergence properties of GENESIS-CBA, and found that GENESIS results in a stable network in most cases. The observed cases of instability occur due to connectivity constraints in the Tier-1 clique. GENESIS-CBA shows both path dependence and dependence on initial conditions, i.e., it can produce different equilibria depending on the initial topology and playing order of nodes which we believe is also true of the Internet. We found that while these equilibria are distinct in terms of network topology, it is possible to predict certain properties of the network or of certain classes of nodes with statistical significance. In ongoing work, we use GENESIS-CBA to understand the effect of the scheme on the fitness distribution of different classes of autonomous systems in the Internet.

In future, we intend to extend GENESIS-CBA to turn it into a practical tool capable of being deployed by the operators managing the ASes in the interdomain network. Our objective is to align the information requirements of CBA with the information available to ASes in the real world. CBA uses the following information for analysis: topological information, e.g., customer-provider relationships, traffic flows between ASes and geographic co-location. While ASes can form estimates of the previously mentioned information from data available to them, their accuracy in the context of CBA is yet to be ascertained, and this is one of the goals of our future work. We envision GENESIS-CBA being deployed in the real world at a microscopic scale, i.e., at the AS level, to carry out what-if analysis of peering decisions. In this context, CBA strategy can be deployed by ASes in the Internet to aid in their analysis to determine which ASes to peer with. They can use CBA to analyze the effect of existing and potential peering links on their economic fitness.

Endnote

[1]There is another type of peering which is known as "paid peering" in which one peer pays the other. However, "paid peering" is not the focus of this work.

Competing interests
The authors declare that they have no competing interests.

Authors' contributions

AD and CD supervised the design of the model and the study. The development work for the model and simulations were carried out by AL. All authors actively participated in data analysis and editing of the draft. All authors read and approved the final draft.

Acknowledgements

This research was financially supported by Google and the National Science Foundation (grant CNS-1017139). Its contents are solely the responsibility of the authors and do not necessarily represent the official views of Google or NSF.

Author details

[1]School of Computer Science, Georgia Institute of Technology, Atlanta, USA. [2]The Cooperative Association for Internet Data Analysis, University of California, San Diego, USA.

References

Akella, A, Seshan S, Shaikh A: **An empirical evaluation of wide-area internet bottlenecks.** In *Proceedings of ACM SIGCOMM IMC*; 2003.

Amsterdam Internet Exchange Pricing 2012. [https://www.ams-ix.net/services-pricing/pricing]

Augustin, B, Krishnamurthy B, Willinger W: **IXPs: mapped?** In *Proceedings of the 9th ACM SIGCOMM Conference on Internet Measurement Conference*, IMC '09. Chicago: ACM; 2009.

Chang, H, Jamin S, Mao Z, Willinger W: **An empirical approach to modeling Inter-AS traffic matrices.** In *Proceedings of ACM SIGCOMM IMC*; 2005.

Chang, H, Jamin S, Willinger W: **Internet connectivity at the AS-level: an optimization-driven modeling approach.** In *Proceedings of ACM SIGCOMM MoMeTools Workshop*; 2003.

Chang, H, Jamin S, Willinger W: **To peer or not to peer: modeling the evolution of the internets AS-level topology.** In *Proceedings of IEEE INFOCOM*; 2006.

Corbo, J, Jain S, Mitzenmacher M, Parkes DC: **An Economically-Principled Generative Model of AS Graph Connectivity.** In *Proceedings of IEEE INFOCOM Mini-Conference*; 2009.

Dhamdhere, A, Dovrolis C: **The internet is flat: modeling the transition from a transit hierarchy to a peering mesh.** In *Proceedings of ACM CoNEXT*; 2010.

Dhamdhere, A, Dovrolis C: **Twelve years in the evolution of the internet ecosystem.** *IEEE/ACM Trans Netw* 2011, **19**(5):1420–1433.

DE-CIX: **Frankfurt internet exchange point global peering pricing.** [http://www.de-cix.net/products-services/de-cix-frankfurt/globepeer/]

Fabrikant, A, Koutsoupias E, Papadimitriou CH: **Heuristically optimized trade-offs: a new paradigm for power laws in the internet.** In *Proceedings of ICALP*; 2002.

Faloutsos, M, Faloutsos P, Faloutsos C: **On power-law relationships of the internet topology.** In *Proceedings of ACM SIGCOMM*; 1999.

Feldmann, A, Kammenhuber N, Maennel O, Maggs B, De Prisco R, Sundaram R: **A methodology for estimating interdomain web traffic demand.** In *Proceedings of ACM SIGCOMM IMC*; 2004.

Gill, P, Arlitt M, Li Z, Mahanti A: **The flattening internet topology: natural evolution, unsightly barnacles or contrived collapse?** In *Proceedings of the Passive and Active Measurement Conference (PAM)*; 2008.

Goyal, S: *Connections: An Introduction to the Economics of Networks*. New Jersey: Princeton University Press; 2007.

Holme, P, Karlin J, Forrest S: **An integrated model of traffic, geography and economy in the internet.** *ACM SIGCOMM CCR* 2008.

Internet Assigned Numbers Authority 2012. [www.iana.org]

Internet Transit Prices: Historical and Projected 2010. [http://drpeering.net/white-papers/Internet-Transit-Pricing-Historical-And-Projected.php]

Jackson, MO: *Social and Economic Networks*. New Jersey: Princeton University Press; 2008.

Labovitz, C, Iekel-Johnson S, McPherson D, Oberheide J, Jahanian F: **Internet inter-domain traffic.** In *Proceedings of ACM SIGCOMM*; 2010.

Li, L, Alderson D, Doyle JC, Willinger W: **Towards a theory of scale-free graphs: definition, properties, and implications.** *Internet Math* 2005, **2**(4).

Lodhi, A, Dovrolis C: **A network formation model for internet transit relations.** In *Proceedings of the Workshop on Economics of 2010 Networks, Systems, and Computation (NetEcon)*; 2010.

Lodhi, A, Dhamdhere A, Dovrolis C: **Analysis of peering strategy adoption by transit providers in the internet.** In *Proceedings of the Workshop on the Economics of Networks, Systems, and Computation*; 2012a.

Lodhi, A, Dhamdhere A, Dovrolis C: **GENESIS: An agent-based model of interdomain network formation, traffic flow and economics.** In *Proceedings of IEEE INFOCOM*; 2012b.

Lodhi, A, Dhamdhere A, Dovrolis C: **Peering strategy adoption by transit providers in the internet: a game theoretic approach?** *SIGMETRICS Perform Eval Rev* 2012c, **40**(2):38–41.

Norton, W: **The great (public vs. private) peering debate: peering at 10Gbps.** 2011.

Peering Strategy Survey 2011. [http://drpeering.net/white-papers/Peering-Policies/A-Study-of-28-Peering-Policies.html]

PeeringDB 2012. [http://www.peeringdb.com]

Telegeography: **IP tranist pricing service.** 2012. [http://www.telegeography.com]
Wang, X, Loguinov D: **Wealth-based evolution model for the internet AS-Level topology.** In *Proceedings of IEEE INFOCOM*; 2006.
Wang X, Loguinov D: **Understanding and modeling the internet topology economics and evolution perspective.** *IEEE/ACM Trans Netw* 2010, **18**(1):257–270.
Yook, SH, Jeong H, Barabasi AL: **Modeling the internet's large-scale topology.** *Proc Natl Acad Sci* 2002, **99**(21):13382–13386.

3

Large-scale global optimization through consensus of opinions over complex networks

Omid Askari-Sichani and Mahdi Jalili*

* Correspondence: mjalili@sharif.edu
Department of Computer Engineering, Sharif University of Technology, Tehran, Iran

Abstract

Purpose: Large-scale optimization tasks have many applications in science and engineering. There are many algorithms to perform such optimization tasks. In this manuscript, we aim at using consensus in multi-agent systems as a tool for solving large-scale optimization tasks.

Method: The model is based on consensus of opinions among agents interacting over a complex networked structure. For each optimization task, a number of agents are considered, each with an opinion value. These agents interact over a networked structure and update their opinions based on their best-matching neighbor in the network. A neighbor with the best value of the objective function (of the optimization task) is referred to as the best-matching neighbor for an agent. We use structures such as pure random, small-world and scale-free networks as interaction graph. The optimization algorithm is applied on a number of benchmark problems and its performance is compared with a number of classic methods including genetic algorithms, differential evolution and particle swarm optimization.

Results: We show that the agents could solve various large-scale optimization tasks through collaborating with each other and getting into consensus in their opinions. Furthermore, we find pure random topology better than small-world and scale-free topologies in that it leads to faster convergence to the optimal solution. Our experiments show that the proposed consensus-based optimization method outperforms the classic optimization algorithms.

Conclusion: Consensus in multi-agents systems can be efficiently used for large-scale optimization problems. Connectivity structure of the consensus network is effective in the convergence to the optimum solution where random structures show better performance as compared to heterogeneous networks.

Keywords: Large-scale optimization, Complex networks, Continuous opinion formation, Consensus, Scale-free networks, Small-world networks
AMS subject classification: 15A04, 54A20, 60J20, 92D25

Background

Networks are everywhere and we confront many networks in our daily life; they are practically present where any kind of information is transmitted or exchanged. Networks such as the Internet, the World Wide Web, engineering, social, biological and economical networks have been subject to heavy studies in the last decade and many applications have been developed based on network science Albert & Barabasi (2002, 1999; Barabasi & Albert 1999; Boccaletti et al. 2006; Newman & Watts 2006; Newman

2003; Newman & Park 2003; Strogatz 2001; Watts & Strogatz 1998). The progress in network science accelerated after the seminal work of Watts & Strogatz (1998) on collective behavior of small-world networks Watts & Strogatz (1998) and Barabasi & Albert (1999) on scale-free graphs (Barabasi & Albert 1999). Watts and Strogatz discovered that many real-world networks have small-world property in that their characteristics path length scales logarithmically with network size (Watts & Strogatz 1998) – a property that is observed in random networks. At the same time, these networks show high levels of transitivity (clustering coefficient) (Watts & Strogatz 1998) – much higher than corresponding random networks. Furthermore, many real networks from different disciplines were shown to have a power-law degree distribution (Barabasi & Albert 1999); the probability of having a node with degree k is $k^{-\gamma}$ with γ being in the range 2–3. Real networks have been shown to have more complex properties such as motifs (Milo et al. 2002) and community structure (Girvan & Newman 2002). These structural features influence dynamics and functionality of networks. For example, synchronization and consensus properties of networks largely depend on their structure (Belykh et al. 2005; Lu et al. 2004).

The most striking pattern of networked structures appear when a number of agents (each with simple behavior) interact leading to complex behaviors as a result of collective motion. Synchronization of interacting agents - as the most striking form of collective behavior - has many applications in science and engineering. For example, techniques available in network theory can be used for efficient distrusted inference in sensor networks (2007; Scutari et al. 2008). In this work we used tools available in network science to perform a numerical optimization task. Optimization is an approach that iteratively improves the performance of a system, which is formulated as a single standard measurement equation called cost (or objective) function.

In order to use network theory for solving an optimization task, we used the concept of consensus formation in the opinions of multi-agent systems. Let us consider a network of agents where a (discrete or continuous) opinion value is associated to each agent. Agents can influence each others' opinions through the connections existing between them, i.e., the edges of the network. Considering some simple update rules and if certain conditions are met, the agents can reach a consensus in their opinions through a number of opinion updates (Kozma & Barrat 2008; Carletti et al. 2006). In this work we considered the evolution of continuous opinions based on the modified version of bounded confidence model (Deffuant et al. 2001), which has been extensively studied in recent years (Gandica et al. 2010; Weisbuch 2004; Urbig et al. 2008). The previous studies of continuous opinion formation have been mainly performed considering uniform agents (Gandica et al. 2010; Weisbuch 2004; Urbig et al. 2008). However, in reality the agents are diverse in their wealth and social status, and hence, have diverse influence on others (Holyst et al. 2001; Lewenstein et al. 1992; Jalili 2013a; Jalili 2013b). Therefore, we associated specific weight for each agent resulting in faster consensus.

The paradigm proposed in this manuscript was applied on a number of benchmark problems. We first considered a simple function with many local optima and showed that the proposed optimization strategy could successfully find the optimum. We then applied the method on a number of benchmark problems from CEC 2010 competition benchmark set (Tang et al. 2009). We compared the performance of the proposed consensus-based optimization approach with that of a number of classic optimization methods including genetic algorithms, differential evolution and particle swarm optimization.

Methods

Optimization through consensus in the network

The paradigm we have proposed for large-scale optimization task is based on consensus in networked structures. In opinion formation models, there is a population of agents, each with a (discrete or continuous) opinion value representing its information about a subject (Deffuant et al. 2001; Gandica et al. 2010; Weisbuch 2004; Urbig et al. 2008). The term opinion is not easy to define in reality; however, it can be considered as a discrete or continuous value expressing the individuals' degree of desire or preference. This opinion is often represented as a real number when the model is unimodal or as a vector of real numbers when the model is multimodal. In this paper we aimed at optimizing an objective function, and therefore, each agent will have an opinion value containing all the input parameters of desired objective function, i.e., a multimodal model.

Opinion formation in multi-agent systems

The agents update their opinions as a result of interactions with their neighboring agents. Consider two neighboring agents i and j with opinions as x_i and x_j, respectively. Their opinions at time $n + 1$ will be a function of their previous opinions, i.e. $x_i(n + 1) = f_1(x_i(n), x_j(n))$, $x_j(n + 1) = f_2(x_i(n), x_j(n))$. If certain conditions are met, after a number of updates in these values, the agents can reach a consensus in their opinions (Kozma & Barrat 2008; Carletti et al. 2006). The collective behavior of the agents over complex networks largely depends on the structural properties of the networks (Amblard & Deffuant 2004), and minor modification in the structure of the network can have drastic effects on the behavior of opinion formation (Nardini et al. 2008).

There are a number of rules for modelling opinion formation in complex networks. For example, considering discrete opinions, in the voter model, randomly selected agents exchange their opinions by that of one of their neighbours (Krapivsky & Redner 2003). The agents might influence their neighbouring agents to change their opinions based on their strength and the neighbours' threshold (Leskovec et al. 2006). In the evolution of continuous opinions on a network, the opinions of two connected agents are updated if their difference is less than a threshold, i.e. the agents have evolving opinions (Deffuant et al. 2001; Amblard & Deffuant 2004; Lorenz 2007; Kurmyshev et al. 2011; Hegselmann & Krause 2002; Guo & Cai 2009).

In this work we considered a specific form of continuous bounded confidence model in which each agent has an opinion in the range [−1,1] - denoted by opinion space – and update its opinions based on a specific rule (Deffuant et al. 2001; Fortunato et al. 2005). First, each agent takes a random value from opinion space. Then, at each proceeding step, each agent finds its best-matching neighbour, i.e., the one that optimizes the objective function furthest among the neighbours, and then updates its opinion value with this best-matching adjacent. The update rule for agent i is as

$$\begin{cases} x_i(n+1) = x_i(n) + \mu\left[x_j(n) - x_i(n)\right] \\ \text{if } f(x_i(n)) > f(x_j(n)) \end{cases} , \quad i = 1, 2, \ldots, N, \qquad (1)$$

$$j = \arg_{\min,k} f(x_k(n)) \qquad , \quad k \in N_i$$

where f is the desired cost (or objective) function to be optimized, N is the network size and N_i is the set of neighbours of agent i. μ is the convergence (or influence) parameter, which often takes a value between 0 and 1. This parameter controls the speed of convergence in such a way that small values of μ corresponds to slow but smooth convergence, while the large values of μ corresponds to faster but wavy convergence.

To some extent, the above model for opinion formation imitates the behavior of agents in real social networks. A person may know many individuals in the society; however, he/she is only influenced by his/her closest friends (i.e., neighbors in the network). In many cases, individuals get the maximum influence through their best (closest) friends and try to make themselves similar to them, i.e., making their opinion closer to their closest friends. People try to behave like their best friends for establishing and maintaining their friendships and they influenced by them more than the others in their life. Sometimes these changes will happen because people want to preserve their connections and friendships and they will act or behave like their close friends (Barry & Wentzel 2006). They project their own attitudes and habits to their friends. Furthermore, research showed that, in general, the influence of the very best friend approximately is equal or comparable to the influence of multiple friends (Berndt & Murphy 2003).

In our model, each agent finds one of its neighboring agents that have the best value in the objective function – which is denoted by best-matching neighbor. For example, if the objective function is an energy function; the neighbor with minimal energy function is selected. The agents then update their opinions using equation (1). It has been shown that considering proper connection weights can enhance the consensus properties of the network, i.e., the network reaches to consensus in a shorter time (Jalili 2013a; Jalili 2013b; Yang et al. 2009; Brunetti et al. 2012). Therefore, we also took proper weights while updating the opinions. The update equations read

$$\begin{cases} x_i(n+1) = x_i(n) + \mu \dfrac{f(x_j(n)) + \varepsilon}{f(x_i(n)) + \varepsilon}\left[x_j(n) - x_i(n)\right] & , \quad i = 1, 2, \ldots, N, \\ \text{if } f(x_i(n)) > f(x_j(n)) & \end{cases} \qquad (2)$$
$$j = \arg_{\min,k} f(x_k(n)) \qquad , \quad k \in N_i$$

where ε is a small value (in order to make the denominator non-zero). The above weighted update rules can be justified as follows. Let us suppose that an objective functions is to be minimized. As the best-matching neighbor is found for each agent, it influences the agent according to its fitness, i.e., its value in the objective function. To this end, the weight for the update rule of an agent gets as the fitness function at that agent divided by the fitness value of the best-matching neighbor, often resulting in a value in the range 0–1 (note that the opinions are updated only when the fitness of best-matching neighbor is better than that of the agent). It is worth mentioning that in some cases, the opinions are in multi dimensions, i.e., x is vector, in which the best matching agent is obtained separately for each dimension.

The method largely depends on the diffusion of good opinions (i.e., those that are good in terms of the objective function) in the network. Agents with opinion values close to the optimal objective function disseminate their opinions through communicating with their neighbors, i.e., getting into consensus with them. Indeed, influence of opinions is in two folds. Indeed, closer opinions to the optimal value of the objective function have a better chance to be selected as best-matching neighbors.

The above rule for opinion formation is somehow inspired by communication in human societies. Our friends influence our behavior in daily life; however, we are usually affected only when our friends are better than us. Here, similarly, for each agent, first, the best matching agent is found, and then, its opinion is updated (using equation (2)) if the fitness of the best-matching neighbor is better (i.e., it results in a lower value in the objective function) than that of the agent.

It is worth mentioning that the consensus (or synchronization) properties of dynamical networks largely depend on their structure and some topologies are favored for fast consensus (Belykh et al. 2005; Ajdari Rad et al. 2008). Network topology plays also important role in the evolution of other dynamical phenomena over complex networks, such as evolution of cooperative behavior among interacting agents (Perc & Szolnoki 2010; Perc 2009).

A pseudo-code of the proposed consensus-based optimization algorithm is illustrated in section Pseudo-Code as follows.

Pseudo-Code for the proposed consensus-based optimization method

Function CBO

N: number of agents in the population (network size)

M: number of attributes of opinion vector

Boundaries: the range of the opinions

F: desired objective function which is needed to be optimized (minimized in this case)

Begin

- Initialize $N * M$ matrix X by a random normal distribution for the opinion values in *Boundaries*;
- *net* = Create a structured network;
- Repeat
- **for** each agent i in population **do**
- **for** each attribute a **do**
- *neighbors_opinion* = mask other attributes of the opinions x in neighbors of agent i in network *net* by a dummy value;
- *self_opinion* = mask other attributes of the opinions x agent i;
- j = find the best agent in *neighbors_opinion* resulting in the best value for F;
- **if** *neighbor_opinion* of agent j optimizes F better than *self_opinion* **then**

$$weight = \frac{F\left(neighbors\_opinion_j\right) + \varepsilon}{F(self\_opinion) + \varepsilon};$$
$$x[i,a] = x[i,a] + \mu \cdot weight \cdot (x[j,a] - x[i,a]);$$
$$x[i,a] = \mathrm{mod}(x[i,a], Boundaries);$$

- end if
- end for
- end for
- Until stopping condition(s) has/have been met
- End

In the beginning of the process, the agents are initialized by some random values in a range acceptable by their opinion values. As indicated by Watts and Dodds (2007) "a minority of individuals who influence an exceptional number of their peers" (Watts & Dodds 2007), there is often a minority of agents that have a significant influence on others, which is mainly due to their specific position in the network The hypothesis of influential agents demonstrated that initiating influential individuals will be explicitly different from initiating non-influential ones in the size and likelihood of a cascade (Watts & Dodds 2007). This means that initial opinions for influential agents probably would bias the result of the consensus. This phenomenon will not happen in proposed method, since CBO is not based on the bounded confidence model. Every agent selects his/her best-matching neighbor regardless of its great social power and degree.

Consensus of opinion values

In this section, we provide a mathematical proof that the update rule expressed in equation (2) leads the opinions to converge. To this end, let us rewrite it as follows (Hegselmann & Krause 2002):

$$x(t+1) = A(t, x(t)).x(t), \tag{3}$$

where $x(t) = [x_1(t), x_2(t), \ldots, x_n(t)]$ is the opinion vector at time t and A is a time-dependence state transition matrix that also depends on the opinion vector. We would like to verify that starting from an initial opinion values $x(0)$, whether or not all opinions converges to a single value, that is $\lim_{t\to\infty} x_i(t) = x^*$ for $i = 1, 2, \ldots, N$. Let us define diameter d of the opinions as

$$d(x) = \max_{1\le i,j\le N} (x_i - x_j), \tag{4}$$

Lemma 1 (Krause 2000): Consider a stochastic matrix A (i.e., a nonnegative matrix with row-sums equal to 1 is defined as a stochastic matrix), then, one has

$$\max_{i,j} \left| x_i(t+1) - x_j(t+1) \right| \le d(A). \max_{i,j} \left| x_i(t) - x_j(t) \right|, \tag{5}$$

or equivalently,

$$d(x(t+1)) = d(A(t, x(t)).x(t)) \le \left(1 - \min_{1\le i,j\le N} \sum_{k=1}^{n} \min\{a_{ik}, a_{jk}\} \right) d(x(t)), \tag{6}$$

The above lemma was proved in (Seneta 1981); however, we also give another proof using a simpler method.

Proof: Expression (6) can be written as

$$d(Ax) = \max_{1i,j\leq N}(A_i x - A_j x) = \max_{1\leq i,j\leq N}\sum_{k=1}^{n}(a_{ik}x_k - a_{jk}x_k)$$
$$= \max_{1\leq i,j\leq N}\sum_{k=1}^{n}(a_{ik} - \min\{a_{ik}, a_{jk}\} + \min\{a_{ik}, a_{jk}\} - a_{jk})x_k$$
$$= \max_{1\leq i,j\leq N}\sum_{k=1}^{n}(a_{ik} - \min\{a_{ik}, a_{jk}\})x_k - \min_{1\leq i,j\leq n}\sum_{k=1}^{n}(a_{jk} - \min\{a_{ik}, a_{jk}\})x_k$$
$$\leq\left(1 - \min_{1\leq i,j\leq N}\sum_{k=1}^{n}\min\{a_{ik}, a_{jk}\}\right)\left(\max_{1\leq i\leq N}x_i - \min_{1\leq j\leq n}x_j\right)$$
$$= \left(1 - \min_{1\leq i,j\leq N}\sum_{k=1}^{n}\min\{a_{ik}, a_{jk}\}\right)\max_{1\leq i,j\leq N}(x_i - x_j)$$
$$= \left(1 - \min_{1\leq i,j\leq N}\sum_{k=1}^{n}\min\{a_{ik}, a_{jk}\}\right)d(x)$$

Our proposed weighted update rule for opinion formation, as expressed by equation (2), can be rewritten as

$$x_i(t+1) = x_i(t) + \mu w_j(x_j(t) - x_i(t)) = (1 - \mu w_j)x_i(t) + \mu w_j x_j(t) = A\ .x(t), \qquad (7)$$

It is clear that in the above representation, matrix A is a stochastic matrix.

Theorem 1: The product of two stochastic matrixes is a stochastic matrix.

Proof: Let A and B are two stochastic matrices, and $C = A.B$ is their product. Entries of C are multiplications of entries of A and B. Since A and B are stochastic, their elements are non-negative and, thus, the entries of C are also non-negative. The row-sum of C (C_i for i = 1, 2, ..., N) can be obtained as

$$C_i = \sum_j C_{ij} = \sum_{j=1}^{N}\sum_{k=1}^{N}A_{ik}B_{kj} = \sum_{j=1}^{N}A_{ik}\sum_{k=1}^{N}B_{kj} = \sum_{j=1}^{N}A_{ik}{*}1 = 1, \qquad (8)$$

Therefore, C is a matrix with nonnegative entries and row-sums of equal to 1, and thus, it is a stochastic matrix.

Let t_1 and t_2 represent time steps ($t_1 < t_2$) and $B(t_1,t_2) = A(t_1\text{-}1)A(t_1\text{-}2)A(t_1\text{-}3)...A(t_2)$, which models the accumulated weights between time t_1 and t_2 (Hegselmann & Krause 2002). It can be simply shown that for any $r \geq 0$, $1 - r \leq e^{-r}$.

Theorem 2 (Convergence Theorem): Considering opinion update rule (2), suppose we have a matrix $B(t_1,t_2) = [b_{ij}(t_1,t_2)]$, which is a stochastic matrix and models accumulated weights where b_{ij} is an element of matrix B the sequences $0 = t_0 < t_1 < t_2 < \ldots \leq T$ and δ_1, δ_2, ..., δ_i, ... are such that $0 \leq \delta_t \leq 1$ and $\Sigma_{t=0}^{\infty}\delta_t = \infty$. . If $\Sigma_{k=1}^{\infty}min\{b_{ik}(t_m, tm-1), b_{jk}(t_m, t_{m-1})\}\geq\delta_m$ for all $m \geq 1$ and $1 \leq i, j \leq N$, then for any initial condition, there exists a consensus, i.e., $\lim_{t\to\infty}x_i(t) = x^*$ for $i = 1, \ldots, N$.

Proof: see the Appendix section.

Optimization tasks

We applied our optimization procedure on a number of benchmark problems and compared its performance with some well-known methods including genetic algorithms (GA), particle swarm optimization (PSO), differential evolution (DE) and distributed dual averaging (DDA) algorithms. GA has been successfully applied to many optimization problems and was used as a basic paradigm in this work. It starts with a population of some random solutions – denoted by chromosomes. Therefore, the first step is to encode the initial solutions from phenotype to genotype. The objective function is then used for ranking the chromosomes. GA works iteratively and, in each step, uses some operators such as parent selection, recombination or cross over and mutation (Holland 1975). There are a number of parameters that should be tuned in order for a GA to work well. These include crossover probability, mutation probability, population model and parent selection models. Crossover probability P_c indicates the probability of creating a new chromosome from two parents. Mutation probability P_m indicates the portion of the population that undergoes mutation in each iteration of the algorithm.

DE is one of the best-performing evolutionary algorithms frequently used for optimization tasks, which often results in the optimal solution in shorter steps as compared to other optimization algorithms. DE uses the difference of a randomly selected pair of chromosomes – indicating diversity of the population – and adds it to one of the chromosomes in the population. Then, it uses crossover operators such as binomial and exponential crossover to combine the chromosomes (Storn & Price 1997). The parameters of the algorithm are as follows. β is a real value showing the coefficient of the difference between two selected chromosomes and controls the amplification of differential variation. P_r indicates the probability of using the mutant (trial) vector. N_v is an integer number indicating the number of couple chromosomes in calculating the mutant vector.

We also compared our algorithm with PSO that is a well-known optimization algorithm based on swarm intelligent (Kennedy & Eberhart 1995). In this algorithm, there is a population of agents called swarms (or particles) interacting with other agents – like our algorithm. PSO has two components: cognitive and social components. The cognitive component is the experience of each particle while the social component is the experience of the community the agents belong to. PSO has shown high degree of flexibility and acceptable speed in solving many optimization problems. Here we used one of the best extensions of PSO that is PSO with Inertia weights (Eberhart & Shi 2000). This feature plays an important role in balancing the powers of exploration and exploitation and making the algorithm more stable. PSO has a number of control parameters. Let us denote the parameters controlling the cognitive and social power of the algorithm as c_1 and c_2, respectively.

Distributed dual averaging (DDA) algorithm – inspired by Nesterov's dual averaging algorithm (Xiao 2010; Nesterov 2009) – has been proposed for optimizing convex functions (Duchi & Wainwright 2012). Similar to CBO optimization algorithm, DDA is a network-based optimization method in which each node computes sub-differential of a local function while receiving information from its neighboring nodes. There is also a weight matrix to model the weighting process of the method. In any iteration, each node updates its solution vector by multiplying the stochastic weight matrix by the summation of its neighbors' parameters and the sub-gradient of the objective function. DDA is computationally efficient and the convergence time depends on properties of the objective function and underlying network topology. Expander graphs have been proposed as efficient connection topology for DDA optimization algorithm (Duchi & Wainwright 2012). Alternating direction method of multipliers (ADMM) is another optimization method which uses properties of dual decomposition and augmented Lagrangian methods simultaneously (Boyd & Vandenberghe 2004). The Lagrange dual function is obtained by convex conjugate definition and the dual problem is solved using gradient ascent.

Benchmark problems

We evaluated the performance of the proposed optimization strategy on a number of benchmark problems. As the first problem, we considered the following cost function

$$F_1(x) = -e^{-2(\ln(2))\left(\frac{x-0.1}{0.8}\right)^2}.\sin^6(5\pi x), \tag{9}$$

which is a function with many local optima. The optimal point for which the minimum 1 is achieved for this function is at $x^* = 0.1$.

We used a number of multiple competitive functions which have been introduced as benchmarks in optimization problems (Tang et al. 2009). The first function of this type is Shifted Rastrigin's function that is defined as

$$F_2(x) = F_{rastrigin}(x) = \sum_{i=1}^{D}\left[x_i^2 - 10\cos(2\pi x_i) + 10\right], \tag{10}$$

which is a multimodal, shifted, separable and scalable function. The other function of this type considered here is Shifted Ackley's function, which is defined as

$$\begin{aligned} F_3(x) &= F_{ackley}(x) \\ &= -20\exp\left(-0.2\sqrt{\frac{1}{D}\sum_{i=1}^{D}x_i^2}\right) - \exp\left(\frac{1}{D}\sum_{i=1}^{D}\cos(2\pi x_i)\right) + 20 + e. \end{aligned} \tag{11}$$

We also considered Shifted Schwefel's function, defined as

$$F_4(x) = F_{schwefel}(x) = \sum_{i=1}^{n}\left(\sum_{j=1}^{i}x_i\right)^2, \tag{12}$$

and Shifted Elliptic function, defined as

$$F_5(x) = F_{elliptic}(x) = \sum_{i=1}^{D}\left(10^6\right)^{\frac{i-1}{D-1}}x_i^2. \tag{13}$$

In all above functions except F_1, $x \in [-5, 5]^D$. Furthermore, the global optimum - which is $F_2^* = F_3^* = F_4^* = F_5^* = 0$ - is achieved at x^* which is a random and different vector of real numbers in each run.

Network structures

One of the key ingredients of the proposed optimization algorithm is the graph structure used for connecting the agents, which is kept unchanged during the optimization process. In other words, the set of neighbours are not changed for the agents. In this work, we used a number of well-known graph structures including, Erdős-Rényi random, Watts-Strogatz small-world and Barabasi-Albert scale-free networks.

We used the model introduced by Erdős and Rényi for construction of pure random networks (Erdős & Rényi 1960). In this model, N nodes are considered and each pair is connected with probability P. Research showed that real networks are neither random nor regular but somewhere in between; they are indeed small-world. In order to construct small-world networks, we used the original model proposed by Watts and Strogatz, as follows (Watts & Strogatz 1998). Starting with a regular ring graph in which each node is connected to its k-nearest neighbours, each edge is rewired with probability P, provided that self-loops and duplication of edges are prohibited. They showed that for some intermediate values of the rewiring probability P, we obtain a network with low characteristic path length, comparable to that of random networks, and high clustering coefficient (i.e., transitivity) that is much higher than corresponding random networks.

Erdős-Rényi and Watts-Strogatz models result in networks with almost homogeneous degree distribution. However, it was shown that many real networks have heterogeneous degree distribution; there are many low-degree nodes in the network, while a few nodes are hubs with high degrees (Albert & Barabasi 2002; Barabasi & Albert 1999; Barabási 2009). Barabasi and Albert proposed a preferential attachment growth model for constructing such networks, which is used in this work (Barabasi & Albert 1999). The model starts with a $k + 1$ all-to-all connected nodes. In each step, a new node with k links is added to the network. This node tips to the old nodes with probability that is proportional to their degree, i.e., the higher is the degree of an old node in the network, the higher the probability of the making connection with the new node. The model results in scale-free networks whose degree distribution obeys a power-law (Barabasi & Albert 1999).

Results and discussion

In this section, we report the performance of the proposed Consensus-Based Optimization (CBO) method as compared to other tools including DE, GA and PSO. In order to assess the performance of the methods, they were applied to five benchmark functions as expressed by equations (9)-(13). Function F_1 is a simple function consisting of many local optima and with a minimum of $F_1(x^*) = -1$ at $x^* = 0.1$ (Figure 1a). All classic optimization tools can solve such a simple task. We investigated whether or not the proposed CBO algorithm could find the optimum for this function. Figure 1b shows the performance of CBO when Barabasi-Albert (BA), Watts-Strogatz (WS) and Erdős-Rényi (ER) networks were used as interaction topology. We set the network parameters as $N = 1000$ and $k = 4$ for BA model; $N = 1000$, $k = 4$, and $P = 0.1$ for WS model; $N = 1000$ and $P = 0.1$ for ER model. As it is seen, the optimization strategy of CBO is efficient and could find the optimum value in all topologies. For this simple problem, GA resulted in the best performance followed by CBO method. However, in order to better compare the performance of the methods, we applied them on more difficult objective functions.

In order to assess the ability of CBO on solving more complex optimization tasks, we considered the connection graphs with structural properties as above. Figures 2 and 3 show the influence of parameters on the performance of CBO algorithm in optimizing the objective function F_2. μ controls the speed of convergence; we obtained the value of $\mu = 0.6$ as optimal for the optimization task (Figure 2). Large values for μ result in zigzag convergence which is clearly indicated in large variance for such cases. Like

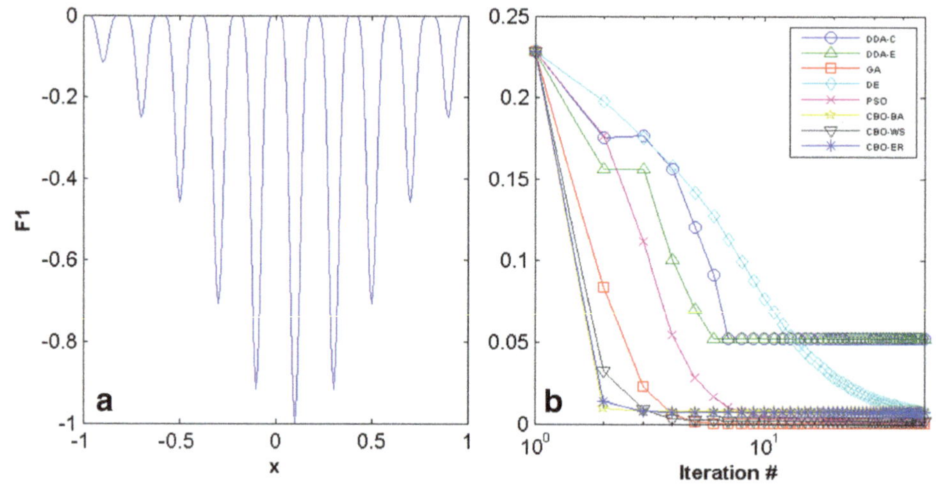

Figure 1 a) Function F_1, b) The optimum value as a function of iteration steps using CBO based on Barabasi-Albert (BA), Watts-Strogatz (WS) and Erdős-Rényi (ER) models, DDA based on Cycle (C) and Expander (E) network model, GA, DE and PSO algorithms. The networks are with N = 1000 and the results show averages over 50 runs.

other optimization algorithms, CBO is sensitive to the number of agents (i.e., network size N), in which as the number of agents increases, the performance gets better (Figure 3). However, increasing the size of the network means increase in the computational complexity of the algorithm. For our simulations, we fixed the network size as $N = 1000$, since for larger networks than this size, the improvement is not significant. We also set the number of agents in other optimization algorithms at $N = 1000$; which makes their results comparable. For the parameters of GA, we set $P_c = 0.85$ and $P_m = 0.05$. Furthermore, we set steady-state form for population model and fitness proportionate

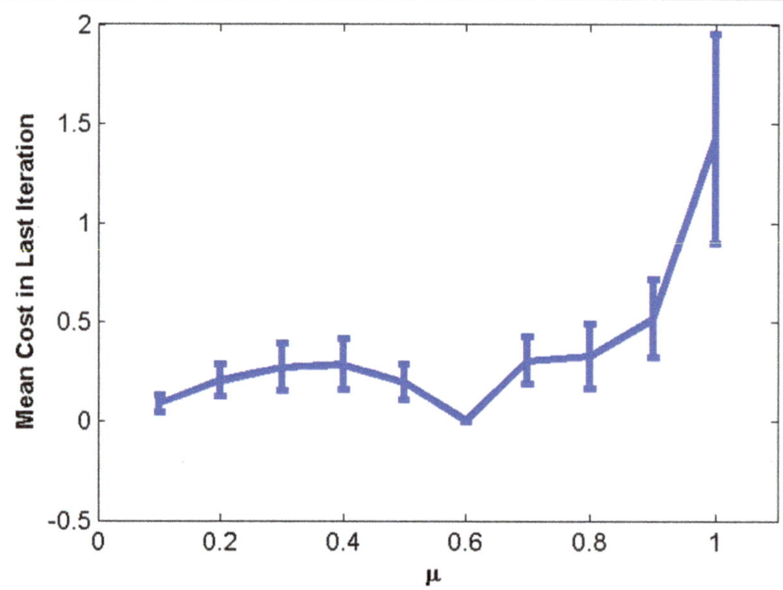

Figure 2 Effect of μ in the performance of CBO; the plot shows the mean and standard error of the objective function F_2 at the end of the convergence (i.e., the last iteration) as a function of μ. The connection graph is Barabasi-Albert (BA) with N = 1000 and data show averages over 50 runs.

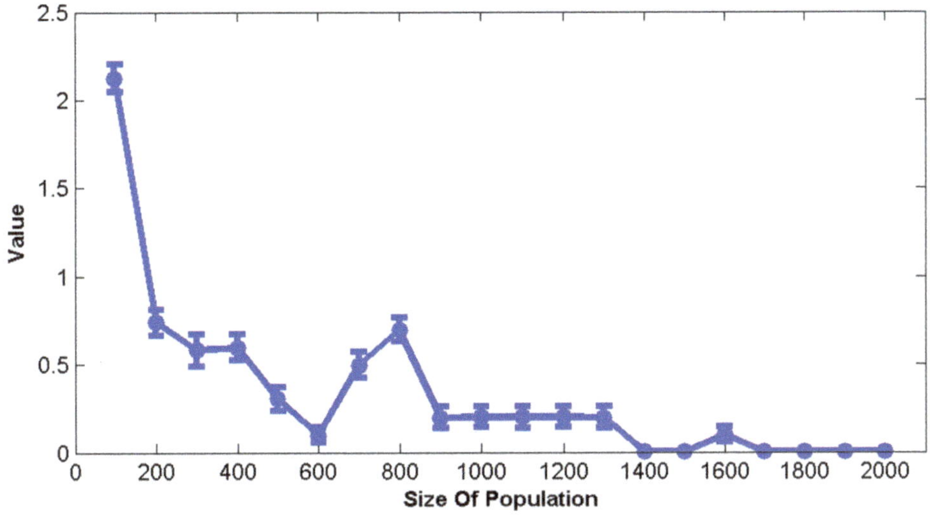

Figure 3 Effect of network size in the performance of CBO; the plot shows the mean and standard error of the objective function F_2 at the end of the convergence (i.e., the last iteration) as a function of the size of Barabasi-Albert (BA) network. Data show averages over 50 runs.

selection policy for parent selection model. The parameters of DE algorithm was set as $\beta = 0.9$, $P_r = 0.9$ and $N_v = 1$ with binomial crossover and random parent selection policy. We set $c_1 = 0.1$ and $c_2 = 0.9$ for PSO resulting in balanced framework for both exploration and exploitation power.

Figures 4,5,6 and 7 show the performance of CBO algorithm using various graph topologies as compared to DE, GA, PSO and DDA methods for optimizing F_2-F_5,

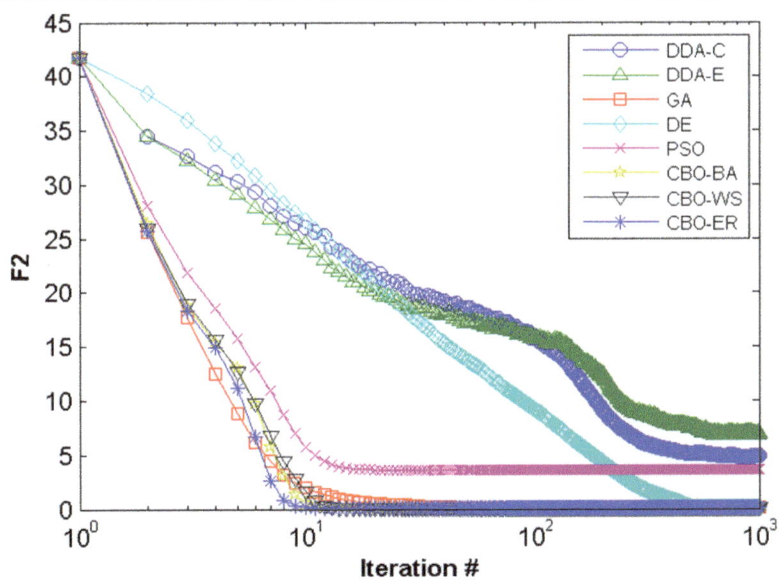

Figure 4 The objective function F_2 as a function of iterations in different optimization algorithms. The optimization methods are distributed dual averaging (DDA) with two network models Cycle (C) and Expander (E), genetic algorithm (GA), differential evolution (DE), particle swarm optimization (PSO) and the proposed consensus-based optimization (CBO) with three schemes for the connection topology between the agents: Barabasi-Albert (BA), Watts-Strogatz (WS), and Erdős-Rényi (ER). All methods have a population of 1000 nodes, the desired function has 5 dimensions and the optimum output is 0. Data show averages over 50 runs.

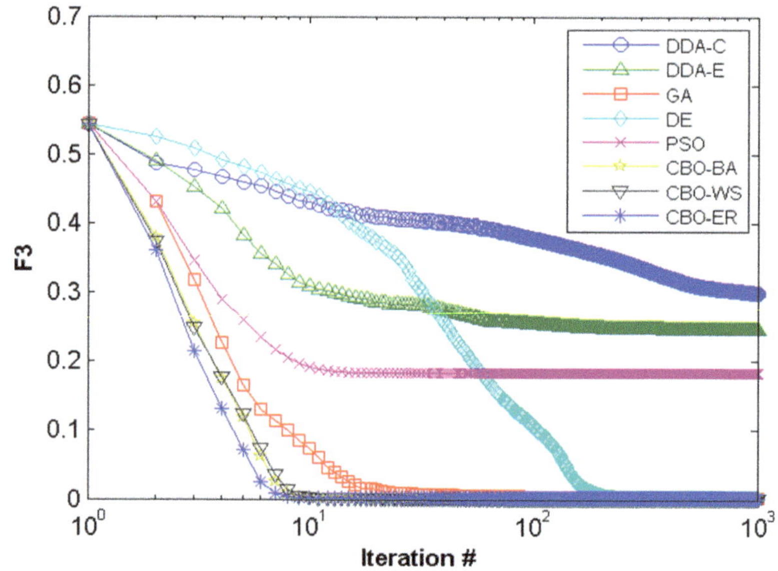

Figure 5 Objective function F_3 as a function of iterations steps averaged over 50 runs. Other designations are as Figure 4.

respectively. The results were averaged over 50 runs. Note that some algorithms could not find the optimum solution in the considered number of iterations in some runs, resulting in a drift from the optimum value in their reported profile (which shows average over 50 runs). Majority of the algorithms could find the optimum; however, they showed significantly different performance. CBO with BA and ER as connection structure among the agents, showed the best performance (i.e., finding the optimum in the least iteration steps) as compared to other algorithms for F_2 (Figure 4). They could find the optimum solution in about 16 steps, which is much less than the steps of GA

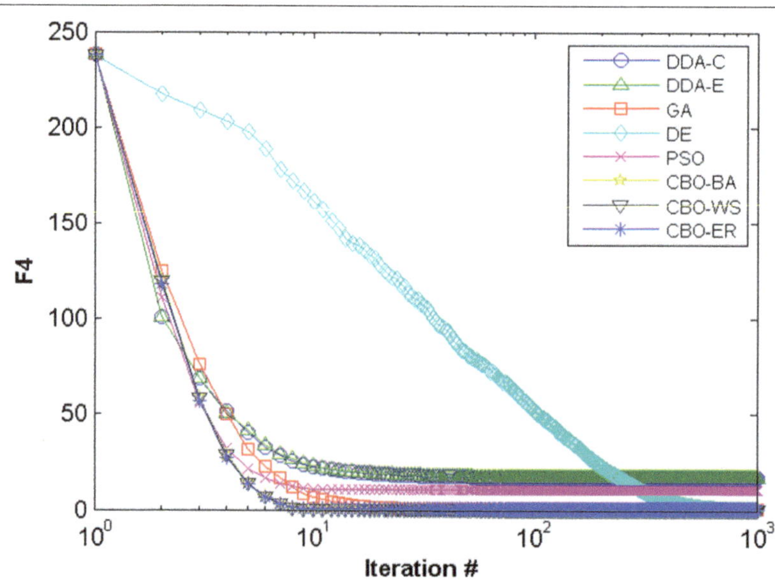

Figure 6 Objective function F_4 as a function of iterations steps averaged over 50 runs. Other designations are as Figure 4.

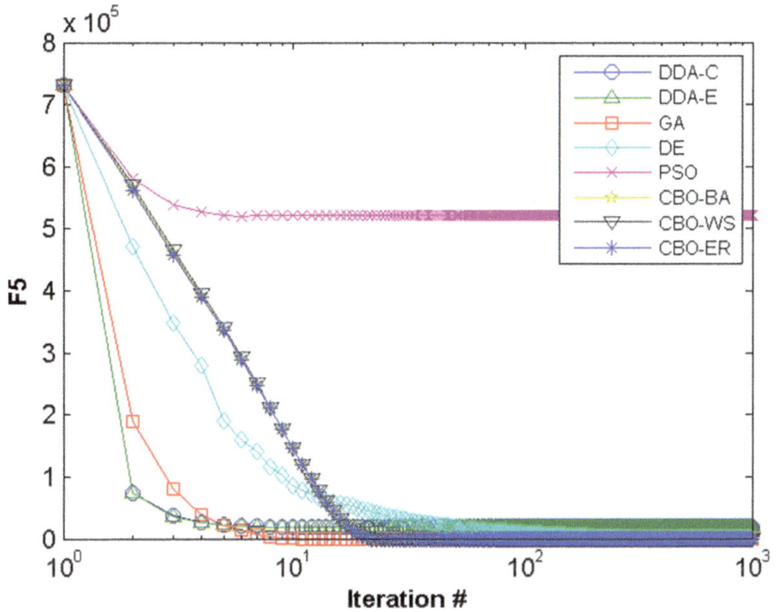

Figure 7 Objective function F_5 as a function of iterations steps averaged over 50 runs. Other designations are as Figure 4.

and DE needed to converge. PSO and CBO with WS topology could not find the solution in some runs resulting in a drift in their performance. Except for F_2, DDA based on Expander network worked better than or equals the one based on Cyclic one. Although DDA has been shown to be well-performing optimizer for convex functions, it did not result in good performance for the objective functions considered in this work. This is due to the fact that these functions are widely non-convex and a method specially designed for convex functions might not properly work on them.

All algorithms could find the optimum solution for objective function F_3 as expressed by equation (11) with the results shown in Figure 5. However, CBO algorithms showed much faster convergence than others. In terms of network topology in CBO, BA and ER topologies worked better than the case when WS model used for constructing interaction topology. CBO was also better than GA, DE, PSO and DDA in optimizing the objective functions F_4 as expressed by equation (12) and F_5 as expressed by equation (13) and the results are shown in Figure 6 and 7, respectively. While, ER topology resulted in a bit faster convergence than BA and WS for F_4, their performance was almost the same for F_5. It is expected that consensus on random networks should be the fastest as compared to the one in BA and WS networks. This is mainly due to the fact that random networks often have shorter average path length compared to other models.

We also compared the computational complexity of these methods by calculating their convergence times for the considered objective functions (Table 1). CBO showed the best performance (i.e., the least number of iterations until convergence) followed by PSO and GA. DE and DDA were the slowest methods among these optimization algorithms.

In sum, our experiments showed that performing an optimization task with a simple consensus network provides the solution with a better performance than a number of

Table 1 Iteration count averaged over 50 runs till convergence happened in optimization methods include distributed dual averaging (DDA), genetic algorithm (GA), differential evolution (DE), particle swarm optimization (PSO) and the proposed consensus-based optimization (CBO)

	DDA	GA	DE	PSO	CBO
F_1	6	6	26	9	**3**
F_2	63	44	535	26	**12**
F_3	13	18	87	13	**9**
F_4	134	49	365	21	**13**
F_5	123	55	121	**24**	31

classic optimization tools including GA, DE, PSO and DDA. Furthermore, we found pure random topologies constructed by Erdős-Rényi model more effective than small-world topologies constructed by Watts-Strogatz model and scale-free topologies obtained through Barabasi-Albert preferential attachment model.

Conclusion

In this paper we introduced a novel application for consensus phenomenon in complex networks. Consensus in networked structures has many applications ranging from engineering (e.g., sensor networks) to sociology (e.g., opinion formation in social networks). In this manuscript, we used network consensus to solve optimization tasks. We considered a number of agents interacting over a networked structure with topology as random, small-world or scale-free. Furthermore, each agent was associated with an opinion value which could change in collaboration with neighboring agents. The agents worked collectively with their friends (which was defined based on the considered network topology and was kept unchanged during the optimization process) to solve an optimization task. To this end, each agent adapted its opinion value based on the best-matching neighbor, i.e., the neighbor with the best value in the objective function. The proposed consensus-based optimization (CBO) method was applied on a number of benchmark problems and its performance was compared with that of a number of classic optimization tools such as genetic algorithms, differential evolution and particle swarm optimization. Our experiments showed that CBO could always find the optimal solution faster and more reliable. We also found Erdős-Rényi random topology better than Watts-Strogatz small-world and Barabasi-Albert scale-free topologies for which it could solve the optimization task faster when used in CBO as connection graph.

Appendix

Proof of the Convergence Theorem: Based on the assumption made in the theorem

$$\sum_{k=1}^{N} \min_{1\le i,j\le N, m\ge 1} \{b_{ik}(t_m, t_{m-1}), b_{jk}(t_m, t_{m-1})\} \ge \delta_m, \tag{14}$$

We have

$$\left(1 - \sum_{k=1}^{N} \min_{1\le i,j\le N, m\ge 1} \{b_{ik}(t_m, t_{m-1}), b_{jk}(t_m, t_{m-1})\}\right). \ d(x(m)) \le (1-\delta_m) \ d(x(m)). \tag{15}$$

Having (15) and the results of Lemma 1, one concludes that

$$\begin{aligned} d(x(m+1)) &= d(B.\ x(m)) \\ &\leq \left(1-\sum_{k=1}^{N} \min_{1\leq i,j\leq n, m\geq 1}\{b_{ik}(t_m, t_{m-1}), b_{jk}(t_m, t_{m-1})\}\right).\ d(x(m)), \\ &\leq (1-\delta_m).\ d(x(m)) \end{aligned} \tag{16}$$

That is

$$d(B.\ x(m)) \leq (1-\delta_m).\ d(x(m)). \tag{17}$$

Having (17) and using the results of Theorem 1, one concludes

$$(1-\delta_m).\ d(x(m)) \leq e^{-\delta_m} d(x(m)). \tag{18}$$

We have

$$\begin{aligned} &e^{-\delta_m} d(x(m)) \leq e^{-\delta_m} e^{-\delta_{m-1}} d(x(m)) d(x(m-1)) \leq \\ &\qquad \mathrm{L} \leq e^{-\delta_m} e^{-\delta_{m-1}} \mathrm{L} e^{-\delta_0} d(x(0)) = e^{-\sum_{i=0}^{\infty} \delta_i} d(x(0)). \end{aligned} \tag{19}$$

Using (17)-(19) and the fact that $\Sigma_{t=0}^{\infty}\delta_t = \infty$, we conclude that

$$d(x(m+1)) = d(B.\ x(m)) \leq e^{-\sum_{i=0}^{\infty} \delta_i} d(x(0)) = 0. \tag{20}$$

Or, in other words

$$\lim_{t\to\infty} d(x(t)) = 0. \tag{21}$$

Since B is a stochastic matrix and $x(t + 1) = B(\mathrm{x}(t),t).x(t)$, we have

$$\min_j x_j(t) \leq\ x_i(t+1) \leq \max_j x_j(t) \quad , \quad i = 1, \ldots, N, \tag{22}$$

Which becomes the following when time approaches to infinity

$$\lim_{t\to\infty} \min_j x_j(t) \leq \lim_{t\to\infty}\ x_i(t+1) \leq \lim_{t\to\infty} \max_j\ x_j(t). \tag{23}$$

Let us denote the lower bound of the opinions by p and their upper bound by q. Then, the above expression becomes

$$p \leq \lim_{t\to\infty} x_i(t+1) \leq q. \tag{24}$$

Now, using equations (21) and (4) we have

$$\lim_{t\to\infty} \max_i x_i(t) - \lim_{t\to\infty} \min_i x_i(t) = p - q = 0 \to p = q. \tag{25}$$

Applying *Squeeze Theorem* (Steinhaus 1938), we have

$$\lim_{t\to\infty} x_i(t+1) = \lim_{t\to\infty} B(x(t), t).x_i(t) = x^* \quad , \quad i = 1, \ldots, N. \tag{26}$$

And this completes the proof.

Competing interests

The authors declare that they have no competing interests.

Authors' contributions

OAS and MJ designed the research and developed the methods. OAS did the numerical simulations. OAS and MJ wrote the manuscript. Both authors read and approved the final manuscript.

Acknowledgment

The authors would like to thank Dr. Hamid Beigy for his insightful comments and discussion on the topic.

References

Ajdari Rad A, Jalili M, Hasler M: **Efficient rewirings for enhancing synchronizability of dynamical networks.** *Chaos* 2008, **18:**037104.

Albert HJR, Barabasi A-L: **"Diameter of the World Wide Web".** *Nature* 1999, **401:**130–131.

Albert R, Barabasi A-L: **Statistical mechanics of complex networks.** *Rev Modern Physics* 2002, **74:**47–97.

Amblard F, Deffuant G: **The role of network topology on extremism propagation with the relative agreement opinion dynamics.** *Physica A* 2004, **343:**725–738.

Barabási AL: **Scale-free networks: a decade and beyond.** *Science* 2009, **325:**412–413.

Barabasi A-L, Albert R: **Emergence of scaling in random networks.** *Science* 1999, **286:**5009–5012.

Barbarossa S, Scutari G: **Decentralized maximum likelihood estimation for sensor networks composed of nonlinearly coupled dynamical systems.** *IEEE Trans Signal Process* 2007, **55:**3456–3470.

Barry CMN, Wentzel KR: **Friend influence on prosocial behavior: The role of motivational factors and friendship characteristics.** *Dev Psychol* 2006, **42:**153–162.

Belykh I, Hasler M, Laurent M, Nijmeijer H: **Synchronization and graph topology.** *Int J Bifurcation Chaos* 2005, **15:**3423–3433.

Berndt TJ, Murphy LM: **Influences of friends and friendships: Myths, truths, and research recommendations.** *Adv Child Dev Behav* 2003, **30:**275–310.

Boccaletti VLS, Moreno Y, Chavez M, Hwang DU: **"Complex networks: structure and dynamics".** *Physics Reports* 2006, **424:**175–308.

Boyd S, Vandenberghe L: *Convex optimization.* Cambridge University Press; 2004.

Brunetti S, Cordasco G, Gargano L, Lodi E, Quattrociocchi W: **"Minimum Weight Dynamo and Fast Opinion Spreading".** In *38th international conference on Graph-Theoretic Concepts in Computer Science.* 2012:249–261.

Carletti T, Fanelli D, Grolli S, Guarino A: **How to make an efficient propaganda.** *Europhysics Letters* 2006, **74:**222–228.

Deffuant G, Neau D, Amblard F, Weisbuch G: **Mixing beliefs among interacting agents.** *Adv Comp Sys* 2001, **3:**87–98.

Duchi AAJC, Wainwright MJ: **Dual averaging for distributed optimization: convergence analysis and network scaling.** *IEEE Trans Automatic Con* 2012, **57:**592–606.

Eberhart RC, Shi Y: **Comparing inertia weights and constriction factors in particle swarm optimization.** *In Proceedings of the Congress on Evolutionary Computation.* 2000, **1:**84–88.

Erdős P, Rényi A: **On the evolution of random graphs.** *Pub Mathematic Ins Hungarian Acad Sci* 1960, **5:**17–61.

Fortunato S, Latora V, Pluchino A, Rapisarda A: **"Vector opinion dynamics in a bounded confidence consensus model".** *Int J Modern Physics C* 2005, **16:**1535–1553.

Gandica Y, Del Castillo-Mussot M, Vázquez GJ, Rojas S: **Continuous opinion model in small-world directed networks.** *Physica A* 2010, **389:**5864–5870.

Girvan M, Newman MEJ: **Community structure in social and biological networks.** *Proc Nat Acad Sci USA* 2002, **99:**7821–7826.

Guo L, Cai X: **Continuous opinion dynamics in complex networks.** *Com Computational Physics* 2009, **5:**1045–1053.

Hegselmann R, Krause U: **Opinion dynamics and bounded confidence models, analysis, and simulation.** *J Art Soc Social Simulation* 2002:5. no. 3.

Holland JH: **Adaptation in Natural and Artificial Systems.** Ann Arbor, MI: The University of Michigan Press; 1975.

Holyst JA, Kacperski K, Schweitzer F: **Social impact models of opinion dynamics.** *Ann Rev Computatio Physics* 2001, **20:**531–535.

Jalili M: **"Effects of leaders and social power on opinion formation in complex networks".** In *Simulation - Transactions of the Society for Modeling and Simulation International.* 2013a. doi:10.1177/0037549712462621.

Jalili M: **Social power and opinion formation in complex networks.** *Physica A* 2013b, **392:**959–966.

Kennedy J, Eberhart R: **"Particle swarm optimization".** *In Proceedings IEEE International Conference Neural Networks* 1995, **4:**1942–1948.

Kozma B, Barrat A: **Consensus formation on adaptive networks.** *Physical Review E* 2008, **77:**016102.

Krapivsky PL, Redner S: **Dynamics of majority rule in two-state interacting spin systems.** *Phys Rev Let* 2003, **90:**238701.

Krause U: **A discrete nonlinear and non-autonomous model of consensus formation.** *Com Diff Eq* 2000, **2000:**227–236.

Kurmyshev E, Juárez HA, González-Silva RA: **Dynamics of bounded confidence opinion in heterogeneous social networks: concord against partial antagonism.** *Physica A* 2011, **390:**2945–2955.

Leskovec J, Adamic LA, Huberman BA: *The dynamics of viral marketing.* ACM Conference on Electronic Commerce. Ann Arbor: Michigan, USA; 2006.

Lewenstein M, Nowak A, Latane B: **Statistical mechanics of social impact.** *Physical Rev A* 1992, **45:**703–716.

Lorenz J: **Continuous opinion dynamics under bounded confidence: a survey.** *Int J Modern Physics C* 2007, **18:**1819–1838.

Lu JH, Yu XH, Chen GR, Cheng DZ: **Characterizing the synchronizability of small-world dynamical networks.** *IEEE Tran Cir Syst I: Fundamental Theory Ad App* 2004, **51**:787–796.
Milo R, Shen-Orr S, Itzkovitz S, Kashtan N, Chklovskii D, Alon U: **Network motifs: simple building blocks of complex networks.** *Science* 2002, **298**:824–827.
Nardini C, Kozma B, Barrat A: **Who's talking first? consensus or lack thereof in coevolving opinion formation models.** *Phys Rev Let* 2008, **100**:158701.
Nesterov Y: **Primal-dual subgradient methods for convex problems.** *Mathematical Programming* 2009, **120**:221–259.
Newman MEJ: **The structure and function of complex networks.** *SIAM Review* 2003, **45**:167–256.
Newman MEJ, Park J: **Why social networks are different from other types of networks.** *Physical Review E* 2003, **68**:036122.
Newman A-LBM, Watts DJ: *The structure and dynamics of networks. 1.* Princeton University Press; 2006.
Perc M: **Evolution of cooperation on scale-free networks subject to error and attack.** *New J Physics* 2009, **11**:033027.
Perc M, Szolnoki A: **Coevolutionary games-A mini review.** *Biosystems* 2010, **99**:109–125.
Scutari G, Barbarossa S, Pescosolido L: **Distributed decision through self-synchronizing sensor networks in the presence of propagation delays and asymmetric channels.** *IEEE Trans Signal Process* 2008, **56**:1667–1684.
Seneta E: *Non-negative matrices and Markov chains. 2nd edition.* Berlin: Springer Verlag, New York; 1981.
Steinhaus H: **A note on the ham sandwich theorem.** *Mathesis Polska* 1938, **9**:26–28.
Storn R, Price K: **Differential evolution - a simple and efficient heuristic for global optimization over continuous spaces.** *J Global Optimization* 1997, **11**:341–359.
Strogatz SH: **Exploring complex networks.** *Nature* 2001, **410**:268–276.
Tang K, Xiaodong L, Suganthan N, Yang PZ, Weise T: **"Benchmark functions for the cec'2010 special session and competition on large-scale global optimization".** *Technical report, University of Science and Technology of China* 2009, **1**:1–23.
Urbig D, Lorenz J, Herzberg H: **Opinion dynamics: the effect of the number of peers met at once.** *J Artific Soc Social Simulation* 2008, **11**:4.
Watts DJ, Dodds PS: **Influentials, networks, and public opinion formation.** *J Consumer Res* 2007, **34**:441–458.
Watts DJ, Strogatz SH: **Collective dynamics of 'small-world' networks.** *Nature* 1998, **393**:440–442.
Weisbuch G: **Bounded confidence and social networks.** *Eur Physic J B* 2004, **38**:339–343.
Xiao L: **Dual averaging methods for regularized stochastic learning and online optimization.** *J Machine Lear Res* 2010, **11**:2543–2596.
Yang H-X, Wu Z-X, Zhou C, Zhou T, Wang B-H: **Effects of social diversity on the emergence of global consensus in opinion dynamics.** *Physical Review E* 2009, **80**:046108.

4

Information theoretical methods for complex network structure reconstruction

Enrique Hernández-Lemus[1,2*] and Jesús M Siqueiros-García[3]

*Correspondence: ehernandez@inmegen.gob.mx
[1]Computational Genomics Department, National Institute of Genomic Medicine, Mexico City, México
[2]Complexity in Systems Biology, Center for Complexity Sciences, National Autonomous University of México, Mexico City, Mexico
Full list of author information is available at the end of the article

Abstract

Purpose: Complex networks seem to be ubiquitous objects in contemporary research, both in the natural and social sciences. An important area of research regarding the applicability and modeling of graph- theoretical-oriented approaches to complex systems, is the probabilistic inference of such networks. There exist different methods and algorithms designed for this purpose, most of them are inspired in statistical mechanics and rely on information theoretical grounds. An important shortcoming for most of these methods, when it comes to disentangle the actual structure of complex networks, is that they fail to distinguish between direct and indirect interactions. Here, we suggest a method to discover and assess for such indirect interactions within the framework of information theory.

Methods: Information-theoretical measures (in particular, Mutual Information) are applied for the probabilistic inference of complex networks. Data Processing Inequality is used to find and assess for direct and indirect interactions impact in complex networks.

Results: We outline the mathematical basis of information-theoretical assessment of complex network structure and discuss some examples of application in the fields of biological systems and social networks.

Conclusions: Information theory provides to the field of complex networks analysis with effective means for structural assessment with a computational burden low enough to be useful in both, Biological and Social network analysis.

Keywords: Complex networks structure, Probabilistic network inference, Feature selection, Information theory

MSC: 94A15, 62B10, 91D30, 05C82

Background

Complex networks, no doubt constitute one of the cornerstones of contemporary research in many branches of science (Barabási 2012; Newman 2003). From systems biology to the study of the role of friendship in the spread of diseases, connections between individuals at different organizational levels are outlined using complex graphs.

Alongside with the statistical analysis of large complex networks, the need for robust methodologies to infer such networks from empirical data has risen. Most of these methods rest on the domain of probabilistic inference and computational learning (Bickel and Doksum 2007; Hernández-Lemus and Rangel-Escareño 2011) and as such, they are subject to expectation errors and asymptotic constraints. Apart from the problem of inferring large networks from noisy data sources (Bansal et al. 2007; de Jong 2002; Hernández-Lemus et al. 2009), complex network researchers in general, are confronted with some subtler structural challenges in network reconstruction. One of such challenges lies in the capacity to assess direct from indirect interactions (Chua et al. 2008; Tresch et al. 2007).

The assessment of direct and indirect interactions may play an important role in understanding network navigability, community structure and information flow, specially when considering the range of influence of given nodes within a network, the strength of the interactions and how these interactions shape the whole correlation structure, as well as the topology and dynamics of the system. For instance, in gene regulatory networks, when it comes to the functional role of subnetworks forming either motifs or pathways, there is an important distinction between genes locally involved in the regulation of a small set of highly specific targets and some other genes that are involved in the transcriptional control of a large number of targets, often by means of a chain of indirect interactions. The first set of genes is responsible for the fine tuning processes involved in environment-specific genomic control, while the second set (known as master regulators) is related with long range, large scale control of genome expression used by the cell mechanisms of growth and proliferation (Baca-López et al. 2012).

In the case of social networks, there is also a growing interest in the role that indirect interactions may play in information and influence flow among nodes (Fowler and Christakis 2007, 2010). In some instances (such as the social epidemiology of obesity) (Fowler and Christakis 2007) it has been shown that indirect connections (i.e. second degree links) within a social network may, under some conditions, exert a greater influence than direct interactions that however shape the global structure of the network. Such structural determination may be one of the keystones to discern between diverse features of influence in social networks such as homophily, social contagion and covariation (Shalizi and Thomas 2011). We may envisage other instances in the realm of complex networks where the distinction between direct and indirect links may be of some relevance, such as ecological or economic-financial networks, etc (Beltrán et al. 2012; Callaway and Howard 2007; Tsatskis 2012).

Methods for direct and indirect interactions assessment found in the reviewed literature were designed *ad hoc* for too dense or too small networks (Nawrath et al. 2010; Yan et al. 2007), and most of them require additional information in order to estimate or tune parameters to differentiate the two kinds of interactions (Chua et al. 2008; Yan et al. 2007). Noteworthy is to say, a great deal of importance was expressed in direct and indirect interactions assessment, regardless the particular field of research (Systems Biology, Economics or Ecology). Most efforts invested in distinguishing between these two kinds of interactions among nodes were either done manually (for instance, see (Beltrán et al. 2012; Callaway and Howard 2007)) or rely on extremely specific issues of the underlying networks (Baldazzi et al. 2010; Nawrath et al. 2010), hence the relevance of the method we introduce in this paper.

We suggest a general method to discover and assess for direct and indirect interactions within the framework of information theory. The method we submit in this paper allows to reconstruct the basic structure of complex networks. Since our method rests on the comparison of Mutual Information (MI) among nodes in a triangle, it is not affected by directionality between links, directionality is detected instead, once the basic structure of the network is already in place. In what follows we will discuss some methods based in probabilistic inference and information theory by means of which researchers may be able to infer and assess complex network structure from the probability distributions of some empirical quantitative features.

Methods

Information theoretical approaches in network inference

Information theory (IT) offers a powerful theoretical foundation that is well-fit to contribute to the development of computational methodologies intended to deal with network inference problems as applied to real data in several branches of complex systems theory (Hernández-Lemus and Rangel-Escareño 2011). IT also provides an analogy with statistical mechanics (SM), that can be useful for inferring network interactions (links) from between-node correlation measures, thus enabling to use (although in a quite non-trivial manner) the huge arsenal of tools of this science. There are, however a number of open questions in the application of IT to the probabilistic complex network inference. The applied algorithms may be able to return *intelligible* models relying on scarce *a priori* information while dealing with a potentially large number of variables. IT methods may also detect non-linear dependencies in highly noisy non-independent probability distributions. The best benchmarking options for such kind of complex network inference, for us, seems to be the use of sequential search algorithms (instead of stochastic search, typically involving the assignment of structures for large constrained datasets, since these procedures have a high computational complexity, even NP-hard" -exponentially large search-space-) and performance measures based on IT, since this makes `feature selection` fast end efficient, and it also provides an easy way to communicate results.

Information theoretical measures have been applied intensively to infer interactions in complex networks, in particular in the field of computational biology (Bansal et al. 2007; de Jong 2002; Fleuret 2004; Hernández-Lemus et al. 2009; Margolin et al. 2006; Peng et al. 2005; van Someren et al. 2002) but also in social network studies (Crowley-Riddey 2009; Dong 2011; Mislove 2009; Mislove et al. 2010; Zhao et al. 2011). A group of correlation measures including mutual information, Markov random fields and Kullback-Liebler divergences, amongst others are considered appropriate to perform probabilistic network inference (Hernández-Lemus and Rangel-Escareño 2011). However, since conditional probabilities obey the so-called *tower property*, a number of *false positives* links may appear as a consequence of indirect correlations (Hernández-Lemus and Rangel-Escareño 2011).

For instance, if node (or agent) A has a high value of conditional correlation (say, mutual information) with node B, and B is also highly correlated with node C, most common algorithms would predict (with a marginal probability p^{ind}) the presence of a (possibly non-existent) link between processes A and C. In order to correct for the presence of indirect links we may implement some methods from IT, such as bounds in the information-theoretical probability measures and the use of the Data Processing Inequality (DPI) (Sehgal et al. 2007). DPI can provide a bound to the extent on which *signal processing* may optimize probabilistic inference. We will discuss these and other ideas in the framework of network inference and structure assessment. We will also discuss some of their implications, and potential applications in the contemporary complex systems scenario.

Some of the essential notions of IT that will be used in this work are: (information-theoretical) entropy, mutual information and other related measures. To do so, let X and Y denote two discrete random variables having the following features:

- Finite alphabet $\mathcal{X}$ and $\mathcal{Y}$ respectively
- Joint probability mass distribution $p(X, Y)$
- Marginal probability mass distributions $p(X)$ and $p(Y)$
- Conditional probability mass distributions $p(X|Y)$ and $p(Y|X)$

Following Shannon (1949), it is possible to define the *information theoretical entropy* H of such distribution as follows

$$H = -K_s \sum_{\nu} p_{\nu}(X) \log p_{\nu}(X) \tag{1}$$

here H is called Shannon-Weaver's entropy, K_s is a constant, useful the determine the units in which entropy is measured (bits, nats, and so on, depending on the base of the log used) and $p_{\nu}(X)$ is the mass probability density for state ν of the random variable given by $X = x$. IT entropy is a measure of the amount of uncertainty associated to the value of X, hence relating the *predictability* of an outcome to the probability distribution. Let us now consider two discrete random variables (Y, X) with a Joint Probability Distribution (JPD) $p(Y, X)$. For these random variables the *joint entropy* $H(Y, X)$ is:

$$H(Y,X) = -\sum_{y \in \mathcal{Y}} \sum_{x \in \mathcal{X}} p(y, x) \log p(y, x) \tag{2}$$

The maximal joint entropy corresponds to independence conditions of the random variables Y and X *i.e.* when the JPD is factorized $p(Y, X) = p(Y)p(X)$. The entropy of the JPD is then just the sum of their respective entropies. An inequality theorem could be stated as an upper bound for the joint entropy:

$$H(Y,X) \leq H(Y) + H(X) \tag{3}$$

Equality holds `iff` X and Y are statistically independent.

Also, given a Conditional Probability Distribution (CPD), the corresponding *conditional entropy* of Y given X is given by:

$$H(Y|X) = -\sum_{y \in \mathcal{Y}} \sum_{x \in \mathcal{X}} p(y, x) \log p(y|x) \tag{4}$$

Conditional entropies measure the uncertainty of a random variable once another one (the conditioner) is known. It can be proved (Cover and Thomas 1991) that:

$$H(Y,X) = H(X) + H(Y|X) \leq H(Y) + H(X) \tag{5}$$

Or:

$$H(Y|X) \leq H(Y) \tag{6}$$

Again, equality holds `iff` X and Y are statistically independent. Equation 6 is useful in the inference/prediction scenario as follows: if Y is a target variable and X is a predictor, adding variables can only decrease the uncertainty on target Y. As it will be shown later, this is essential for network inference when applying IT methods. *Entropy reduction by conditioning* can be accounted if we consider a measure called the *mutual information*, $I(Y, X)$ which is a symmetrical measure (i.e. $I(Y, X) = I(X, Y)$) that is written as:

$$I(Y,X) = H(Y) - H(Y|X) \quad \text{or} \quad I(X,Y) = H(X) - H(X|Y) \tag{7}$$

If we resort to Shannon's definition of entropy (equation 1) (Shannon and Weaver 1949) and substitute it into equation 7 we get:

$$H(Y,X) = -\sum_{y \in \mathcal{Y}} \sum_{x \in \mathcal{X}} p(x,y) \log \frac{p(x,y)}{p(x)p(y)} \tag{8}$$

A comprehensive catalogue of algorithms to calculate diverse information theoretical measures (including mutual information) has been developed for [R], the statistical scientific computing environment (INFOTHEO). We will analyze the special role that MI has in the field of complex networks inference from quantitative feature data. MI has been applied successfully as a measure to infer 2-way interactions in complex networks (quite specially in the field of Gene Regulatory Networks or GRNs) (Andrecut and Kauffman 2006a; Andrecut and Kauffman 2006b; Madni and Andrecut 2007; Margolin et al. 2006). As we have seen, MI quantifies the degree of statistical dependency between two random variables (say α and β). One can see that $MI(\alpha, \beta) = 0$ `iff` α and β are statistically independent.

Hence, if we measure some quantitative-feature of interest ϑ (say expression level of genes in GRNs), by studying its profile (and more specifically the mutual correlation profile for a set of nodes) we may find interactions conforming a network. A pair of agents characterized by feature distributions ϑ_i and ϑ_j for which $MI(\vartheta_i, \vartheta_j) \neq 0$ are said to interact with each other. Since MI is *reparametrization invariant*, one usually calculates the normalized mutual information. In this case $MI(\vartheta_i, \vartheta_j) \in [0, 1]$, $\forall i, j$.

Distinguishing between direct and indirect interactions

With these definitions in mind, let us consider two random variables, X and Y, whose mutual information is $MI(X, Y)$. Now consider a third random variable, Z, that is a (probabilistic) function of Y only. It can be shown that $P_{Z|XY} = P_{Z|Y}$, which in turn implies that $P_{X|YZ} = P_{X|Y}$, as follows from Bayes' theorem.

An information-theoretical theorem called the Data Processing Inequality (DPI) states that Z cannot have more information about X than Y has about X; that is $MI(X;Z) \leq MI(X;Y)$. We can see that $MI(X;Z) = H(X) - H(X|Z) \leq H(X) - H(X|Y,Z) = H(X) - H(X|Y) = MI(X;Y)$. Inequality follows because conditioning on an extra variable (in this case Y as well as Z) can only *decrease* entropy (in a similar way to what occurs in statistical physics when adding constraints to a thermal system), A formal definition of such a theorem would be:

Definition 1. *Three random variables X, Y and Z are said to form a* ***Markov chain*** *(in that order) denoted $X \rightarrow Y \rightarrow Z$, if the conditional distribution of Z depends only on Y and is independent of X. i.e. if we know Y, knowing X doesn't add anything new to what we already know about Z than if we know only Y.*

If X, Y and Z form a Markov chain, then the Joint Probability Distribution can be written as follows:

$$P(X,Y,Z) = P(X)P(Y|X)P(Z|Y) \tag{9}$$

Theorem 1. Data Processing Inequality: If X, Y and Z form a Markov chain, then

$$MI(X;Z) \leq MI(X;Y) \tag{10}$$

Proof. By the chain rule for mutual information we can state that:

$$\begin{aligned} MI(X;Y,Z) &= MI(X;Z) + MI(X;Y|Z) \\ & MI(X;Y) + MI(X;Z|Y) \end{aligned}$$

By the Markov property, since X and Z are independent, given Y, $MI(X;Z|Y) = 0$, then, since $MI(X;Y,Z) \geq 0$ we have: $MI(X;Z) \leq MI(X;Y)$ c.q.d. □

In reference (Margolin et al. 2006), the application of DPI has shown that if nodes ϑ_1 and ϑ_3 interact only through a third node, ϑ_2 within a given network, $MI(\vartheta_1, \vartheta_3) \leq \min[MI(\vartheta_1, \vartheta_2); MI(\vartheta_2, \vartheta_3)]$. Hence, the least of the three MIs values may come from indirect interactions. The proposed algorithm examines each triplet vertex for which all three MIs are measured and compared to some threshold value MI_0. If there is an edge with an MI value below the threshold, then it is removed (see Figure 1). DPI is thus useful to quantify efficiently the dependencies among a large number of nodes. The DPI algorithm is useful in the problem of complex network structure assessment as well, since it eliminates those statistical dependencies that might be of an indirect nature.

In some cases, however, it may happen that the Markov chain structure is not absolutely fulfilled. Say when nodes ϑ_1 and ϑ_3 interact not only through a third node, ϑ_2, but also by means of a direct interaction. Hence ϑ_1 and ϑ_3 may be two-fold connected, in this case pruning-out one of the links may render an inaccurate version of the actual interaction pattern. This scenario can be accounted for by means of establishing a threshold for removing a link, i.e. the link with the lesser MI measure would be removed only if its

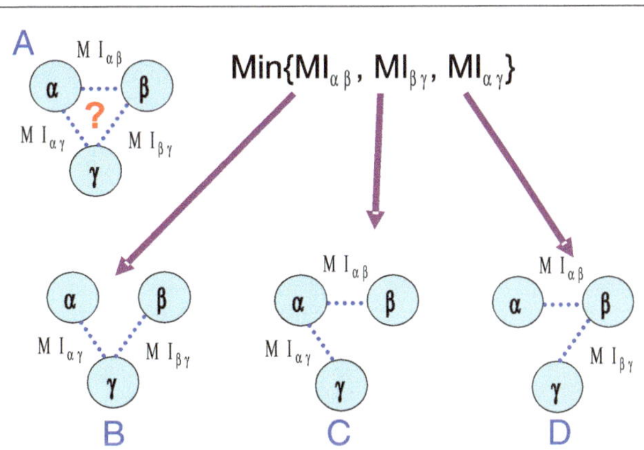

Figure 1 Absolute application of the DPI inequality to assess the direct interaction structure of a three node sub-network. Panel **A** depicts a three node subnetwork composed by nodes α, β and γ. Interactions were inferred by means of mutual information calculations for a selected quantitative feature. The interaction strength between the nodes i, j is given by the respective MI_{ij} $(i, j = \alpha, \beta, \gamma)$ values. Panels **B**, **C**, and **D** result from pruning the subnetwork from the link with the lowest MI_{ij}: $MI_{\alpha,\beta}$ (B), $MI_{\beta,\gamma}$ (C), and $MI_{\alpha,\gamma}$ (D).

value is below a certain threshold (that we hereon call DPI_{tol}) -to be determined in a case-by-case basis by close examination of the network and also by considering its intended applications- and it stays in the network otherwise (Figure 2).

There are similar approaches to the one just presented, for instance the ones in reference (Liang and Wang 2008) and in reference (Zhang et al. 2012). Both approaches are based in *conditional mutual information* (i.e. the degree of information a variable X and a variables Y share, given a third variable (or group of variables) Z). These algorithms try to account for indirect links by means of conditioning the associated mutual information distributions. In reference (Zhang et al. 2012), the authors further improve the performance of the algorithm by using an adaptive computation framework. In addition to these good performance general methods (references (Liang and Wang 2008; Zhang et al. 2012) were developed for gene regulatory networks although with minimal adjustments can ba applied to any other probabilistically inferred networks), there are also more specific approaches based on somehow *Ad Hoc* considerations. We can mention, for instance the MARINa algorithm (Lefebvre et al. 2010; Lefebvre et al. 2007; Mani et al. 2008) developed specifically for the assessment and reconstruction of gene regulatory networks based on *statistical enrichment* of certain signatures (Subramanian et al. 1554), an approach close in philosophy of that of conditioning variables that, however requires for additional information (i.e. the signatures themselves) to be useful, hence is more restricted to its scope and applications as are approaches relying on additional phenotypic information (Wu et al. 2009; Yu et al. 2006).

Data and algorithms

In order to introduce the importance of telling direct network interactions from indirect ones, we performed a topological analysis on the Gene Network of the fruit fly, *Drosophila melanogaster* (D.m.). The fruit fly Gene Network is a paradigmatic system for genetic studies and one of the best annotated organisms in genomics databases. It also presents a high genomic similarity to that of mammals -humans included- (about 61% of disease-associated genes in humans have a D.m. counterpart) and there is open access to its high-throughput inferred biological network (Costello et al. 2009). By discussing

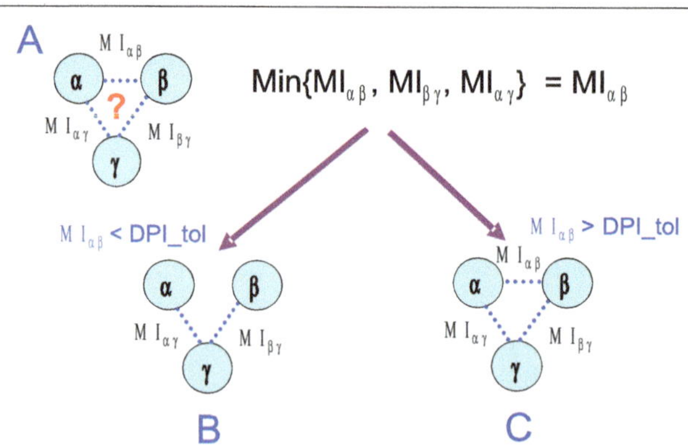

Figure 2 Relative application of the DPI inequality to assess the direct interaction structure of a three node sub-network. Panel **A** depicts a three node subnetwork composed by nodes α, β and γ. Panels **B**, and **C** result from pruning (or not) the subnetwork from the link with the lowest MI_{ij}, in this case $MI_{\alpha,\beta}$. The α-β link is removed only when its respective MI-value is lower than a predetermined threshold DPI_tol and left in the network otherwise.

some features of the network structure of this highly studied species we introduce the problem of finding direct and indirect interactions in complex networks inferred from experimental data. Once this problem has been outlined, we proceed to illustrate how the methods of information theory may be appropriate to distinguish between direct and indirect interactions in order to sketch (at least partially), the network structure on a gene regulatory network inferred from experimental data obtained from 1191 whole genome gene expression experiments in breast tissue from breast cancer patients/controls and on a social network inferred from researchers at Mexico's National Institute of Genomic Medicine coauthorship collaborations data, retrieved from the *PubMed* database.

As explained before, the methods of information theory used here correspond to the implementation of MI calculations and DPI to infer and prune respectively such networks. There is a number of different methods for computing this quantities in the literature (Hernández-Lemus and Rangel-Escareño 2011) and most of them are quite functional and almost equivalent in performance. Here we used the C++ implementation of the `aracne` algorithm (in particular we resort to `aracne 1.0` even if there is a new version `2.0` with an improved algorithmic complexity performance, because version `2.0` uses a bootstrapping method that we have found to be still a little bit unstable) (Margolin et al. 2006) for Biological Networks and `Python` scripts (some customized and others from the `NetworkX` library) for the Social Networks. The `aracne 1.0` algorithm is useful for our purposes since it is based on crystal clear MI calculations (Hernández-Lemus and Rangel-Escareño 2011), it is possible to implement DPI thresholds and its algorithmic complexity and performance are quite good (we have benchmarked `aracne 1.0` against other information-theoretical methodologies such as Information Based-Similarity (`ibs`) and linear correlation predictors in the past (Hernández-Lemus et al. 2009) with very acceptable results). `Cytoscape` and `Python`'s library `NetworkX` were used to depict and analyze the networks (Assenov et al. 2008).

Microarray pre-processing of the data was performed by using the `affy` library in `BioConductor` running under [R] on a 128 Gb RAM 8-Power5+ dual core-processor, symmetric multiprocessing (SMP) unit by IBM. All statistical tests were performed on a Dell Precision Series 16 Gb RAM QuadCore Workstation by using `limma` package in [R]/BioConductor. Information theoretical measure calculations for biological systems were performed by the `aracne` v 1.0 program in the IBM SMP machine. `Python` scripts were used instead for Social network calculations. Graphical depiction and network analyses were performed on a MacBook Pro 8 Gb i7.

The Drosophila melanogaster GRN used to highlight the presence of hierarchical structure was not further used here to demonstrate IT methods of network assessment. The reason for this is that it was inferred (Costello et al. 2009) by using Pearson correlation metric, which is a *linear* measure, thus unable to capture the whole statistical dependency spectrum. Let us recall that for two statistically independent random variables Pearson correlation coefficient is 0. However, the converse is not always true, because Pearson correlation coefficient detects only linear dependencies between two variables. Null Pearson correlation coefficients only implies statistical independence for the special case of *jointly normal* distributions. Since this is not the general case in gene expression distributions, values of linear correlations are not enough to determine statistical dependency (Hernández-Lemus et al. 2009).

Results and discussion

Network assessment in biological networks

In order to introduce the need for complex network assessment of indirect interactions, let us consider the case of the gene regulatory network of the fruit fly (*Drosophila melanogaster*, D.m.), which is one of the best curated biological networks that has been constructed ever. By means of computational integration in a probabilistic model of high throughput data sets from DNA, RNA and protein assays, combined with data mining for individual genes to phenotypes treats, such network has been assembled with the goal of providing a *...meaningful, functional gene network and to draw new and unforeseen connections...*(Costello et al. 2009). D.m. is of course a model organism for too many instances in genetics and genomics. One particular feature that is long known for the genetic structure of D.m., is the fact that gene regulation is largely determined by the action of a relatively small set of molecules that are able to exert transcriptional control in the highly active stages of development and proliferation (Baker 2001; Brennecke et al. 2007; Harrison et al. 2011; Wells 2009). For instance, aspects of regeneration are often regulated by complex mechanisms of activation for several growth regulatory pathways in damaged tissue. It has been proved that every pathway involved is being regulated by the *p53* molecule in D.m., (dP53). dP53 is thus a critical master regulator in the GRN of D.m., in particular with regards to cell growth, proliferation and differentiation. Interestingly enough, the human homolog of this very molecule is known to play a quite important role in most human cancers. dP53 is by no means the only master regulator in drosophila; an extremely important gene that acts as a universal master regulator in D.m. is the molecule called eyeless (*ey*). This gene was first study in relation to eye development (hence the name, larvae with knocked-up eye does not develop eyes). However, the role of *ey* is not restricted to eye development. Abnormal expression of the gene is able to convert other tissues into eye-cells, including legs, wings and antennae. *ey* is also involved in controlling the processes that coordinate differentiation of several cell types in a very precise way in order to develop the fly eye. The *ey*-homolog *Pax6* and homologs of other eye determination genes from D.m., are also required for tissue development in vertebrates. It has been shown that *ey* becomes a master regulator of development deep until later stages by means of *regulation by signaling* through the Notch and EGFR signaling pathways (two outstanding important generalistic signaling pathways) (Baker 2001).

Other master regulator genes are known to play special roles in D.m. Such is the case of the zinc-finger protein *Zelda* (*ZLD*) that plays a key role in transcriptional activation and as a master regulator of genome activation in the earliest stages of D.m. development (Harrison et al. 2011). The ZLD transcription factor bounds to thousands of sites across the genome at all developmental stages of D.m., with relatively small changes in binding between stages. The number and range of ZLD targets - around 2,000 genes have ZLD bound to their promoters and/or enhancers - demonstrate that it plays a major role in maternal-to-zygotic transition activation. Hence Zelda *should* be an important hub in the transcriptional regulation network for D.m. However, probably the most important group of master regulators in the fruit fly GRN is formed by the so-called *piwi* family of transcription factors. The piwi class of genes was identified long ago as encoding regulatory proteins that are responsible for the maintenance of incomplete differentiation in stem cells and also in preserving the stability of cell division rates in germ line cells, a quite important set of processes related again to all stages of development, growth and differentiation (Brennecke et al. 2007). There are plenty of other examples (*Drosophila melanogaster* is known to have some 50 or more master regulators in its GRN (Costello et al. 2009) but we think that these may be sufficient to establish the point that within the GRN for D.m. the presence of such highly connected nodes (hubs) should be evident.

If we refer to Figure 3, in panel A we can see a rendering of the whole D.m., genome GRN (Costello et al. 2009). A clustering algorithm (Bader and Hogue 2003 was used to color-code genes (white-to-red) according to their degree of clustering (red genes are strongly clustered whereas white ones are largely isolated). The relative importance of master regulator genes -that may be measured, for instance through their associated centrality degree- is not evident from the network structure. These genes are also important for biological reasons, since they determine to a large extent the regulation patterns of the entire network. In panel B we present the same network but we color-coded individual nodes according to their degree distribution. Some important *clusters* seem to appear,

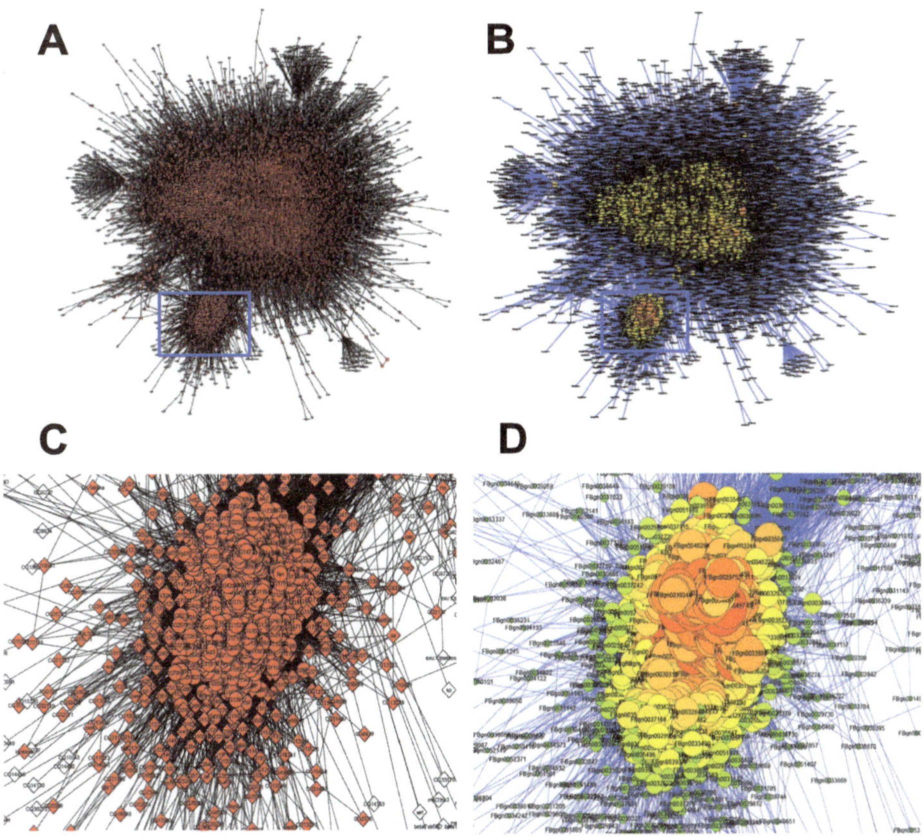

Figure 3 Largest island in the gene regulatory network of drosophila melanogaster. Gene Network constructed from top 200,000 gene pairs (Costello et al. 2009). Panel **A**: Nodes are color coded (white-to-red) according with their MCode scores (Bader and Hogue 2003) for clustering. Red nodes are genes tightly clustered. It can be seen that there is strong clustering between nearly all genes. However, no definite structural features are evident. Panel **B**: Same gene network as in panel A but with node-coloring and sizing determined dy connectivity degree: Big red nodes are highly connected genes whereas small green ones are lowly connected genes. Panels **C** and **D** present zoomed-in renderings of the boxes in panels **A** and **B** respectively. As in panel A, panel C does not permit to observe network structure. Panel **D**, in the other hand, shows the presence of some hyper-connected genes (or hubs, here in red) that are densely linked to highly connected nodes (dark orange) passing through less connected ones (light orange) until reaching barely connected genes (yellow and finally green) suggesting some kind of hierarchical structure in the network *wiring*.

however it is difficult to asses the relevance of individual nodes. In panels C and D we can see zoomed-in renderings of the rectangular regions in panels A and B respectively. As in panel A, panel C is inadequate to highlight the relative importance in the very same aspect as in panel A of individual nodes. In panel D, highly connected genes are highlighted (big red nodes), however the complex entanglement of direct and indirect interactions makes impossible to detect the actual relative importance of these, since they are surrounded by many medium to medium-high connected nodes. This situation is precisely the one calling for a method to assess for direct and indirect interactions. In what follows we will show a proposal for such method, based in the tenets of information theory as applied to both a biological network (a GRN for primary breast cancer) and (in the next subsection) a social network (a scientific collaboration network based in co-authorship probabilities).

Gene regulatory network for primary breast cancer

To show the application of IT methods to infer and assess complex biological networks, let us consider the GRN for primary breast cancer as inferred by means of MI-calculations (Margolin et al. 2006) (see Methods) in the whole genome gene expression levels (for differentially expressed genes between biopsy-captured primary breast cancer and healthy breast tissues as controls) (Baca-López et al. 2012). In Figure 4 we present the aforementioned network, where a complex entangled structure reflecting the intricate regulatory relationships driving the cancer phenotype is displayed. As in the D.m. case, in panel B we show a color- and size-coded (see Figure 4 caption) rendering based in individual degree values for every node. The relative importance of a number of genes becomes more evident, but still is not clear. It is now established that there are some (few) genes in such network acting as master regulators (Baca-López et al. 2012). This is still not evident from the topology/visualization in panel B. If we analyze the network topological structure we can see some reasons behind this. In panel C we plot the degree distribution that show an almost homogeneous behavior for two-plus orders of magnitude in the degree: from genes with a few interactions to nodes with more than a hundred connections. This scenario may be consistent with a highly structured hierarchic structure instead of the dominion of a few master regulator genes. The average clustering coefficient distribution in panel D follows a power-law-like behavior, indicating the relative importance of second neighbor and higher order interactions in the structure of the network as opposed to a neighbor dominated by a few hubs. Panel E displays the Shortest-Path length distribution that turned-out to be a short-tailed bimodal (with maxima at distances 2 and 4), indicating high network navigability.

If we now consider the MI-inference of a GRN network with the same gene expression profiling data as the one in Figure 4, but constrained by the absolute (i.e. $DPI_{tol} = 0$) application of the Data Processing Inequality (as in Figure 1), the network depicted in Figure 5 is generated. Panels A-E are the same as in Figure 4. Panel A highlights the relative importance genes as master regulators. This is even more evident in panel B where the role of *MEF2C* (big red node) and *MNDA* (big dark-orange node) (known master regulators (Baca-López et al. 2012) becomes evident, with important but bounded interrelationships with other genes. Panel C displays the degree distribution, a monotonic-decaying power-law like function highlighting the importance of a few highly connected nodes, whereas the vast majority of genes present low connectivity. Panel D shows the average clustering coefficient distribution that has obviously dropped to zero since all

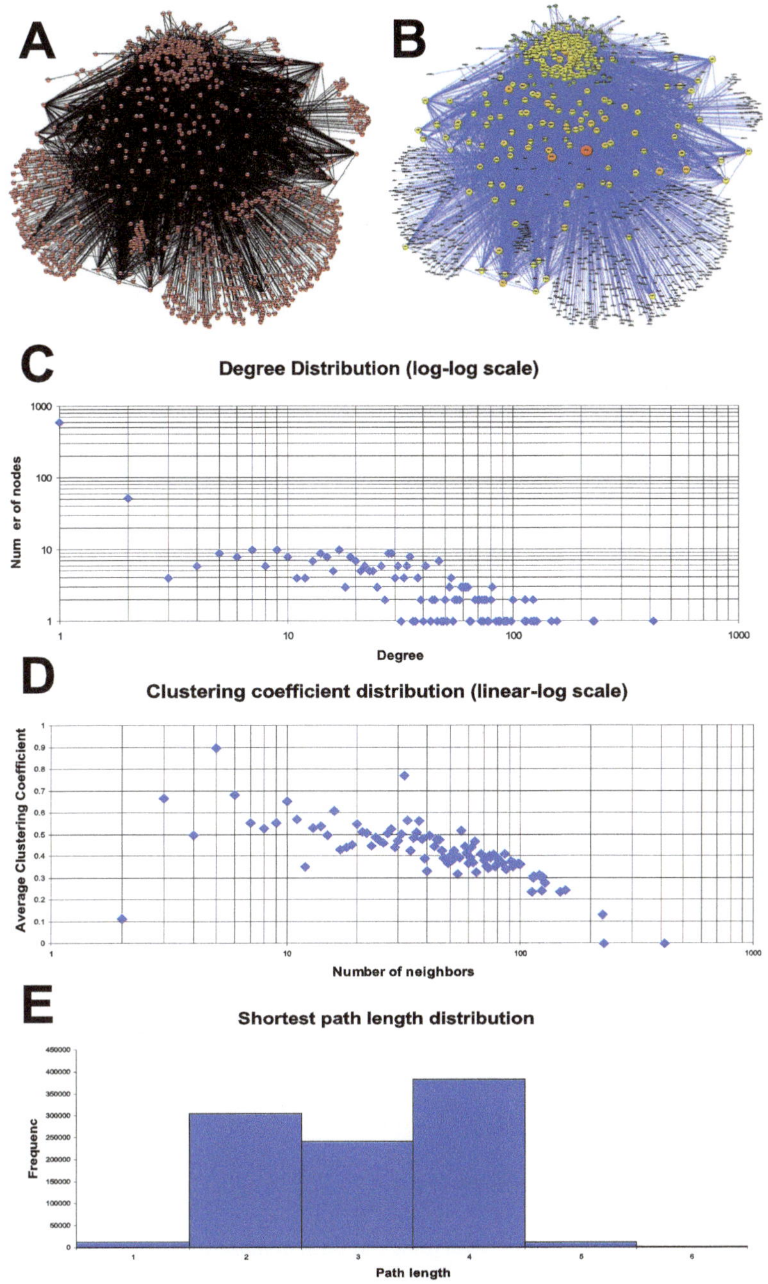

Figure 4 Gene regulatory network associated with proliferation in primary breast cancer. Gene Regulatory Network (GRN) inferred from differential gene expression profiling in 1191 whole genome expression experiments for biopsy samples from breast cancer patients/controls (Baca-López et al. 2012). Panel **A** depicts the associated GRN. The relative importance of highly connected genes is not evident. Panel **B** depicts the same network, nodes are size-coded and color-coded according with their connectivity degree (big red nodes correspond with highly connected genes, whereas small green nodes are lowly connected genes). Some genes apparently stand-out as *relevant*, however the intricate network structure does not permit to tell indirect connections from direct ones, also a number of medium-level connected nodes add complexity to the analysis. In panel **C** we can see the connectivity degree distribution: no definite trend is evident -e.g. a power law, a stretched exponential, etc.- but the distribution remains somehow homogeneous for the range between a few connections and more than a hundred connections. In panel **D** we can see how the average clustering coefficient almost follow a power-law ($R^2 = 0.85$ for the power-law fit) indicating that there may be a hierarchy related with how nodes associate. This could be an indication that a number of not-so-strong interactions are present. Panel **E** presents the short path length distribution that in this case is a bimodal (with maxima at distances of 2 and 4 links) with a short tail.

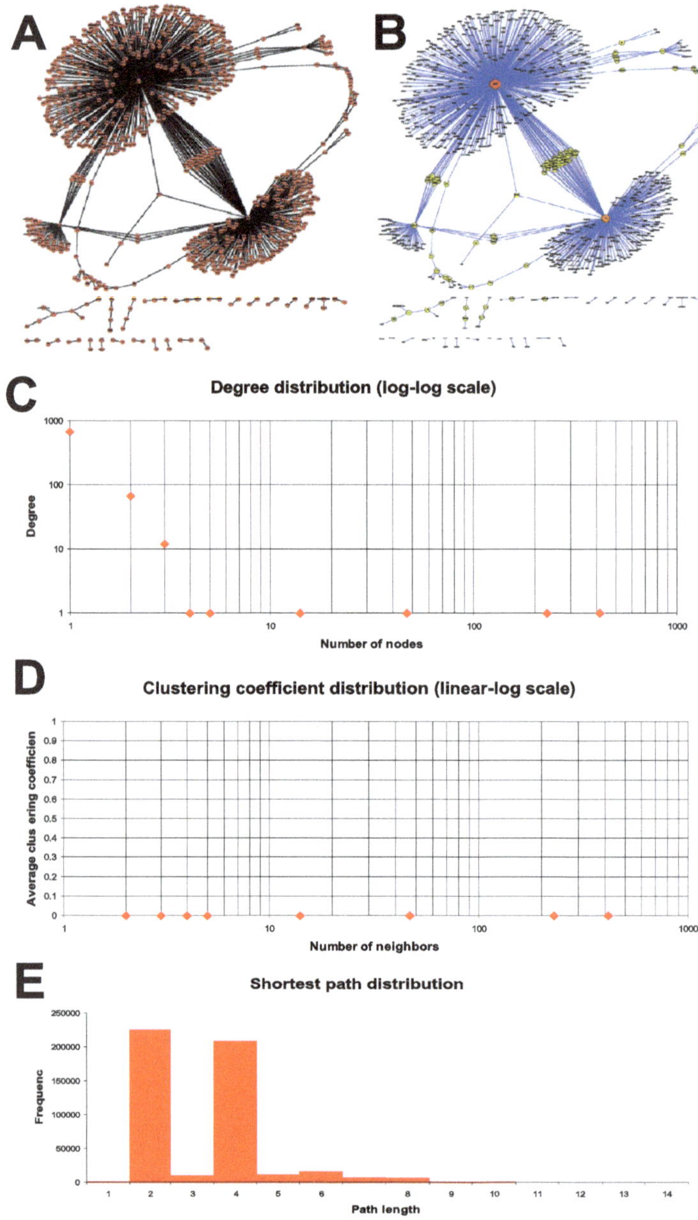

Figure 5 Gene regulatory network associated with proliferation in primary breast cancer after DPI pruning of indirect interactions. Panel **A** depicts the Breast Cancer associated GRN (Baca-López et al. 2012) (same as in Figure 4)after eliminating all indirect interactions by means of the application of the Data Processing Inequality (DPI). The relative importance of highly connected genes is now more evident. The network is basically founded on the role of two major hubs (that are also connected by means of intermediate nodes) and a couple of medium-high connected nodes. Panel **B** depicts the same network, nodes are size-coded and color-coded according with their connectivity degree. Two major hubs appear corresponding to *MEF2C* (red node) and *MNDA* (dark orange node). These two genes have been recognized as *transcriptional master regulators* in breast cancer (Baca-López et al. 2012). In panel **C** we can see the connectivity degree distribution that now resembles a power law distribution (too few nodes, however to have a reliable statistic for the fit) indicating a relatively high importance of few nodes and a low importance for most nodes. In panel **D** we can see how the average clustering coefficient has drop to zero. This is a clear effect of pruning for indirect interactions, since the relative importance of a node is now given in terms of its direct connections and not because of neighbor-by-neighbor influence. Panel **E** presents the short path length distribution that again is a bimodal (with maxima at the same distances of 2 and 4 links) that however shows a somehow larger tail than in Figure 4. This may be due to the fact that navigability in the network was easier in the presence of indirect interactions.

indirect interactions were pruned-out of the network. Interestingly enough, the Shortest-Path length distribution grossly remains a bimodal (with the same maxima at distances 2 and 4) that however exhibits a long tail, a clear indicator of diminished network navigability that results from eliminating shortcuts given by indirect interactions.

Network assessment in social networks

Scientific collaboration networks

In order to test the value of IT methods for community inference in the context of social networks, we describe a Scientific Collaboration Network (SCN from now on) based on the coauthorship history of the researchers of the National Institute of Genomic Medicine of Mexico (referred as INMEGEN). INMEGEN is one of the National Institutes of Health in Mexico and it was created in 2005, being so the second youngest institute of all 13.

The study of SCN dates back to the Erdös Number Project (Cardillo et al. 2006) and due to the accessability of data through the Internet, it has become lately a common place for those interested in social networks (Newman 2004). In this context, a very intuitive way of understanding scientific collaboration is by means of coauthorship, that is, when two scientists have worked together in one or more publications. In order to build the network, we used data of the publications reported by INMEGEN (as retrieved from *PubMed*) from 2005 to the beginning of 2012. The network includes collaborations among researchers of INMEGEN with themselves and with scientists from other institutions, as well as those collaborations between the latter as long as they have also coauthored publications with researchers from INMEGEN within the same network. The edges of the network are weighted according with their corresponding co-authorship probability (the value of every link depends on the strength of collaboration between two scientists -a MI-like function inferred from the number of collaborations with a maximum value of 11 and a minimum of 1-). This network is called *Network 1* (Figure 6). From *Network 1* we obtained a subnetwork (*Network 2* in Figure 7). *Network 2* is also a collaboration network of the researchers of INMEGEN but in which collaborators external to INMEGEN that otherwise are not connected with each other have been left out. Finally, by applying IT methods for network assessment to *Network 2*, and eliminating the weakest link between the nodes of a triangle, we obtained one last network, (*Network 3* shown in Figure 8). From *Network 3* we were able to identify four research groups and we built the collaboration network for each one of them. Groups were defined by the fact that all their members have coauthored at least two papers.

We paid special attention to general values such as number of nodes N, clustering coefficient C, centrality degree, network centralization, and characteristic pathways $\langle l \rangle$. *Network 1* has $N = 847$, a very high clustering coefficient $\langle C \rangle = 0.925$ (see Figure 6C) and a network centralization of 0.221. The characteristic path length $\langle l \rangle$ is 3.09 (Figure 6E), and the average number of neighbors is 24.8. *Network 2* has $N = 535$, a $\langle C \rangle = 0.411$ (Figure 7C), and a network centralization of 0.199. The characteristic path length $\langle l \rangle$ is 3.36 (Figure 7E), and the average number of neighbors is 4.3. *Network 3* displayed the following values, $N = 535$, $\langle C \rangle = 0.341$ (Figure 8C), and a network centralization of 0.197. The characteristic path length $\langle l \rangle$ is 3.43 (Figure 8E), and the average number of neighbors is 4.1.

Due to the shortness of the characteristic path length and its clustering coefficient, the structure of *Network 1* is close to that of a *small-world* network, which is not at all strange

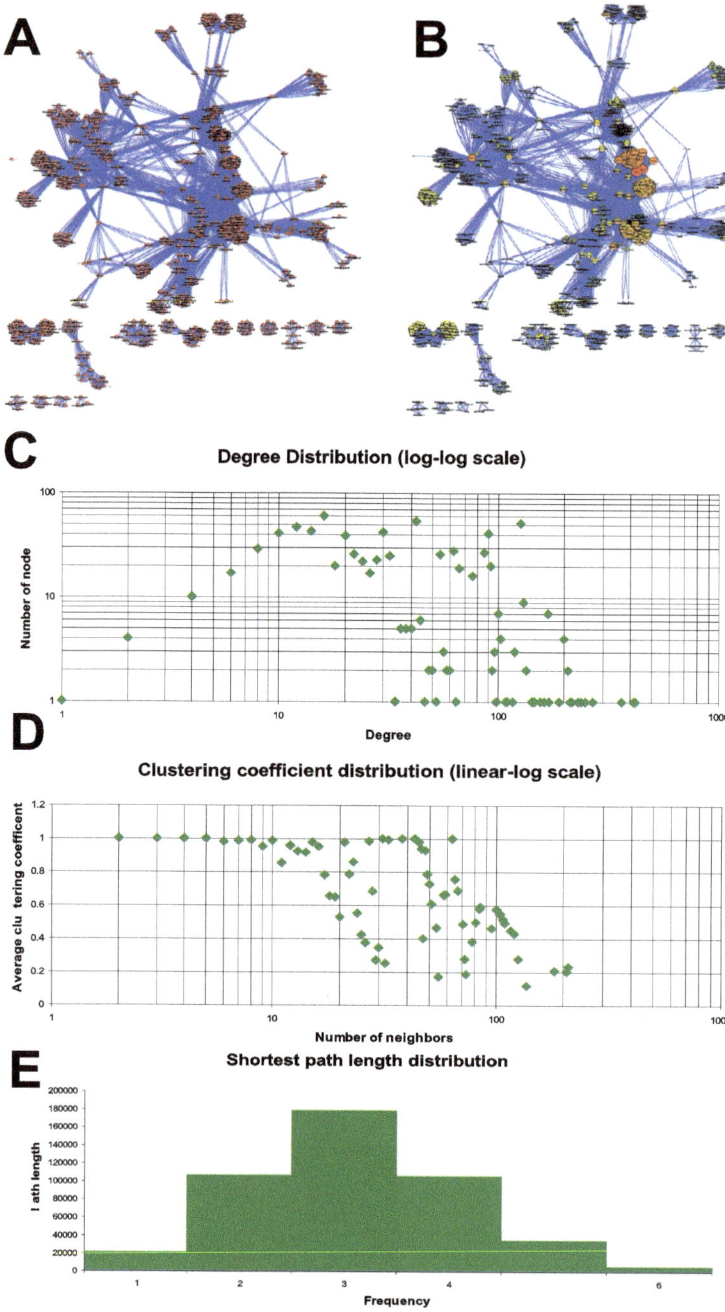

Figure 6 Scientific collaboration network of researchers at the National Institute of Genomic Medicine (INMEGEN) and their extended partners - network 1. Panel **A** presents a SCN that includes collaboration links among researchers of INMEGEN with each other and also with scientists from other institutions, also collaborations between the latter as long as they have also coauthored publications with researchers from INMEGEN within the same network. Panel **B** depicts the same network, nodes are size-coded and color-coded according with their connectivity degree. The presence of a couple of well-connected individuals (bigger red nodes) as well as a number of medium-high connected ones (orange mid-sized nodes) points out to the existence of some kind of hierarchic structure. In panel **C** we can see the connectivity degree distribution that shows a somehow anomalous behavior in the very low degree region, and then displays a typical power-law behavior. This anomaly (a very low number of barely connected nodes) may be due to incidental collaboration. In panel **D** we can see the average clustering coefficient that also presents a *left-hand* tail, most likely also due to incidental collaboration and after this a power-law like behavior. Panel **E** presents the short path length distribution, which is a quasi-symmetric unimodal with an average length of three steps.

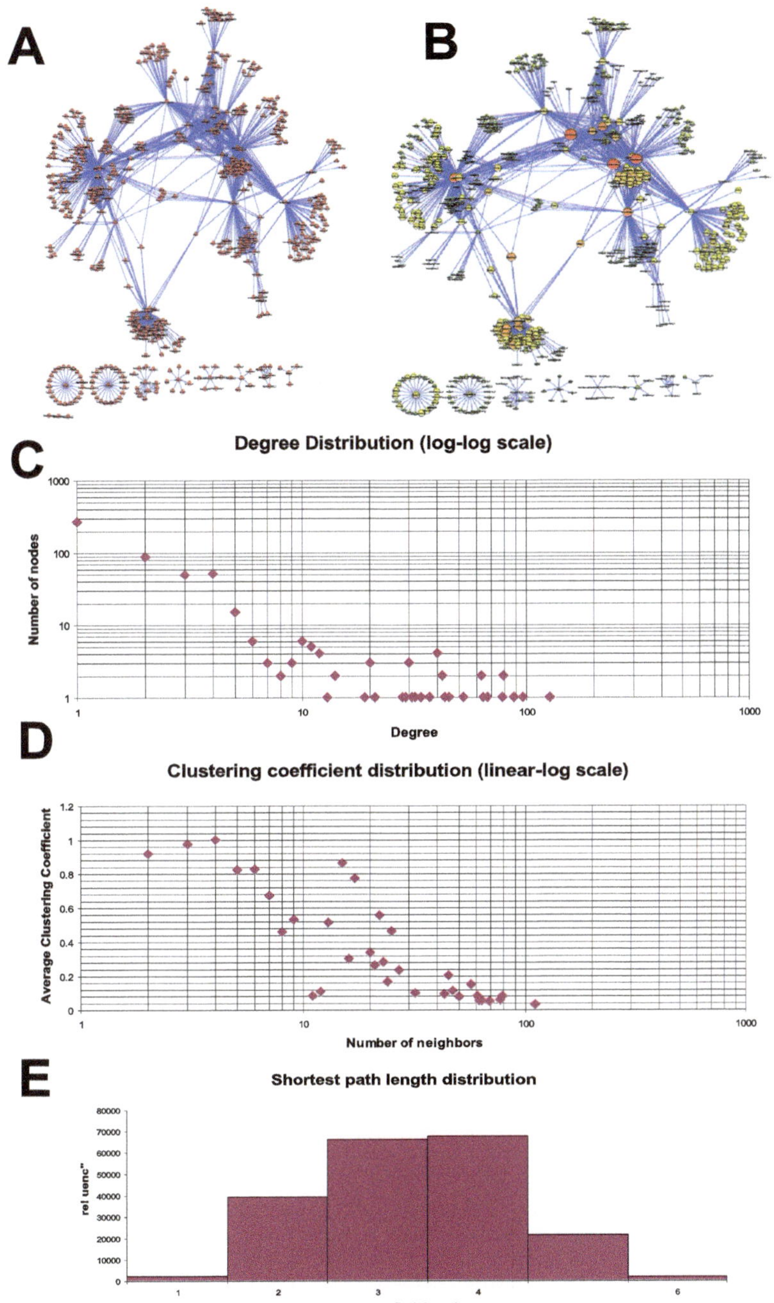

Figure 7 Scientific collaboration network of researchers at the National Institute of Genomic Medicine (INMEGEN) and their extended partners - network 1. Panel **A** presents a SCN that includes collaboration of INMEGEN researchers but in which collaborators external to INMEGEN -otherwise not connected with each other- have been left out. We can see that this network presents a topology resembling that of Network 1 but decimated in the number of links. Panel **B** depicts the same network, nodes are size-coded and color-coded according with their connectivity degree. The presence of a couple of well-connected individuals (bigger red nodes) as well as a number of medium-high connected ones (orange mid-sized nodes) points out to the existence of some kind of hierarchic structure. In this Network, the presence of localized hubs (that we may later identify as group leaders) is more evident. In panel **C** we can see the connectivity degree distribution that shows a power-law behavior with no further appearance of the incidental collaboration anomaly. In panel **D** we can see the average clustering coefficient displaying again a power-law like behavior. Panel **E** presents the short path length distribution, which is also a quasi-symmetric unimodal but with a less-defined expected value (between three and four steps) for the separation length.

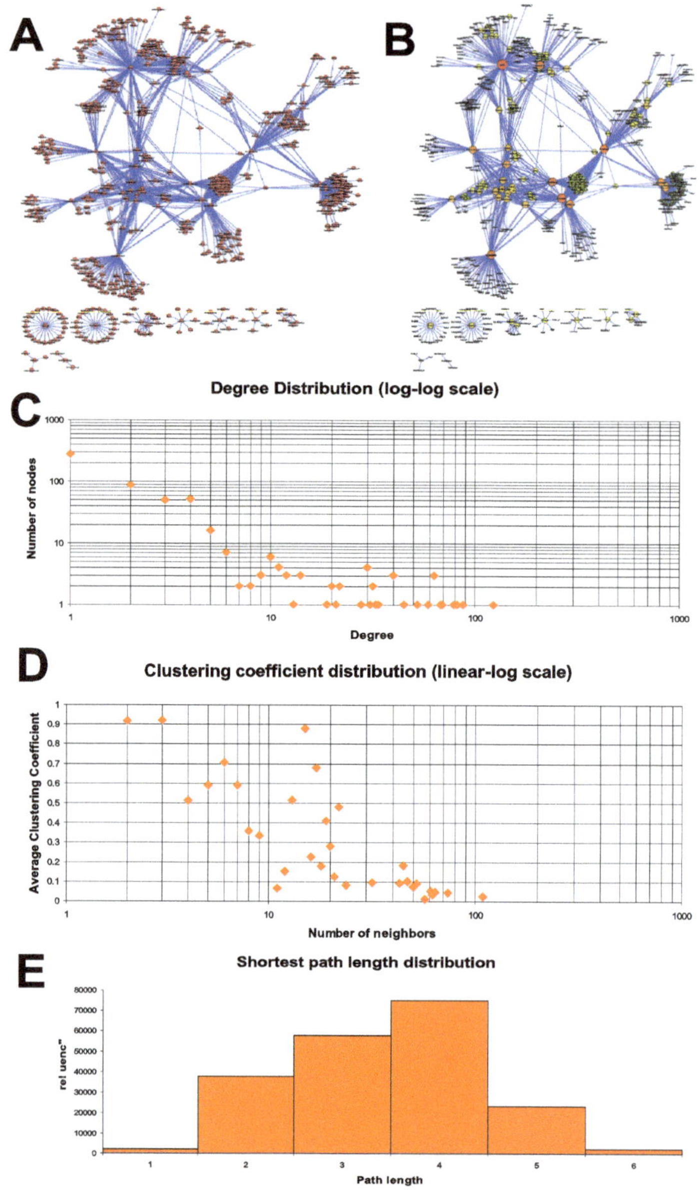

Figure 8 Scientific collaboration network of researchers at INMEGEN after DPI pruning of indirect interactions - network 3. Panel **A** presents a SCN similar to Network 2 but indirect interactions have been pruned-out, (same as in Figure 7) after eliminating all indirect interactions by means of the application of the DPI. Network 3 is similar to Network 2, i.e. the effect of eliminating *indirect* interactions has a small impact. In many triadic interactions (triangles in the network) two-out of three interactions presented the same value so that no edge was eliminated. Panel **B** depicts the same network, nodes are size and color-coded according to their connectivity degree. The presence of a couple of well-connected individuals (big red nodes) as well as a number of medium-high connected ones (orange mid-sized nodes) points out to some kind of hierarchic structure. Research groups are easily identified, yet a certain degree of collaboration between most large groups is found. In panel **C** the connectivity degree distribution shows a power-law like behavior but with a smaller high-degree tail than in Network 2. In panel **D** the average clustering coefficient displays again a power-law like behavior. The reason for finite (non-zero) clustering coefficients is that a number of triangles remained in the network for two of its edges presented equal strength of interaction. Panel **E** presents the short path length distribution, which is unimodal but with a sharp expected separation length of 4. It is noticeable that eventhough DPI-pruning did not eliminated all the triadic connections in the network, the effect of network assessment to knock-out clear indirect interactions affected network navigability (average path length went from a clear value of three in Network 1, to between three and four in Network 2 and then to a clear value of three in Network 3).

to collaboration networks and to SCN, as has been noted before (Yousefi-Nooraie et al. 2008). When compared to *Network 2*, its clustering coefficient decreases abruptly compared to its value in *Network 1*, and the characteristic path length increased slightly. Given the difference between the clustering coefficients of these two networks, it lead us to suggest that external collaboration plays a great deal in defining the small-world structure of INMEGEN's collaboration network, and most of all, it gives it a closure that makes it easily navigable. By closure we mean that INMEGEN, being such a young institution, it may still depend, to a degree, on external collaborators. Such dependency might be fueled by the fact that INMEGEN doesn't provide clinical services which means that it doesn't have the possibility to systematically recruit research subjects from where to get biological samples, depending so, on other institutions for this purpose. This idea still remains to be tested and is part of our future work.

The differences between *Network 2* and *Network 3*, are minimal. *Network 3* was inferred by IT Methods for network assessment, even though, its clustering coefficient remained almost unaffected -decreased from $C=0.411$ in *Network 2*, to $C=0.341$ in *Network 3*- and the characteristic path length was also basically the same. This is an important result because it means that most of the clustering was due to external collaboration as it was noted already, and for *Networks 2* and *3*, triangle formation depends on having at least two nodes from INMEGEN.

IT Methods applied to *Network 2*, eliminated what was left from the 'unbalanced' triangles and the weakest link between two nodes in a triangle was deleted, generating *Network 3*. This method made possible to identify more clearly the existence of four communities that we could recognize as INMEGEN's main research groups, with a leader (laboratory or researcher leader) easily detectable. It is important to mention that there were other, rather small groups, that have not been articulated to the main network and with a small weight value in their collaborations links. These are hypothesized to be groups in process of consolidation.

In order to analyze the structure of the four groups, we created a subnetwork for each one, based on researchers that have collaborated in at least two publications. The subnetworks made also possible to recognize different collaborative strategies among them. Group 1 and 2, with an $N = 79$ and $N = 139$ respectively, were very similar, both displayed a high value for network centralization, over 0.750 and low clustering coefficient, having Group 2 the highest $\langle C \rangle = 0.356$. For these groups, there was a leader with a very high k, and one or two very close collaborators. The difference between first and second order collaborators was large, with the highest weight (corresponding approximately to $k = 9$) between the leader and its first order collaborator and a lesser weight (implying $k = 2$) with the lowest, with an average connectivity difference of 7. The structure of these two networks could be interpreted as having a tendency towards a star-shaped structure.

In group 3, with $N= 106$, network centralization was not as high as in groups 1 and 2, although it remained important (Network centralization= 0.681). Noteworthy was the clustering coefficient over $\langle C \rangle = 0.500$. According to these numbers, group 3 behaves more like a group and the weight of collaborations is similar among its members.

Group 4 had the highest network centralization 0.938 as well as the highest clustering coefficient $\langle C \rangle = 0.746$. The network of group 4 depends on two members of INMEGEN on which all collaborations are fixed. The situation for group 4 is that from all, this is the extreme case of group behavior since there is a minimal connectivity degree difference of 1 between first and second order collaborators, that is, for most nodes, the weight of collaboration among each other is the same (weight 2) and only few have a weight of 1, and none of weight 3 or higher. It is important to mention that this group is different from the other three because is the newest and the smallest ($N = 64$).

We created these networks -i.e. *Networks 1, 2, and 3*- in such a way because we wanted to have a complete image of the collaborative network of INMEGEN, and to assess INMEGEN's internal community structure. Overall, there are some emerging groups and senior researchers are still quite scarce. This may be the consequence of genomics being a new field (as compared, for instance to other biomedical disciplines), and INMEGEN is still in its infancy. If this is true, we will be able to see it in our future work when we compared INMEGEN's SCN with the networks of other National Institutes of Health, some of them with more than 50 years of history. We would also like to model the collaboration strategies followed by the different groups using agent based modeling.

Conclusions

When reconstructing the basic structure of a network, to being able to assess direct and indirect interactions among nodes can be very useful and informative. One important aspect when analyzing complex networks is to being able to distinguish and assess direct from indirect interactions (even more, with regards to local interaction levels that may shape the large scale structure and the functional features of such networks). In this paper we have shown how an application of simple theorems of information theory (Hernández-Lemus and Rangel-Escareño 2011) makes possible for researchers to grasp the nature of the links in a clear-cut way. We discuss this in the context of both biological and social networks. However, we believe that this same general method arguments may apply to any network inferred by means of mutual information measures, and to some extent to other networks inferred by other quantitative interaction measures. What is more, such a general method may serve to the general purpose of unveiling similarities and differences between networks that are as different as it is a transcriptional network from a collaborative one. It is important to recall that DPI-pruned networks may be considered along with non-DPI pruned (and even with relatives degrees of pruning as given by different values of DPI_{tol}) in order to assess for structural features of the network. This is specially relevant if one is to consider the *de-Novo* functional discovery of the role of specific individuals or the reassurance of already envisioned hypothesis along the same lines.

The material presented here is intended to show some useful features of the application of IT methods for network inference and assessment. The biological and social networks chosen here were of different nature: different in size and, more important, in structure. This may be seen by the fact that DPI-pruning of the biological network (that was much larger than the social one) resulted in a tree-like structure (with no triangles and hence with a null clustering coefficient), whereas in the social network, DPI-pruning uncover different structural properties due to the presence of undecidable cases where three edges were equally weighted. However, in both cases one important outcome of direct-vs-indirect interaction analysis resulted, i.e. the presence (or better the *visibility* of the presence) of *key players*: in one case transcription factor genes acting as master regulators and in the other researchers identified as *group leaders*. Both studies revealed some surprising elements -like the presence of novel master regulators, or some *in-consolidation* researchers that are really emerging as *emerging group leaders*- given the structure of the direct interaction networks.

The study of the relationship between topological structure and functional organization in complex networks is, of course, still in a very early stage of development. The problems and challenges that arise include not only the determination of direct and indirect interactions and the role that some privileged nodes may play in network structure and navigability. A thorough study may also considered the issues of community structure and local connectedness. In the case of networks probabilistically inferred from experimental data one may also take into account the role that inference errors, noise and asymptotics may play. Further on, since complex networks are often dynamically adaptive systems, driven both by their inner structure and their environmental constraints, the role that dynamic evolution, fluctuations and adaptability may play will most surely also be fundamental to understand their behavior. We can only envisage what lies in the future: too many challenges should still be tackled before reaching a complete understanding of the behavior complex networks. However, we believe that information theoretical concepts and tools may play a fundamental role when facing such complexities.

Competing interests

The author declare that he have no competing interests.

Authors' contributions

EHL conceived the study, EHL and JMSG performed simulations, analyses and calculations. EHL and JMSG wrote the paper. All authors read and approved the final manuscript.

Authors' information

EHL is a theoretical physicist trained in chemical physics, probability theory and non-equilibrium statistical mechanics; his main area of research is the interface of systems biology, statistical physics and complex system's theory. JSG is an anthropolgist/ethnologist trained in theoretical biology and philosophy; his current research is in social networks; science, technology and society- studies; and genome-bioethics.

Acknowledgements

The authors gratefully acknowledge support by grants: 179431/2012 (CONACYT) and PIUTE10-92 (ICyT-DF) [Contract 281-2010], as well as federal funding from the National Institute of Genomic Medicine (México).

Author details

[1]Computational Genomics Department, National Institute of Genomic Medicine, Mexico City, México. [2]Complexity in Systems Biology, Center for Complexity Sciences, National Autonomous University of México, Mexico City, Mexico. [3]Ethics, Legal and Social Studies Department, National Institute Genomic Medicine, Mexico City, México.

References

Andrecut, M, Kauffman SA: **A simple method for reverse engineering causal networks.** *J Phys Math Gen* 2006, **39:**L647–L655.

Andrecut, M, Kauffman SA: **Mean-field model of genetic regulatory networks.** *New J Phys* 2006, **8:**148.

Assenov, Y, Ramírez F, Schelhorn S, Lengauer T, Albrecht M: **Computing topological parameters of biological networks.** *Bioinformatics* 2008, **24:**282–284.

Barabási, AL: **The network takeover.** *Nat Phys* 2012, **8:**14–16.

Baca-López, K, Hidalgo-Miranda A, Mayorga M, Gutiérrez-Nájera N, Hernández-Lemus E: **The role of master regulators in the metabolic/transcriptional coupling in breast carcinomas.** *PLoS ONE* 2012, **7**(8):e42678.

Bader, GD, Hogue CW: **An automated method for finding molecular complexes in large protein interaction networks.** *BMC Bioinformatics* 2003, **4**(1):2.

Baker, NE: **Master regulatory genes; telling them what to do.** *Bioessays* 2001, **23**(9):763–766.

Baldazzi, V, Ropers D, Markowicz Y, Kahn D, Geiselmann J, de Jong H: **The carbon assimilation network in Escherichia coli is densely connected and largely sign-determined by directions of metabolic fluxes.** *PLoS Comput Biol* 2010, **6**(6):e1000812. doi:10.1371/journal.pcbi.1000812.

Bansal, M, Belcastro V, Ambesi-Impiombato A, di Bernardo D: **How to infer gene networks from expression profiles.** *Mol Syst Biol* 2007, **3:**78.

Beltrán, E, Valiente-Banuet A, Verdú M: **Trait divergence and indirect interactions allow facilitation of congeneric species.** *Ann Bot* 2012, **110:**1369–1376.

Bickel, PJ, Doksum KA: *Mathematical Statistics: Basic, Ideas and Selected Topics* Vol. 1. 2nd. New Jersey: Pearson-Prentice Hall; 2007.

Brennecke, J, Aravin AA, Stark A, Dus M, Kellis M, Sachidanandam R, Hannon GJ: **Discrete small RNA-generating loci as master regulators of transposon activity in Drosophila.** *Cell* 2007, **128**(6):1089–1103.

Cardillo, A, Scellato S, Latora V: **A topological analysis of scientific coauthorship networks.** *Physica A* 2006, **372**(2):333–339.

Chua, HN, Ning K, Sung W-K, Leong HW, Wong L: **Using indirect protein-protein interactions for protein complex prediction.** *J Bioinform Comput Biol* 2008, **6**(3):435–466.

Callaway, RM, Howard TG: **Competitive networks, indirect interactions, and allelopathy: a microbial viewpoint on plant communities.** *Prog Bot* 2007, **68:**317–335.

Costello, JC, Dalkilic MM, Beason SM, Gehlhausen JR, Patwardhan R, Middha S, Eads BD, Andrews JR: **Gene networks in Drosophila melanogaster: integrating experimental data to predict gene function.** *Genome Biol* 2009, **10**(9):R97.

Cover, TM, Thomas JA: *Elements of, Information Theory.* New York: John Wiley & Sons; 1991.

Crowley-Riddey, L: **An information-theoretic approach to finding community structure in networks.** *Ph.D. Dissertation* 2009. Trinity College Dublin, School of Mathematics.

de Jong, H: **Modelling and simulation of genetic regulatory systems: a literature review.** *J Comp Biol* 2002, **9**(1):67–103.

Dong, W: **Mutual Information: inferring tie strength and proximity in bipartite social network data with non-metric associations.** 2011. M.Sc. Dissertation, University of Illinois at Urbana-Champaign, USA.

Fowler, JH, Christakis NA: **The spread of obesity in a large social network over 32 Years.** *N Engl J Med* 2007, **357**(4):370–379.

Fowler, JH, Christakis NA: **Cooperative behavior cascades in human social networks.** *Proc Natl Acad Sci USA* 2010, **107**(12):5334–5338.

Fleuret, F: **Fast binary feature selection with conditional mutual information.** *J Mach Learn Res* 2004, **5:**1531–1555.

Harrison, MM, Li X-Y, Kaplan T, Botchan MR, Eisen MB: **Zelda binding in the early drosophila melanogaster embryo marks regions subsequently activated at the maternal-to-zygotic transition.** *PLoS Genet* 2011, **7**(10):e1002266.

Hernández-Lemus, E, Rangel-Escareño C: **The role of information theory in gene regulatory network inference.** In *Information Theory: New Research.* Edited by Deloumeaux, P, Gorzalka JD: Mathematics Research, Developments Series, Nova Publishing; 2011:109–144.

Hernández-Lemus, E, Velázquez-Fernández D, Estrada-Gil JK, Silva-Zolezzi I, Herrera-Hernández MF, Jiménez-Sánchez G: **Information theoretical methods to deconvolute genetic regulatory networks applied to thyroid neoplasms.** *Physica A* 2009, **388:**5057–5069.

INFOTHEO: A collection of information theoretical tools based on several entropy estimators. [http://cran.r-project.org/web/packages/infotheo/index.html] Published 2012-12-10, Last accesed 2013-03-17.

Lefebvre, C, Lim WK, Basso K, Dalla-Favera R, Califano A: **A context-specific network of protein-DNA and protein-protein interactions reveals new regulatory motifs in human B cells.** *Lect Notes Bioinform* 2007, **4532:**42–56.

Lefebvre, C, Rajbhandari P, Alvarez MJ, Bandaru P, Lim WK, Sato M, Wang K, Sumazin P, Kustagi M, Bisikirska BC, Basso K, Beltrao P, Krogan N, Gautier J, Dalla-Favera R, Califano A: **A human B-cell interactome identifies MYB and FOXM1 as master regulators of proliferation in germinal centers.** *Mol Syst Biol* 2010, **6:**377. doi:10.1038/msb.2010.31.

Liang, KC, Wang X: **Gene regulatory network reconstruction using conditional mutual information.** *EURASIP J Bioinform Syst Biol* 2008:253894. doi:10.1155/2008/253894.

Madni, AM, Andrecut M: **Design and implementation of a gene network reverse engineering method based on mutual information.** *J Integr Des Process Sci* 2007, **11**(3):55–68.

Mani, KM, Lefebvre C, Wang K, Lim WK, Basso K, Dalla-Favera R, Califano A: **A systems biology approach to prediction of oncogenes and molecular perturbation targets in B-cell lymphomas.** *Mol Syst Biol* 2008, **4:**169.

Margolin, AA, Nemenman I, Basso K, Wiggins C, Stolovitzky G, Dalla Favera R, Califano A: **ARACNe: An Algorithm for the reconstruction of gene regulatory networks in a mammalian cellular context.** *BMC Bioinformatics* 2006, **7**(Suppl I):S7.

Mislove, AE: **Online social networks: measurement, analysis, and applications to distributed information systems.** *Ph.D. Dissertation* 2009. Rice University, Department of Computer Science.

Mislove, AE, Viswanath B, Gummadi KP, Druschel P: **You are who you know: Inferring user profiles in online social networks.** In *Proceedings of the 3rd ACM International Conference on Web Search and Data Mining WSDM'10*; 2010:251–260.

Nawrath, J, Romano MC, Thiel M, Kiss IZ, Wickramasinghe M, Timmer J, Kurths J, Schelter B: **Distinguishing direct from indirect interactions in oscillatory networks with multiple time scales.** *Phys Rev Lett* 2010, **104:**038701.

Newman, MEJ: **The structure and function of complex networks.** *SIAM Rev* 2003, **45:**167–256.

Newman, MEJ: **Coauthorship networks and patterns of scientific collaboration.** *Proc Nat Acad Sci* 2004, **101**(Supl1):5200–5205.

Peng, H, Long F, Ding C: **Feature selection based on mutual information: criteria for max-dependency, max-relevance and min-redundancy.** *IEEE Trans Pattern Anal Mach Intell* 2005, **27**(8):1226–1238.

Sehgal, MSB, Gondal I, Dooley L, Coppel R, Mok GK: **Transcriptional gene regulatory network reconstruction through cross platform gene network fusion.** In *Pattern Recognition in Bioinformatics, Lecture Notes in Computer Science* 2007, **4774:**274–285.

Shalizi, CR, Thomas AC: **Homophily and contagion are generically confounded in observational social network studies.** *Sociol Meth Res* 2011, **40**(2):211–239.

Shannon, CE, Weaver W: *The Mathematical, Theory of Communication.* Urbana: The University of, Illinois Press; 1949.

Subramanian, A, Tamayo P, Mootha VK, Mukherjee S, Ebert BL, Gillette MA, Paulovich A, Pomeroy SL, Golub TR, Lander ES, Mesirov JP: **Gene set enrichment analysis: a knowledge-based approachfor interpreting genome-wide expression profiles.** *Proc Natl Acad Sci USA* 1554, **102:**5–15550.

Tsatskis, I: **Systemic losses in banking networks: indirect interaction of nodes via asset prices.** *SSRN 2062174* 2012. [http://papers.ssrn.com/sol3/papers.cfm?abstract_id=2062174]

Tresch, A, Beissbarth T, Saltmann H, Kuner R, Poustka A, Buness A: **Discrimination of direct and indirect interactions in a network of regulatory effects.** *J Comput Biol: J Comput Mol Cell Biol* 2007, **14**(9):1217–1228.

van Someren, EP, Wessels LFA, Backer E, Reinders MTJ: **Genetic network modelling.** *Pharmacogenomics* 2002, **3**(4):507–525.

Wells, BS: **Drosophila p53: A master regulator of DNA and tissue damage repair.** 2009. Ph.D. Dissertation, Columbia University, Genetics, Department.

Wu, X, Liu Q, Jiang R: **Align human interactome with phenome to identify causative genes and networks underlying disease families.** *Bioinformatics* 2009, **25:**98–104.

Yan, KK, Maslov S, Mazo I, Yuryev A: **Prediction and verification of indirect interactions in densely interconnected regulatory networks.** *arXiv preprint arXiv:0710.0892* 2007. [http://arxiv.org/abs/0710.0892]

Yousefi-Nooraie, R, Akbari-Kamrani M, Hanneman RA, Etemadi A: **Association between co-authorship network and scientific productivity and impact indicators in academic medical research centers: a case study in Iran.** *Health Res Policy Syst* 2008, **6:**9.

Yu, H, Xia Y, Trifonov V, Gerstein M: **Design principles of molecular networks revealed by global comparisons and composite motifs.** *Genome Biol* 2006, **7:**R55.

Zhang, X, Zhao XM, He K, Lu L, Cao Y, Liu J, Hao JK, Liu ZP, Chen L: **Inferring gene regulatory networks from gene expression data by path consistency algorithm based on conditional mutual information.** *Bioinformatics* 2012, **28**(1):98–104. doi:10.1093/bioinformatics/btr626.

Zhao, K, Karsai M, Bianconi G: **Entropy of dynamical social networks.** *PLoS ONE* 2011, **6**(12):e28116.

5

Cloud identity management security issues & solutions: a taxonomy

Umme Habiba[1*], Rahat Masood[1], Muhammad Awais Shibli[1*] and Muaz A Niazi[2]

*Correspondence:
umme.habiba@seecs.edu.pk;
awais.shibli@seecs.edu.pk
[1] Department of Computing, School of Electrical Engineering & Computer Science, National University of Sciences & Technology (NUST), Sector H-12, Islamabad, Pakistan
Full list of author information is available at the end of the article

Abstract

Purpose: Cloud computing systems represent one of the most complex computing systems currently in existence. Current applications of Cloud involve extensive use of distributed systems with varying degree of connectivity and usage. With a recent focus on large-scale proliferation of Cloud computing, identity management in Cloud based systems is a critical issue for the sustainability of any Cloud-based service. This area has also received considerable attention from the research community as well as the IT industry. Numerous Cloud Identity Management Systems (IDMSs) have been proposed so far; however, most of those systems are neither widely accepted nor considered highly reliable due to their constraints in terms of scope, applicability and security. In order to achieve reliability and effectiveness in IDMs for Cloud, further extensive research needs to be carried out to critically examine Cloud based IDMSs and their level of security.

Methods: In this work, we have holistically analyzed Cloud IDMSs to better understand the general as well as the security aspects of this domain. From the security perspective, we present a comprehensive list of attacks that occur frequently in Cloud based IDMSs. In order to alleviate those attacks, we present a well-organized taxonomy tree covering the most desired features essential for any Cloud-based IDMSs. Additionally, we have specified various mechanisms of realization (such as access control polices, encryption, self-service) against each of the features of Cloud IDMSs. We have further used the proposed taxonomy as an assessment criterion for the evaluation of Cloud based IDMSs.

Results: Our in-depth analysis of various Cloud based IDMSs reveals that most of the systems do not offer support to all the essential features of Cloud IDMS and the ones that do, have their own certain weaknesses. None of the discussed techniques heuristically covers all the security features; moreover, they lack compliance to international standards which, understandably, undermines their credibility.

Conclusion: Presented work will help Cloud subscribers and providers in understanding the available solutions as well as the involved risks, allowing them to make more knowledgeable decisions while selecting potential Cloud IDMSs that best suits their functional and security requirements.

Keywords: Cloud computing security; Identity management; Assessment criteria; IDMS taxonomy

Background

Cloud computing has emerged as a relatively new and influential paradigm for managing and delivering internet-based services and is considered to be an evolution of grid computing, which itself is based on traditional distributed system concepts (Youseff et al. 2008). Cloud computing offers many benefits to the IT industry by offering them unlimited storage and computing capacity. In addition, Cloud is based on pay-as-you-use model that allows organizations to outsource their data and IT services, offering on-demand self-service, broad network access and rapid elasticity at low cost (Mahmood 2011; Wang and Mu 2011). Cloud, being a service oriented computing architecture, is capable of providing anything-as-a-service, including but not limited to Software-as-a-service (SaaS), Platform-as-a-service (PaaS), Infrastructure-as-a-service (IaaS), Database-as-a-service (DbaaS) or Identity-as-a-service (IDaaS). Google (PaaS), Amazon (PaaS and IaaS) and Salesforce (SaaS) are few examples of major Cloud Service Providers (CSP) that offer on-demand and low-cost services/applications to the Cloud Service Consumers (CSC).

Another aspect of the Cloud systems is complexity. The problem in understanding cloud systems stems from the fact that it is simply quite difficult to model them. Cloud is a very dynamic system with numerous users, devices and networks, connecting and disconnecting simultaneously with the cloud. This complexity is to such an extent that it can perhaps be likened to the complexity of a human brain where neurons connect and change their synaptic structure continuously to store information. However, what is a problem here is the fact that unlike the brain, where the connecting neurons are already authenticated, cloud systems require extensive authentication as well as identity management systems. Still, these are simply not enough to cater for the ever-growing requirements of novel paradigms such as the Internet of Things (IoT) in relation to its connectivity with the cloud.

Despite the very attractive features that Cloud promises, the rate of migration to Cloud is rather slow, mainly due to the inherent security challenges associated with the technology. These challenges include data privacy, transparency, risk management, compliance and information security (Fox et al. 2009; Jansen 2011; Subashini and Kavitha 2011). Consequently, the security of Cloud paradigm has become a hot research area, which is being explored by both academic and industrial research communities. In this regard, issues related to the handling and management of sensitive identity credentials have garnered a lot of interest amongst the research communities (Albeshri and Caelli 2010; Chen and Zhao 2012; Gunjan et al. 2012). Storage and processing of identity information by a third party (CSP), outside the organizational boundary, brings in loss of control and transparency issues. This contributes to the reluctance of organizations to move their critical identity information to Cloud.

Cloud based Identity Management Systems (IDMSs) differ from the traditional IDMSs in that they require dynamic governance of provisioning, de-provisioning, synchronization, entitlement, scalability and access control (Gopalakrishnan 2009; Jansen 2011). In addition to this, Cloud IDMSs are required to have updated and synchronized identity information to avoid any conflicts caused by the usage of old user data. Management of sensitive identity information in the Cloud environment raises many privacy and confidentiality concerns. Moreover, security requirements vary from CSC to CSC; therefore, it is very important to choose the most suitable identity management system that best supports the CSC's security requirements.

An extensive survey on existing Cloud identity management related literature reveals that it is quite dispersed and mainly covers only few specific security features at a time; for instance authentication, authorization and access right delegation etc. Several Cloud identity management systems (Ates et al. 2011; Chowdhury and Noll 2007; Choudhury et al. 2011; Ranchal et al. 2010) have been proposed so far; however, none of those systems holistically meet the requirements of the IT industry, which necessitate a comprehensive identity management system for Cloud. Additionally, to be effective in Cloud, IDMSs are required to incorporate all the essential functional and security features of this paradigm. Furthermore, there exists no benchmark against which one can evaluate existing and newly proposed Cloud identity management systems. As a result, whenever a CSC or CSP needs to opt for an identity management system, they are left unclear about what an IDMS is offering (in terms of their services) and how their security or privacy requirements will be fulfilled.

After reviewing many state-of-the-art Cloud based IDMSs, we present a comprehensive list of attacks that involve identity either as an attack tool or as a target. Keeping in view the potential attacks and security requirements of Cloud IDMSs, we have identified attacks and proposed key features that should be a part of every Cloud IDMS. Presented taxonomy would help IT professionals and researchers understand the importance of these features that are worth considering when implementing or selecting an IDMS for Cloud. We have presented those identified features in the form of a well-organized taxonomy, and have suggested ideal mechanisms for providing said features. As a test case, we have applied the proposed taxonomy as an assessment criterion for the evaluation of existing Cloud IDMSs.

This paper presents a holistic view of identity management domain: a brief introduction about the evolution of IDMSs followed by identity lifecycle management, categories of IDMSs, list of attacks that can be launched against an IDMS, features pertaining to the security of IDMSs in the form of well-informed taxonomy etc. The paper is further organized as follows: Section 'Identity lifecycle management' describes the identity lifecycle management. Further, the novel concept of identity-as-a-service is presented in Section 'Cloud Identity-as-a-Service (IDaaS)'. Evolution of identity management systems is discussed in Section 'Classification of identity management systems', whereas Section 'Open-source cloud platforms & identity management' presents a comprehensive list of well-known open-source Cloud computing platforms and highlight their identity management services and properties. Section 'Cloud identity management: security challenges' highlights many open IDM related security issues and challenges. Well-known attacks against Cloud IDMSs are presented in Section 'Cloud identity management: known attack matrix'. Taxonomy based on the proposed security features is elaborated in Section 'Methods'. Section 'Results' comprises of the evaluation of Cloud identity management systems using the proposed taxonomy. Section 'Discussion' highlights current gaps, challenges and future key-trends related to Cloud identity management systems. Finally, the Section 'Conclusion' highlights the key-points derived out of this study.

Identity lifecycle management

Identity management systems are primarily responsible for the storage, maintenance and retrieval of CSC credentials for either authentication, authorization or some other business functions. The process of identity lifecycle management is the same for Cloud based and conventional systems. Identity lifecycle management encompasses the whole process of identity creation, management of account changes, password management and deletion or de-activation of CSC account in a synchronized manner (Gopalakrishnan 2009), as shown in Figure 1. We have explained each phase of identity lifecycle management in the following subsections.

User Provisioning: Provisioning is also referred to as on-boarding or creation of CSC account by an identity management system (Gopalakrishnan 2009; Meier et al. 2009; Slone 2004). In a Cloud environment, IDMS typically ensure CSC's provisioning via *Just-in-time* or *On-demand* user provisioning techniques. Cloud based IDMSs support provisioning via *Service Provisioning Markup Language (SPML)* or *Security Assertion Markup Language (SAML)* (Maler et al. 2003; Sodhi 2004). When new CSC subscribes for any Cloud service at CSP, IDMS creates his account and stores all the required information. Cloud IDMS assigns identity credentials to the CSC that he uses while acquiring/accessing various Cloud resources. IDMS further defines CSC roles and associates them with certain services to ensure authorized access to the Cloud services, resources and data.

Account changes/management: Over the course of time, CSC might subscribe for other Cloud services or resources; such an activity requires consistent adjustment in the CSC's account information (Gopalakrishnan 2009; Meier et al. 2009; Slone 2004). For instance, if CSC's title or role is changed from Assistant Manager to Manager then the identity management system is required to change the account details accordingly and timely. Other reasons of account modification include change in access privileges or attribute values. Thus, Cloud identity management system is required to accommodate all the changes to CSC account across all the systems in a consistent and synchronized manner. It helps avoid any potential conflicts or security breaches such as unauthorized or illegal access to Cloud resources.

User De-Provisioning: De-provisioning or account deactivation is another important phase of identity life-cycle management (Gopalakrishnan 2009; Meier et al. 2009; Slone

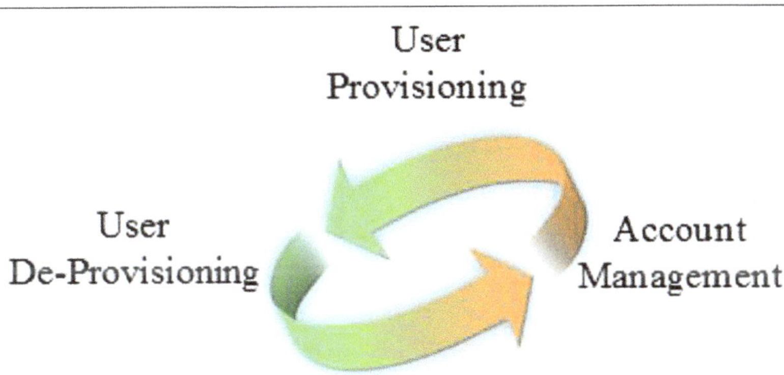

Figure 1 Identity life-cycle management.

2004). In a Cloud environment, de-provisioning means real time revocation that involves a synchronized deletion or suspension of CSC's account along with immediate termination of his access rights from all the organizational services and resources at Cloud. Any delay in de-provisioning could lead to many security risks like malicious use of privileges.

Cloud Identity-as-a-Service (IDaaS)

Cloud Identity as a Service (IDaaS) is essentially the management of identities in the cloud, outside the organizational boundary and applications that use them. The service is provided as third party management of identity functions, including user life cycle management and single sign-on. The term IDaaS is quite broad, and encompasses all three service layers of Cloud computing paradigm including software, platform, or infrastructure; and for both public and private clouds. Hybrid solutions may also exist, whereby identities can still be managed internally within an organization, while other components such as authentication, authorization etc. are externalized through *Service Oriented Architecture (SOA)*. IDaaS besides providing desired identity management services offers all of the Cloud benefits as well, including reduced hardware cost, easy management with wide range of integration options etc. (Rimal et al. 2009; Subashini and Kavitha 2011). For that reason, most of the organizations are moving their existing enterprise IDMSs to Cloud based services.

However, externalizing any portion of identity management functions to third-party provided IDaaS provider may raise several security and privacy challenges as well, which mainly includes *identity data locality, confidentiality, trust establishment, availability* etc. IDaaS may provide a level of benefit to an organization when it comes to functions like account management for an enterprise's SaaS partners, but in the short-term only. Identity management services are still best when managed internally, since, identity management represents the keys to the kingdom and IDaaS vendors don't take on the risk associated with losing critical identity information. Besides that, from an identity management perspective, there are a number of other uncertainties that have arisen with the concept of IDaaS, such as a clear definition of what exactly identity services are and what type of functionality is to be expected, application developers must adhere to SOA requirements, interoperability must be satisfied along with defining an *Application Programming Interface (API)* model that facilitates IDaaS development among many others.

Classification of identity management systems

Identity management systems have actively followed the IT evolution, initiating from Cluster computing followed by Grid and Peer-to-Peer systems that have now transformed into Cloud computing paradigm (Slone 2004; Youseff et al. 2008). Cloud Identity management is a broad, fascinating and continuously evolving domain with some misconceptions regarding its features and services, thus, requiring further investigation. Various Cloud identity management solutions exist and in order to highlight their strengths, weaknesses and suitability for Cloud, we have characterized them on the basis of their *deployment architecture* and *functional behavior*. Figure 2 presents the classification of identity management systems followed by a brief description for each of these systems.

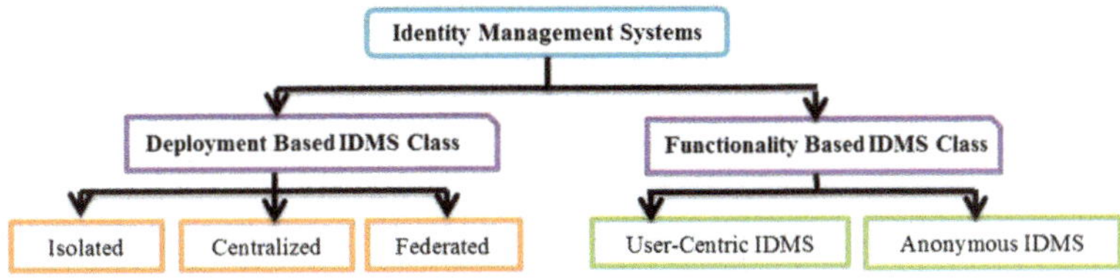

Figure 2 Classification of Cloud Identity Management Systems.

Deployment based classification

Deployment based classification includes the *Isolated, Centralized* or *Federated* identity management architectures. This classification mainly deals with underlying architecture for the storage, management and flow of identity information. Identity information can be stored on a single storage server or it could be distributed across various servers, considering the requirements of CSCs and CSPs.

1. ***Isolated Cloud IDMS*** Isolated Cloud Identity management system is based on the common deployment model used by the small or medium organizations. In an isolated Cloud IDMS, single server acts as a *Service Provider (SP)* as well as the *Identity Provider (IdP)* and is responsible for the storage of identity information and user operations (Alrodhan and Mitchell 2010; Cao and Yang 2010; Jøsang et al. 2005). A common use case is depicted in Figure 3, prior to the service acquisition, (1.1) CSCs are required to perform authentication at the CSP. Here, CSP redirects the user's authentication request to its own IdP for further processing. After successful authentication, (1.2) an authentication response is generated and returned to the corresponding user. This identity management system does not rely on a Trusted Third Party (TTP) for the credential issuance and verification. However, Isolated IDMS becomes unmanageable with the increase in services and resources, since each service needs to know the credentials of authorized users (Cao and Yang 2010; Jøsang et al. 2005).
2. ***Centralized Cloud IDMS*** Centralized Cloud identity management system is slightly different from the isolated IDMS, since it separates the functions of SP and IdP. In a centralized IDMS, a single IdP (a trusted third party) is responsible for the issuance,

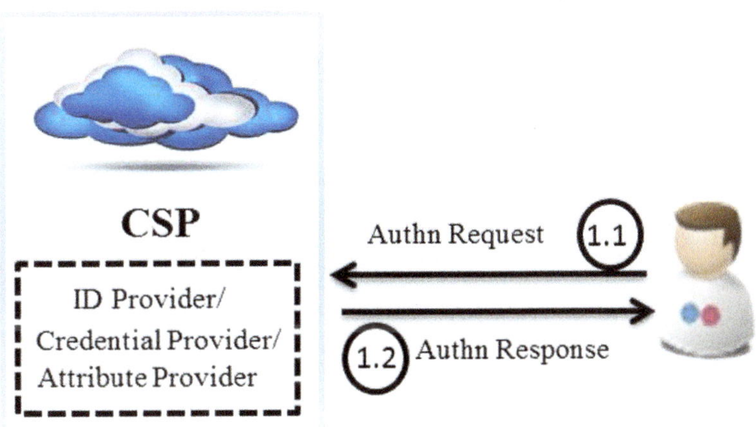

Figure 3 Isolated IDMS.

storage and management of identity data (Cao and Yang 2010; Jøsang et al. 2005; Windley 2005). As a first step, IdP collects all the identity information from CSPs to manage centrally. Later, (1.1) CSC sends an authentication request to CSP, (2.1) Authn request is redirected to the concerned IdP, (2.2) Authn response is sent back to the CSP (1.2) CSC will receive the Authn response (either successful or an error message), as depicted in Figure 4. Typically, single CSC may avail the services of different CSPs that may have a common IdP. In this scenario, CSPs and CSCs are required to have a common trusted IdP since it is responsible for the handling of sensitive identity credentials. An obvious drawback of the centralized IDMS model is single point of failure.

3. ***Federated Cloud IDMS*** Federated Cloud identity management system is the realization of federated identity management model that enables the subscribers of multiple organizations to use the same identification information for acquiring access to all the networks within any particular trusted group of enterprises (Cao and Yang 2010; Chen et al. 2012; Jøsang et al. 2005). Federated Cloud Identity management system has received significant attention from the IT industry because of its design agility that inherently allows cross-domain access to its users by eliminating the need of creating additional user accounts for external parties (Arias-Cabarcos et al. 2012; Shin et al. 2009; Suriadi et al. 2009). Federated IDMS follows the distributed storage architecture, where identity information is stored at multiple locations. The workflow of user request and service provider's response is depicted in Figure 5, where (1) CSC forwards an authentication request to the CSP1, (1.1), the CSP1 being a federated IDMS, forwards the authentication request to the CSP2 for the collection of CSC's identity credentials. As a next step, (1.2) CSP2 forwards the authentication request to the next IdP and retrieves the required attributes from its Identity data store. Finally, an authentication response is created and sent back to the requesting CSP. This process continues until it collects all the attributes required for authentication. In a federated IDMS, CSC's authentication request results in the linking of their information across multiple IdPs, so as to enhance security.

Feature based classification

The functional behavior class includes *Anonymous* and *User-centric* identity management systems. These systems are independent of underlying architecture for instance, a

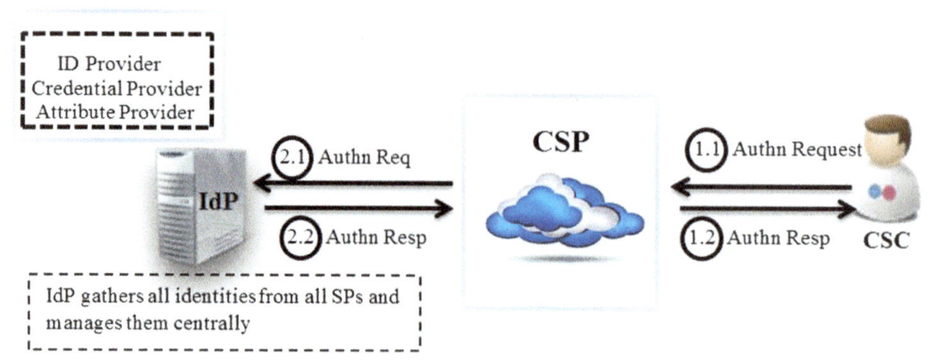

Figure 4 Centralized IDMS.

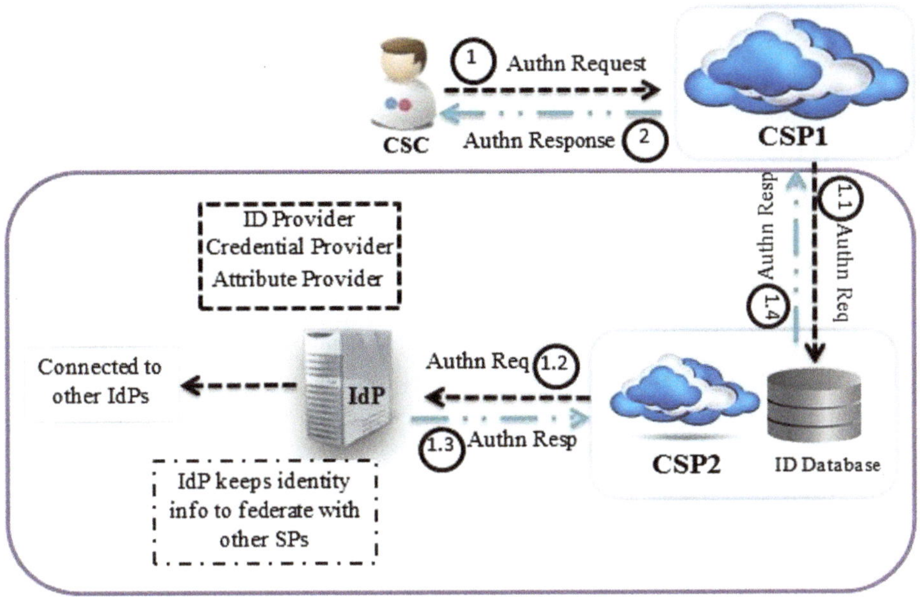

Figure 5 Federated IDMS.

User-Centric IDMS might be based on federated identity management architecture, or it may follow the centralized identity management approach for the management and storage of identity credentials. In the functional behavior class, the key emphasis is on the functionality of the identity management system such as *User-centricity* and *Anonymity*.

1. ***User-Centric Cloud IDMS*** User-centric Cloud identity management system involves user in every identity provisioning transaction. As shown in Figure 6, in order to acquire any service, (1.1) CSCs sends a credential request to the concerned IdP, 1.2) the IdP responses back with all the required credential that are needed to perform authentication. In a User-centric identity management system, CSCs are responsible for the storage, management and retrieval of their personal identity information. It further requires the CSCs to take decisions about the exchange of their identity credentials with other trusted entities such as CSPs, IdPs or users (Alrodhan and Mitchell 2010; Bhargav-Spantzel et al. 2007; Cao and Yang 2010; Suriadi et al. 2009). In

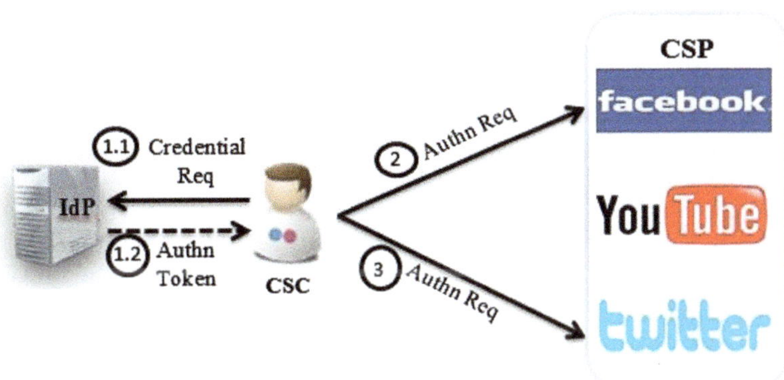

Figure 6 User-centricIDMS.

addition, user-centric IDMSs improve privacy by considering user preferences before disclosing the identity information to the SPs. However, in a service-oriented architecture like Cloud, each application/service requires the CSC to perform authentication and authorization, which is rather cumbersome for the CSC.

2. ***Anonymous Cloud IDMS*** An identity management system that offers anonymity as a feature is termed as an anonymous identity management system. An Anonymous Cloud identity management system is capable of keeping its entity (owner) secret from everyone else (Bhargav-Spantzel et al. 2007; Conrado et al. 2003; McCallister 2010). Anonymous identity should be strong enough to make it hard, if not impractical, to reveal actual identity because data inferred eventually may be connected with other information and can be republished (Bhargav-Spantzel et al. 2007; Conrado et al. 2003). However, anonymous identity management also has some shortcomings, such as lack of trust between the CSCs, and CSPs. It further seems to be negating the purpose of logging and monitoring, since the service consumer is performing actions using some temporary identity.

Open-source cloud platforms & identity management

In this section we present a comprehensive list of well-known open-source Cloud computing platforms and highlight their identity management services and properties.

Eucalyptus - identity & access management

EUCALYPTUS is the acronym for *Elastic Utility Computing Architecture for Linking Your Program to Useful System*, which is an open source private cloud software for building private or hybrid cloud resources for compute, network, and storage that are compatible with *Amazon Web Service (AWS)* APIs (Kumar et al. 2014). I*dentity and Access Management (IAM)* is an authentication, authorization, and accounting system feature within the Eucalyptus private cloud software that is responsible for the management of user identities, enforcement of access controls over resources and providing reports on resource usage as a basis for auditing and managing cloud activities. Eucalyptus by default stores all of the identities and policies in the local *Cloud Controller (CLC)* database (Eucalyptus Identity and Access Management IAM 2012). Identity data can also be pulled from LDAP or Active Directory. The user identity organizational model and the scheme of authorizations are compatible with the AWS Identity and Access Management system with some Eucalyptus extensions that support private clouds. Some of its key benefits include secure credentials management along with efficient and flexible policy based resource access and utilization management.

OpenStack identity service - keystone

OpenStack is an open source cloud computing project that has rapidly become very popular since its first release on October 21st, 2010 (Kumar et al. 2014). OpenStack mainly consist of three core software project which are OpenStack Compute Infrastructure - *Nova*, Object Storage Infrastructure - *Swift* and Image Service Infrastructure - *Glance*. The default identity management system for OpenStack is the OpenStack Identity Service, named as Keystone. Keystone integrates the OpenStack functions for authentication, policy management, and catalog services, including registering all tenants (customer, account, or any organizational unit) and users (person, system, or service), authenticating users and granting tokens for authorization, creating policies that span all users and services, and managing a catalog of service endpoints (Rhoton 2013). The existing Keystone implementation is centralized, where all users are required to be enrolled in its database, either manually by the OpenStack administrator (via a command line interface or Horizon), or via bulk loading from a corporate database such as LDAP, prior to their access to any of the services. There are a number of well-known limitations with this design, such as users having accounts in multiple systems have to remember multiple credentials for each one.

OpenNebula - users & group management

OpenNebula was first established as a research project back in 2005 and was used by many enterprises as an open and flexible Cloud virtualization infrastructure on their VMware-based data center enabling highly scalable hosting environments (Kumar et al. 2014). OpenNebula includes a complete user & group management system. Users in an OpenNebula installation are classified into four types, *Administrators*, (user belongs to an admin group), *Regular users* (may access most OpenNebula functionality), *Public users* (may access only public interfaces) and *Service users* (user with OpenNebula service account). The resources a user may access in OpenNebula are typically controlled by a permissions system that resembles the typical UNIX one. By default, only the owner of a resource (VM/image) can use and manage it. Users can easily share the resources by granting use or manage permissions to other users in their group or to any other user in the system. Furthermore, OpenNebula comes with an internal user/password authentication system; however, an external authentication driver may also be enabled if required. OpenNebula further supports three customizable authentication configurations; namely, *Command Line (CLI)* (basic user/password, X509, LDAP), *Sunstone* (SSL proxy - Apache) and *Servers* authentication.

Cloud identity management: security challenges

Identity management is a broad domain that involves many open security issues and challenges including *Openness, Identity Theft, Least Privileges, Elevated Privilege Control, Availability, Confidentiality, Integrity, Trust Management* etc (Cloud Security Alliance SecaaS Guidance, Category 1: Identity and Access Management, 2012 by Cloud Security Alliance 2013). Since, technology area such as Cloud Identity Management is still evolving, thus, no well-established standards have been developed so far, resulting in issues like vendor lock-in and lack of openness (availability of specifications/data formats/protocols). *Openness* is critical, so that converters can be developed in the future to ensure interoperability and scalability. Further, since identity information could be accessed by unauthorized or malicious users/intruders. Therefore, adequate security *(encryption, authentication access control, monitoring, etc.)* mechanisms are required to be in place for accessing identity and administration interfaces; otherwise, unauthorized access over the network *(eavesdropping)* would be a key risk factor (Kumaraswamy et al. 2010). In addition to this, Cloud IDMS must subscribe to the principle of *Least Privilege* and proceed using a work-flow mechanism for additional approvals. However, since least privilege not only involves static roles and resources, but also complex context, and dynamic changes (both technical and non-technical). Thus, many Cloud IDMS deployments today enormously over-provision effective access rights to users who even do not require those access rights. Such excessive access provisioning leads to many critical security issues including unauthorized disclosure, fraud, accidental access and identity theft.

Another most critical aspect of identity management is *Privilege Access Management.* Besides traditional audit and logging requirements, privilege access needs to be a governance mechanism as well. Since, most enterprise applications suffer from mismanagement of roles, violation of segregation of duties, therefore reporting (both success and failure events) on this aspect becomes crucial. Moreover, attacks on identity services or network connectivity, such as *DDoS attacks or resource hogging*, could risk the availability or degrade the performance of an IDMS. If high availability and/or performance are required, redundancy and fail-over options must be considered. Aside from this, due to the ubiquitous nature of Cloud, it is accessible through various devices and applications resulting in an increased number of access points, which sooner or later, adds to the threat of unauthorized disclosure *(confidentiality)* of identity credentials from both insiders (staff) and outsiders (users/intruders) (Mather et al. 2009). Similarly, considering the increased number of access points and system entities, *integrity* of identity data and information stored at Cloud is another important concern which needs immediate attention (Halpert 2011; Lang 2010). In addition to the above-mentioned issues, *management of trust* between the Cloud identity provider and the subscriber is one of the main issues that today's Cloud IDMSs is dealing with. Furthermore, trust is a subjective and context sensitive term which makes it even more difficult to select a Cloud identity provider with fully trusted identity services (Pearson and Benameur 2010).

Cloud identity management: known attack matrix

Currently, a lot of research is focused on Cloud identity management systems; however, security of Cloud IDMSs is an aspect that is still in nascent stages, and requires further exploration. Existing IDMSs are susceptible to various security and performance bottlenecks, which limits their widespread adoption as a potential solution for dynamic Cloud environment. Therefore, we have contributed to this part by conducting a survey on state-of-the-art Cloud based identity management systems and security issues. We present a list of attacks that are either launched against IDMSs or use identity as a principal tool for attack. A brief description of all the identified attacks is presented in Table 1. In addition to this, we have assigned a unique label to each identified attack that we use later in this paper to refer to these attacks. The compiled list aided in determining the key security features that a Cloud based IDMS must provide to ensure the protection and security of Identity credentials in Cloud.

In addition to the attacks, we have also identified the features for Cloud IDMSs that are capable of alleviating the impact of most if not all of the attacks mentioned here. Along with the list of eminent Cloud IDMS features, we have also provided a list of mechanisms that help achieve the identified features. Strengths and weaknesses of each mechanism are also specified wherever required. Cloud IDMS features along with the mechanisms to implement these features is presented in the form of a well-organized taxonomy in the following subsection.

Table 1 Known Attacks against Cloud IDMSs

Label	Attack	Description
A1	Brute-force attack	Brute-force attack generally allows the attacker to gain unauthorized access to sensitive identity credentials of CSCs stored in an identity management server using different possible combinations for user ID and password. Dictionary attack is one example of brute force attack that might be launched against an IDMS if it fails to comply with international standards of strong password settings. Once successful, attacker intensifies their attack in an attempt to uncover the security holes or vulnerabilities of an IDMS. They analyze the server responses and manipulate them to achieve their malicious purposes (Almorsy et al. 2010; Brute Force Attack; Kumaraswamy et al. 2010; Meier et al. 2009; O'Gorman 2003; Ratha et al. 2001; Yassin et al. 2012).
A2	Cookie-replay attack	Here the attacker steals a cookie containing valid session information along with the CSC's identity credentials and reuses it to trick the identity management server into believing that a previously authenticated session is still ongoing and authentic. Through this attack method, attacker may get unauthorized access to victim's (person whose credentials are stolen) confidential information other than Cloud services and resources (Meier et al. 2009).
A3	Data Tampering Attack	It refers to the unauthorized modification of data related to identification of CSC in an identity data-store at Cloud. These modifications may provide the attacker with an opportunity to transgress and damage the Cloud services and resources. This is the attack on the integrity of identity information stored at Cloud mainly due to the loopholes in access control systems (Angin et al. 2010; Meier et al. 2009; Ranchal et al. 2010; Subashini and Kavitha 2011; Thompson et al. 2006).
A4	Denial of Service (DOS) Attack	DoS attack can be launched against an IDMS if it does not provide mechanisms for logging user activities. Since in a DoS attack, attacker overwhelms the Cloud identity management server with false authentication or authorization requests (malformed input data) and tries to either stop the service or consume all of its available resources so that it may not be able to process the legitimate user requests (Almorsy et al. 2010; Meier et al. 2009; Thompson et al. 2006). Therefore, proper logging mechanisms are required to be ensured, so as to make the IDMS intelligent enough to detect and prevent such attacks.
A5	Eavesdropping	This is the attack at communication level, when the Cloud identity management server and CSC exchange the identity credentials for authentication or authorization purpose. It refers to the unauthorized real-time interception and stealing of sensitive consumer information by the attacker either through listening or reading the un-encrypted sensitive data off the network (Bhadauria et al. 2011; Jansen 2011; Jansen and Grance 2011; Jensen et al. 2009; Meier et al. 2009).
A6	Elevation of Privilege	Privilege escalation attack involves legitimate subscribers of the IDMS with limited set of privileges. They illegitimately escalate their access rights by impersonating other CSC that has higher privileges than theirs in order to achieve their illicit objectives and may cause severe damage to the stored information (Meier et al. 2009; Saripalli and Walters 2010; Subashini and Kavitha 2011; Thompson et al. 2006).
A7	Identity Forgery/Cloning/Spoofing Attack	It refers to the unauthorized copying or manipulation of identity tokens or credentials issued from the trustworthy authorities (such as CSP or government), with the intent to deceive or mislead the investigation if followed. Cloud based IDMS should be able to detect the forged identity by implying strict (two-factor) authentication mechanisms. Forged identities further help in committing fraud and identity theft and requires expert knowledge, exceptional skill-set and sometimes much greater effort than the benefits achieved (Chang 2003; Choudhury et al. 2011; Jensen et al. 2009; Kumaraswamy et al. 2010; Meier et al. 2009; Nabeel et al. 2011; Saripalli and Walters 2010; Subashini and Kavitha 2011; Thompson et al. 2006; Zissis and Lekkas 2012).

Table 1 Known Attacks against Cloud IDMSs *(Continued)*

A8	Identity Theft	Identity theft refers to the stealing of someone's identity (such as their name, personally identifiable information, or credit card number), with the intent to acquire Cloud resources or other financial benefits in that victim's name. The victim of identity theft may undergo adverse consequences if they are held responsible for the actions of actual delinquent. In addition to this, identity theft further paves the way for many other crimes such as fraud and forgery (Angin et al. 2010; Ranchal et al. 2010).
A9	Luring Attack	An IDMS that neither ensures user-centricity (such as consistent user experience) nor provides logging & reporting mechanisms is considered to be more prone to luring attack. It is a more specialized form of privilege escalation attack, where the authorized service consumer unknowingly executes the attacker's code fragment in a more privileged security context. More precisely, the adversary targets and 'lures' a high-privileged CSC to perform some illegal activities on their behalf (Angin et al. 2010; Meier et al. 2009).
A10	Phishing Attack	IDMS that offers no support to user-centricity, strong password schemes and privacy preservation considerations is more vulnerable to phishing attack. Phishing is an act of acquiring CSC's information such as name, passwords, social security number, bank account numbers and credit card details by redirecting the CSC to enter his particulars to some replica website whose look and feel is almost identical to the authentic one. Attacker manipulates the communication so that they may appear to be from a legitimate IdP to successfully lure the unsuspicious CSC (Angin et al. 2010; Jansen 2011; Jansen and Grance 2011; Jensen et al. 2009; Kumaraswamy et al. 2010; Olden 2011).
A11	Replay Attack	Replay attack occurs when an IDMS fails to ensure the security of identity credentials during their transmission. In a replay attack, adversary captures the valid identification information and retransmits it, possibly as part of impersonation attack. Unless mitigated, the IDMS subject to the attack, processes user request as an authentic message, resulting in a range of bad consequences, such as unauthorized disclosure of information followed by fraud, forgery and impersonation (Almorsy et al. 2010; Choudhury et al. 2011; Jansen 2011; Jansen and Grance 2011).
A12	Repudiation	Repudiation attack occurs when the Cloud service consumer denies an action. In addition to this, the Cloud IDMS does not implement any controls to maintain service consumer activity logs so no proof exists to prove him accountable for his actions. Due to the absence of real-time tracking and activity logging mechanisms, service consumers can easily repudiate their malicious activities that they have actually performed on Cloud servers, such as unauthorized manipulation of data and forgery of identity credentials (Bertino and Takahashi 2010; Thompson et al. 2006; Yan et al. 2009; Zissis and Lekkas 2012).
A13	Side-Channel Attack	An IDMS may fall victim to side channel attack if it does not follow the principle of federation and access control. As in the side channel attack, attacker steals the information (like session identifiers, timing information, OAuth tokens and electromagnetic leaks) from the physical implementation of a security system. Therefore, it is recommended to deploy a federated IDMS that stores the sensitive identity information in fragments across multiple servers in order to make it hard for the attacker to achieve his malicious objectives (Angin et al. 2010; Bhadauria et al. 2011; Jansen 2011; Jensen et al. 2009; Ranchal et al. 2010; Zhou and Feng 2005).
A14	Skimming Attack	Since skimming is an attack method where criminals steal the sensitive information from authentication tokens (smart card). IDMS should be capable of ensuring strong encryption and secure distribution of identity credentials across multiple servers (Jacobs and Poll 2011; Zissis and Lekkas 2012).
A15	Snooping	Snooping attack permits the illegitimate collection of sensitive information such as identities, available services and network topology from an identity server in Cloud environment. Snooping is slightly different from eavesdropping since it includes more sophisticated surveillance techniques to intercept secret communications such as through remote activity and key-stroke (key-loggers) monitors (Donevski et al. 2013; Salsano et al. 2002).

Methods

Cloud IDMS features & mechanisms: a taxonomy

Having compiled a list of potential attacks against IDMSs, we move on to identifying the most imperative Cloud IDMS features that are required to ensure the security of CSC's credentials. Most of the identified features do exist for traditional IDMSs, but as Cloud brings in new security challenges, we need to revisit them from the perspective of Cloud environment. In this respect, Ferdous et al. in (Ferdous and Poet 2012) have presented a very comprehensive taxonomy covering almost all the functional and non-functional requirements of an identity management system including *usability, trustworthiness, affordability, security* and *privacy*, as shown in Figure 7. However, among all these requirements, security is one key factor that is categorically considered as the most important requirement, and has not been covered in detail thus requires further investigation and exploration. Therefore, in this paper, we have extended their work and provided a well-informed taxonomy (as shown in Figure 8) - that thoroughly covers all the required Cloud IDMS security features and their corresponding mechanisms of realization. Whereas, for other aspects such as *usability, affordability, liability & law enforcement* etc., we affirm the taxonomy presented in Ferdous and Poet (2012) to be reliable and adequate, since it is inclusive of all the crucial requirements. Finally, detailed description of all the identified Cloud IDMS features, along with their mechanisms, is presented in the following subsections.

Authentication

An identity management system has a primary objective of authenticating users or subscribers registered with it. Authentication is the process of verification that ensures whether the person or the application is actually the one or who it claims to be (Jansen 2011; Pashalidis and Mitchell 2003; Shin et al. 2009; Slone 2004; Subashini and Kavitha 2011; Windley 2005; You et al. 2012). In the multi-tenant Cloud environment, users and computer programs are required to prove their identity/legitimacy to the authentication service whenever they attempt to access some application or while acquiring a web resource. Authentication can be provided as a part of an IDMS or it may exist as a separate

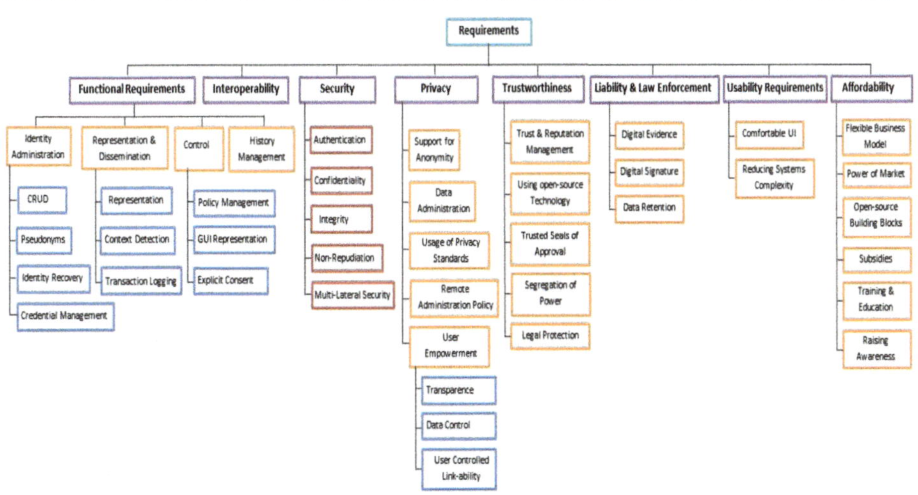

Figure 7 Taxonomy of Identity Management Requirements.

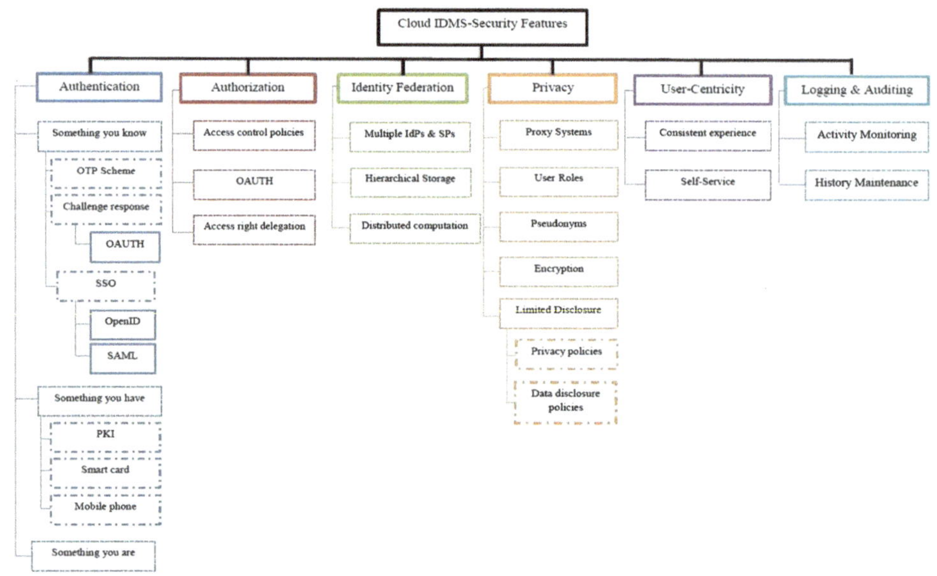

Figure 8 Taxonomy tree of cloud IDMS Components and Mechanisms.

service/system. Authentication commonly relies upon at least one of the factors described below considering the security requirements of the service, since highly secure services require stronger authentication mechanism.

1. ***Something you know*** The conventional and the easiest way of authentication is via password or Personal Identification Number (PIN), something that is shared between the user and the authentication service and is, ideally, both secret and hard to guess. Identity management systems are responsible for the storage, management and secure transmission of user ID and password to the authentication service for verification. In a common Cloud hosted web-service access scenario, user enters his user ID and password/PIN to prove his identity to the authentication server. However, this mechanism is considered to be the least secure authentication mechanism for the dynamic Cloud environment that is susceptible to replay attack and identity theft since user passwords/PIN can easily be stolen. Other reasons might include sharing of passwords or usage of overly simplistic passwords among many others. Password based authentication mechanisms can be enhanced if applied as One-Time-Password (OTP) scheme (Luo et al. 2009; Olden 2011), Challenge-Response mechanism (Choudhury et al. 2011) and Single-Sign-On (SSO) (Ates et al. 2011; Chadwick and Casenove 2011; Choudhury et al. 2011; Kim et al. 2010), each offering different levels of security.

   OTP Scheme: Cloud based IDMS can authenticate users via OTP scheme also (Olden 2011), where user authentication is performed each time with new password that is randomly generated and has no likability to the one used for previous transaction or resource acquisition (Luo et al. 2009). Thus, improves the security and privacy of user during the entire authentication process; however, it decreases the efficiency of the overall system since the authentication system has to generate the OTP and transmit it to the respective user in response to every authentication request.

   Challenge Response Mechanism: In order to make conventional passwords non-reusable and secure, Cloud IDMSs can offer authentication via challenge-response mechanism (Feng et al. 2010; Ratha et al. 2001; Saroiu and Wolman 2009). Authenticator maintains a list of challenges that are later presented to the user in response to every authentication request. User provides response to those challenges and on the basis of the verification results, user is allowed or denied by the service provider. Challenge-response mechanism also has certain weaknesses as the system is prone to playback/replay attacks and password guessing attacks. Moreover, authentication server is required to have a wide range of challenges each with a unique response, which is hard to achieve practically.

OAUTH: It is an algorithm developed by Initiative for Open authentication and is based on a challenge-response algorithm (Almorsy et al. 2010; Gopalakrishnan 2009; Olden 2011). This algorithm offers one-way authentication as well as mutual authentication, and has digital signatures capability also, so is widely adopted as an authentication mechanism by the Cloud IDMS providers and is widely trusted by CSPs and CSCs.

SSO: In order to facilitate the users, Cloud IDMSs offer SSO capability that does not require its users to remember a large number of passwords. SSO is a onetime assertion or authentication per session/per credential where users can access multiple trusted services or resources at Cloud using a single user ID and password (Chadwick and Casenove 2011; Kim et al. 2010; Pashalidis and Mitchell 2003; Suriadi et al. 2009). Organizations may maintain their independent IDMSs but authentication is provided via unique user credentials. An individual's identity information stored across various identity management systems is linked to those unique credentials while performing authentication. SSO controls access among many interrelated, but autonomous systems (Jansen 2011) and can be achieved via any of the following mechanisms.

(a) **OpenID:** OpenID is an authentication protocol whose main objective is to minimize the number of credentials that a user needs to maintain for the purpose of accessing multiple Cloud services (Angin et al. 2010; Celesti et al. 2010; Gopalakrishnan 2009; Olden 2011; Ranchal et al. 2010). It is a type of federated identity which enables the user to access multiple services using OpenID credentials instead of service specific user name and password. OpenID, on behalf of user, requests the service provider for the required service. Moreover, it eliminates the chances of password disclosure, since no OpenID-based Cloud service actually stores user's password information.

(b) **Security Assertion Markup Language (SAML):** SAML is an open standard framed by OASIS for the representation and exchange of identity information among various trusted CSPs while performing user authentication or identification (Celesti et al. 2010; Gopalakrishnan 2009; Jensen et al. 2009; Maler et al. 2003; Olden 2011). SAML version 2.0 offers support to web SSO using token, containing user identity information, that is issued by the IdP. These tokens are recognizable to various trusted CSPs while performing user authentication in response to the user's SSO request.

Since, it is impractical to eliminate passwords from the authentication mechanisms, their security can be further enhanced if they are used in conjunction with other authentication factors and technologies such as biometrics or smart cards.

2. ***Something you have*** Identity management systems can also offer authentication on the basis of *"something you have"* that is commonly referred to as token-based authentication and is a comparatively stronger authentication mechanism. The secret user authentication credentials are encoded in hardware or software tokens that are verified against the information stored by the corresponding IDMS prior to the access of every web service/resource. However, authentication-tokens individually are also vulnerable to identity theft attacks since these cards can be lost or stolen and the adversary can use them for their own malicious purposes. The security of this authentication mechanism can be enhanced by combining it with *"something you know"* for instance a PIN code or password. Token-based authentication can be realized via any of the following methods:

   Public Key Infrastructure(PKI): Identity management systems are capable of accepting digital certificates as software authentication tokens issued by the trusted certification authority (CA). Authentication server traces back the certificate to the issuing party in order to ensure the legitimacy of the user presenting the certificate.

   Smart Card: SecureID tokens or smart cards are also accepted as valid authentication tokens by various IDMSs as smart cards are capable of storing long and hard to recover secret user information (Cameron 2005; Ratha et al. 2001). Smart cards are designed to be tamper resistant. Users present the smart card to the authentication service that verifies the encoded information from its identity management database before sanctioning the user to access any web service or resource.

   Mobile Phone: Identity management systems may incorporate cellular devices (mobile phone) having a SIM (Subscriber Identification Module) card to connect the users to the authentication service via mobile network (Ratha et al. 2001). Mobile phone network is used to communicate the authentication information (password) between the user and the authenticator via SMS (Short Message Service). Authentication information can be voice confirmation from the user or OTP send by the server that the user enters afterwards to verify the legitimacy of user.
3. ***Something you are*** The final category of authentication that an IDMS can support is based on biometrics where user verification is performed on the basis of some natural characteristics such as fingerprint, voice patterns, or iris characteristics that are unique for every individual (Jensen et al. 2009; Leandro et al. 2012; Senk and Dotzler 2011).

To make the authentication system stronger, any two of the aforementioned factors can be combined as two-factor authentication, which is considered highly secure. Most commonly *"something you have"* is combined with *"something you know"* e.g. smart card along with a PIN is used to authenticate a user more securely in Cloud IDMSs.

Authorization

Authorization plays an important part in Identity management system; authorization is the process of granting and denying access to any web resource and service (Slone 2004; Subashini and Kavitha 2011; Windley 2005; You et al. 2012). The authorization process decides what a subject (users or applications) is allowed to do on the system and this decision is taken by using subject's identity information. Cloud is a multi-provider environment where one user might have access to multiple services each can be from a different provider having different security levels. Therefore, Cloud service providers are required to ensure effective authorization for their resources and services in every possible situation. Authorization in a Cloud environment is commonly achieved via access control policies, defined and implemented by the CSPs for their subjects (CSCs), so that only authorized subject may accesses the services and resources.

Access Control Policies (ACP): Cloud service providers implement access control policies on top of identity management systems to ensure that only the legitimate users acquire/access the services and resources they offer (Ates et al. 2011; Chadwick and Casenove 2011; Chowdhury and Noll 2007; Hoellrigl et al. 2010). Complete knowledge about 'Who' can access 'What' and 'When', gives a clear picture of the security and protection of CSP's resources/services. Accordingly, CSPs can take effective security measures to make illegitimate use, unauthorized modification and disclosure of their services, relatively hard if not impractical.

Access Right Delegation: In an IDMS, delegator (person who wants to delegate his access rights) may want to delegate his access rights of some Cloud based web-service to some delegatee (person who receives the delegation) as desired by the delegator or delegatee (Gopalakrishnan 2009; Hoellrigl et al. 2010; Li et al. 2010; Zissis and Lekkas 2012). Delegation of access rights is typically based on any of the access control models, such as Role based access control (RBAC) or Attribute based Access Control (ABAC), where each model specifies a different mechanism for delegation. For instance, in RBAC, access permissions are granted to a role that is later linked to the concerned delegatee. However, in the process of delegation, delegator is required to protect his privacy by restricting the delegatee (via access control policies or privacy policies) from accessing any additional identity information of the delegator.

OAUTH: OAUTH is an open standard for authorization; it allows the service consumers to access the CSP's services or resources on the behalf of the resource/service owner (Almorsy et al. 2010; Gopalakrishnan 2009; Olden 2011; Sanchez et al. 2012). OAUTH follows the principle of access right delegation where resource owners delegate access rights to service consumers without sharing their identity credentials (such as user-ID and password pair). In a common OAUTH scenario, the CSC first accesses a service/application that redirects him to the IdP, where he performs authentication and after successful authentication, IdP redirects the service consumer back to the Cloud service along with the identity token. From then on, service consumers access that particular service via that token.

Identity federation

Identity management systems may offer support to identity federation that enhances the security of identity information during the transmission and storage of credentials. Identity federation has two meanings: it can be a method of associating and communicating a person's digital identity and attributes that are stored across multiple Cloud based IDMSs or it could be any standard, contract and technology that make the identity information movable across independent domains (Arias-Cabarcos et al. 2012; Chen and Zhao 2012; Shin et al. 2009; Slone 2004; Suriadi et al. 2009). Identity federation plays the key role in securing user credentials while moving towards the Cloud. Identity federation can be achieved via any of the following mechanisms:

Multiple IdPs and SPs: An IDMS might involve multiple IdPs and CSPs for the storage and processing of service consumer's identity information (Celesti et al. 2010; Chadwick and Casenove 2011; Hoellrigl et al. 2010). In other words, the identity credentials of a single service consumer are stored at multiple Cloud servers in small chunks following the distributed architecture. Consequently, while performing authentication, authenticator collects the identity information from all the relevant IdPs and CSPs and processes them as required.
Hierarchical Storage: Identity information stored in hierarchical fashion either on a single server or on multiple nodes (typical Cloud scenario) is considered to follow the identity federation principle (Yan et al. 2009). Hierarchical storage offers different levels of access and improves the information accessibility and security. In the hierarchical storage, IDMSs store user information with different access levels considering the sensitivity of identity credentials; therefore, in case if any intermediate node gets compromised, the impact of information loss will be minimal.
Distributed Computation: Distributed Computation: IDMSs are said to be following the identity federation principle, if they follow the distributed processing architecture for user authentication. The distributed architecture works by disseminating the identity information among multiple Cloud servers for the computation or verification of user credentials. This approach accelerates the authentication process and improves security (Ates et al. 2011; Chowdhury and Noll 2007; Kim et al. 2010; Ranchal et al. 2010).

Privacy

Identity management systems are meant to store and process sensitive user credentials that the user wants to protect from unauthorized or unwanted disclosure (Bhargav-Spantzel et al. 2007; Cameron 2005; Leskinen 2012; McCallister 2010; Windley 2005). Privacy is among the top most concerns of users while storing or sharing their secret identity information for either e-commerce or other purposes. Moreover, storage of sensitive identity information in the Cloud, which is outside the organizational boundaries and beyond the user control, elevates user's privacy concerns. Listed below are the mechanisms through which identity management systems can guarantee user privacy:

Proxy systems: Proxy can be a Cloud service or server that personifies someone else (user/application) and acts on their behalf. Identity management systems assign a unique credential to almost every entity be it a system, resource or a service consumer, thus raising privacy concerns (Ates et al. 2011; Olden 2011; Wang et al. 2010). Proxy is particularly intended to enhance user's privacy by requesting the desired Cloud resource/service on behalf of its registered users/systems. The identifying information that is forwarded, for either authentication or resource acquisition, is of the Cloud proxy system rather than the actual user who has requested for the service.
User Roles: A single user might have multiple contextual identities such as personal (name, e-mail ID), social, professional (employee-ID, Salary), and citizen (for instance National ID card), each with different security requirements (Ates et al. 2011). In order to ensure user's privacy while accessing multiple Cloud services, only relevant identity attributes must be shared. An IDMS should be capable of understanding and specifying separate user roles so that only required and related information is forwarded to the respective CSP.

Pseudonym: Pseudonym is a temporary name that is generated either by an IdP upon user request or by the users themselves, whenever they need to protect their privacy while accessing a Cloud based web service/resource (Luo et al. 2009; Zhang and Chen 2010). Pseudonyms intend to protect an individual's true identity in an online transaction. It is recommended to use a different pseudonym for every different Cloud service so that they cannot be linked to the user's original identity.
Encryption: Encryption is used to convert the identity information into some unintelligible form so that even if disclosed, information might not be useful to the adversaries (Chadwick and Casenove 2011; Kim et al. 2010; Nabeel et al. 2011). In order to ensure privacy, Cloud based identity management systems must offer encrypted storage and transmission of identity credentials. The key used for the encryption of identity credentials is securely shared between the communicating parties only so that the information will only be disclosed to the authorized user having the right key pair.
Limited Disclosure: The best practices regarding identity management dictates that information should be disclosed on a *"need to know"* basis, and stored on a *"need to retain"* basis, as that may help to ensure the minimal damage in the event of any violation such as theft of identity information (Cameron 2005).

1. **Privacy Policies:** Identity management systems implement privacy policies to protect their service consumer's identity information in accordance with their preferences. Identity disclosure policies are specified and implemented at the service consumer end (Angin et al. 2010; Celesti et al. 2010; Chowdhury and Noll 2007; Choudhury et al. 2011; Ranchal et al. 2010; Yan et al. 2009. CSCs (data owner) demand transparency about 'who' wants to access 'what' part of their identity information. Whenever CSC attempts to access or acquire a Cloud service/resource, CSP needs certain amount of identity information to verify his authenticity. Therefore, before forwarding CSC's credentials to the CSP, identity management system must consult the privacy policies or may ask for the CSC's consent prior to identity information disclosure. It enables the CSC to decide whether they want to acquire that resource by sharing their identity information with the CSP or if they would prefer to protect their privacy so that they may withdraw their resource acquisition request.
2. **Data disclosure policies/preferences:** Commonly, user-centric IDMSs offer consistent user experience by allowing the user to select the credentials and attributes from identity attribute lists issued by the IdP, in advance, to each operation (Hoellrigl et al. 2010; Zhang and Chen 2010). Service consumers may also specify their identity revelation rules prior to actual service usage to ensure the security and privacy (Fox et al. 2009; Hoellrigl et al. 2010; Zhang and Chen 2010) of their information. User centric IDMSs allow its users to choose an IdP of their own choice which they believe is more appropriate for any particular transaction thus, places greater responsibility in the hands of users.

User-centricity

User centricity is an important identity management feature which emphasizes on transferring the control of identity attributes and disclosure preferences in the hands of identity possessor (CSC). In a user-centric IDMS, CSCs are involved in every identity provisioning operation and have full control over their online identity information (Dhamija and

Dusseault 2008; Leskinen 2012; Windley 2005). CSCs are responsible for the management of their identity credentials; however, user-centric systems have problems in terms of security and privacy. Following are the mechanisms through which identity management systems provide user centricity to the Cloud service consumers.

Consistent User Experience: Identity management systems are required to offer support to various technologies (e.g. OpenID, SAML, OAuth) run by multiple IdPs to ensure the interoperability among various systems (Cameron 2005; Celesti et al. 2010; Chowdhury and Noll 2007; Ranchal et al. 2010). An IDMS that supports interoperability among various identity management systems (by either offering support to multiple technologies or by following common schema for the exchange of identity credentials) help to ensure consistent user experience while accessing multiple services.

Self-service: Identity management systems are required to provide a self-service feature that enables the user to alter his sensitive personal information stored by the Cloud service providers (such as postal address, phone number, password etc.) (Ates et al. 2011; Choudhury et al. 2011; Luo et al. 2009). Users must be allowed to modify/update their personal identification information that uniquely identifies them in an IDMS. It also reduces the SP's operational costs and troubleshooting time.

Logging & auditing

Identity management systems incorporate logging and auditing feature that helps to ensure the proper working of identity management system and gain the trust of its customers/users in the system (Bhargav-Spantzel et al. 2007; Windley 2005). In a multi-tenant Cloud environment, it is comparatively harder to identify the person responsible for any security breach or misbehavior. Therefore, Cloud IDMSs are required to provide logging and auditing facility that compliments the security of Cloud applications and identity management system both. Generally, Cloud SPs are more concerned about the maintenance of logs such as recording of all sensitive user activities (authentication requests forwarded to an IDMS). Logs help in identifying the person responsible for any security breach, thus, accordingly the legally responsible person is penalized for his actions. Whereas, auditing is of mutual interest to the CSCs and CSPs, auditing at one end help CSPs in making sure that their systems/services are in line with their business objectives. On the other hand, auditing highlights the security loopholes that the service consumer needs to know and upon which he either makes his decision of electing the IDMS. Logging and auditing can be achieved via any of the mechanisms mentioned below:

Activity Monitoring: Identity management systems may offer auditing and logging via user activity monitoring functionality (Ates et al. 2011). Often system administrators in Cloud come across situations when they need to answer "Who did it?" though that happens fairly less, but those events are critical. In order to get ahead in security and to timely identify system misuse and user misbehavior, monitoring should be a necessity. An IDMS offers monitoring by tracking and storing each user action or web access in the log files that are later used for auditing purposes.

History Maintenance: Identity management systems might incorporate history maintenance module for maintaining the history of all user actions (such as web access) and attribute exchanges among the service consumers, IdPs and CSPs (Angin et al. 2010;

Ranchal et al. 2010). History maintenance module benefits the CSPs by preserving the user activity logs thus in case of any security breach the wrongdoer can be identified appropriately. Whereas, Cloud service consumers are benefited by the storage of attribute sharing history that helps them in making future attribute disclosure decisions whenever they access the same service or resource again.

There is a strong association between the above-mentioned security features and the attacks presented in Section 'Cloud identity management: known attack matrix'. Each of these mechanisms is capable of mitigating one or more attacks that could be launched against a Cloud IDMS. In order to provide better understanding, we have summarized our findings in Table 2, which depicts a clear mapping between the identified features, mechanisms and the attacks they can mitigate. For instance, if a Cloud IDMS offers **Authentication** using *Biometrics*, then it is capable of mitigating attacks such as brute-force (A1), dictionary attack (A6), forgery (A10), impersonation/identity spoofing (A13) and repudiation (A17), since biometrics involve natural characteristics (such as fingerprint, voice patterns etc.) that are unique for every individual and hard to counterfeit. Similarly, if a Cloud IDMS follows **Identity Federation** principle through *Hierarchical Storage*, then it will limit Unauthorized disclosure of confidential data (A7), Elevation of privileges (A9), Forgery (A10), Identity theft (A12) and Impersonation attack (A13), as the information will be stored in fragments across multiple servers. Therefore, although the attacker somehow manages to compromise one Cloud identity management server and accesses its stored information even then they are unable to launch any attack because all they can acquire is partial information that is simply insufficient to perform the attack.

Table 2 Relationship b/w Features, Mechanisms and Attack Matrix

Features	Mechanism	Mitigated attacks
	Something You Know (OTP & CR)	A2, A5, A6, A7, A10, A11, A12
Authentication	Something You Have (Tokens)	A5, A14
	Something You Are (Biometrics)	A1, A4, A5, A12
	Access Control Policies	A3, A6, A13
Authorization	OAuth	A5, A6, A7, A8, A10, A11
	Access Right Delegation	A3, A6
	Smart-card (Encryption)	A5, A12
Identity federation	Multiple IdPs and CSPs	A7, A8, A13, A14
	Hierarchical Storage	A6, A7, A8
	Distributed Computation	A15
	Proxy-systems	A8
	User-roles	A6
Privacy	Pseudonyms	A8, A10
	Encryption	A5, A7, A8, A10, A14, A15
	Limited Disclosure	A6, A7, A8
User-centricity	Consistet Experience	A9, A10
	Data Disclosures Policies	A3, A6, A9, A13
Audit & Logging	Activity Monitoring	A1, A4, A12
	History Maintenance	A9, A12

Results

Analysis of cloud identity management systems

Taxonomy, presented in Section 'Methods', is being used here as an assessment criterion for the evaluation of existing IDMSs. The objective of our analysis is to assist CSCs and CSPs in selecting the most appropriate IDMS that best suit their functional and security requirements. We have critically analyzed various Cloud based identity management systems; Table 3 provides a brief overview of the reviewed systems and their important technical features. In addition to this, identity management systems from each category (Deployment based, Featured based) have been analytically evaluated against the IDMS features discussed above and findings from the analysis are presented at the end of this section in Tables 4 and 5.

Deployment based classification

1. ***Isolated IDMS***

   - ***A Strong User Authentication Framework for Cloud Computing:*** Choudhury et al. (2011) propose a strong user authentication framework that conforms to the Isolated IDMS properties, where single Cloud SP is responsible for the storage, maintenance and verification of identity credentials and does not rely on any trusted third party (IdP). Proposed framework uses *smart card* based on bilinear pairings and *user password* to provide **two-level Authentication**, in order to avoids illegitimate access to Cloud resources and data. The framework offers **User-Centricity** by offering support to *self-service* feature via password change facility specified in the system design. The proposed secure Cloud framework ensures user **Privacy** via *limited disclosure principle* by using two separate channels for the transmission of sensitive user credentials that minimizes the impact of information disclosure and ascertain protection against identity theft attacks. Distribution of credentials is achieved in a way that some information is stored in the user's smart card, whereas *one-time key (password)* is sent by the server to user's mobile via SMS, thus offering support to **Identity federation** as well. However, the system specifies no **Authorization** mechanism. Furthermore, the design of proposed framework falls short, as far as **Auditing and Logging** is concerned.
   - ***Protection of Identity Information in Cloud Computing without Trusted third party:*** Ranchal et al. (2010) has described a security solution for Personal Identity Information (PII) that prevents unauthorized disclosure and usage of user's sensitive identity information. Proposed scheme ensures user **Privacy** via AB, which discloses restricted PII to the SP following the *limited disclosure* principle, and provides security against identity attacks. AB scheme offers **User Centricity** by incorporating 'Disclosure Policy' component through which users can specify policies for their data. *Disclosure policy* component ultimately ensures the security of their sensitive information accumulated at the Cloud SPs and offers a *Consistent user experience* across multiple SPs. Another significant feature of AB scheme is 'Identity Data', which stores encrypted identity information that is later used for secure user **Authentication**. 'Disclosure History' is yet another important feature of AB scheme which offers *history maintenance*, and is later used to carry

Table 3 Key Features of Cloud IDMSs

Type	Cloud IDMSs	Salient features
Deployment Based Cloud IDMS		
Isolated IDMS	A Strong User Authentication Framework for Cloud Computing (Choudhury et al. 2011)	- Conforms to isolated IDMS properties - Offers security and privacy of user by restricting illegal access - Mutual authentication (Challenge Response & OTP scheme) - Secure session key generation and distribution - Multi-factor authentication (Password and smart-card).
	Protection of Identity Information in Cloud Computing without Trusted third party (Ranchal et al. 2010)	- Isolated IDMS, since it does not rely on any trusted third-party - Protects PII against unauthorized disclosure - Computes assertions over encrypted data - Active bundle scheme for un-trusted hosts - Encrypted storage of identity data
Centralized IDMS	An Identity-Centric Internet: Identity in the Cloud, IDaaS and other delights (Ates et al. 2011)	- Realization of centralized Cloud IDMS - Defines the concept of Identity in Cloud Agents (IC-Agents) - IC-Agents as an identity proxy perform identity propagation transactions - Explains the IDaaS module in the context of Personal Data-as-a-Service - Authentication and Authorization as-a-Service module
	Distributed Identity for Secure Service Interaction (Chowdhury and Noll 2007)	- Presents a role based IDMS architecture - Categorizes digital identity as Personal, Corporate and Social identity - Restricted disclosure of identity credentials to the CSPs - Centralized IdP is responsible for the sharing and distribution of user's identity credentials
Federated IDMS	Security and Cloud Computing: ICIMI (Celesti et al. 2010)	- Inter-Cloud Identity Management Infrastructure (ICIMI) is a federated IDMS - Allows for the expansion of virtualization infrastructure - Establishment of trust among CSPs - Offers standardized, scalable & dynamic authentication
	Strengthen Cloud Computing Security with FIM Using HIBC (Yan et al. 2009)	- Allocates unique identities in hierarchal fashion - Mutual authentication for Hybrid Cloud environment - Handles the establishment of secret session keys
	Chord Based Identity Management for e-Healthcare Cloud Applications (Kim et al. 2010)	- SSO service for Cloud based e-Healthcare application - Uses Peer-to-Peer service model for load balancing - Distributes session information in the federated Cloud environment - Limits the number of authentication requests to central IdP
	Security APIs for My Private Cloud (Chadwick and Casenove 2011)	- Federated access rights to Cloud resources - Proposes Authz API for maintaining the identity database and defining the access control mappings - Authn API for authenticating the Cloud users - Delegation API to delegate access rights to anyone at any time

Table 3 Key Features of Cloud IDMSs *(Continued)*

Functionality Based Cloud IDMS		
	An Identity-Based OTP Scheme with Anonymous Authentication (Luo et al. 2009)	- Identity based One-time Password (OTP) authentication scheme - Operates on smart card based bilinear pairings - Generates a temporary identity to protect user's actual identity - Describes Process Setup and User Registration module - Guarantees user's anonymity and privacy throughout the communication process
Anonymous IDMS	UIMM Based on Anonymous Credentials (Zhang and Chen 2010)	- Universal Identity Management Model (UIMM) that operates on anonymous credentials - Allows for access right delegation - Ensures user's privacy preservation via unlikable self-generated pseudonyms - Extend WS-Federation to implement Identity Meta-system model.
	An Entity-centric Approach for Privacy and Identity Management in Cloud Computing (Angin et al. 2010)	- Entity-centric architecture for Identity Management in - Implements Active Bundles (AB) scheme to ensures user's anonymity - AB encapsulates Personal Identity Information (PII), Privacy preserving rules and VM (Virtual Machine) - Implements anonymous identification
	(Enhancing Privacy and Dynamic Federation in IdM for Consumer Cloud Computing (Sanchez et al. 2012)	- Dynamic privacy-enhanced federated identity management solution that defines an enhanced privacy - Introduces a new reputation protocol and implements Enhanced Client Profile (ECP) - Presents Trust aware IDM architecture that mainly comprises of Identity Management (IdM) layer and Trust layer - IdM Layer facilitates user authentication, authorizations and profile management - Trust layer deals with the management, negotiation and distribution of trust related data to other layers.
User-Centric IDMS	User-Controlled Automated Identity Delegation (Hoellrigl et al. 2010)	- Implements Identity Delegate that applies user defined data disclosure policies and resolves the information consistency problem - Allows for the integration of multiple IdPs and SPs - Dissemination of identity credentials is kept under the control of the identity owner

out auditing, thus offering support to **Auditing and Logging** principle. This scheme eliminates the need of a TTP for the handling of identity management functions as it follows the *distributed storage and computation* mechanism by distributing the secret identity information among multiple nodes to ensure security as well as **Identity Federation**. However, **Authorization** is not included in the proposed solution for the assurance of appropriate access control to the appropriate applications.

Table 4 Analysis of Deployment based IDMSs

Categories	Cloud IDMSs	Authn	Authz	Identity Federation	Privacy	User-Centricity	Audit & Logging
	A Strong User Authentication Framework for Cloud Computing (Choudhury et al. 2011)	Smart card + OTP	–	Mobile phone + Smart Card	Limited Disclosure	Self-Service	–
Isolated IDMS	Protection of Identity Info. in CC without TTP (Ranchal et 2010)	–	–	Distributed Computation	Limited Disclosure	Consistent Experience	History Maintenance
	An Identity-Centric Internet: Identity in the Cloud, IDaaS (Ates et al. 2011)	SSO	Access Control Policy	Distributed Computation	User Roles	Self-Service	Activity Monitoring
Centralized IDMS	Distributed Identity for Secure Service Interaction (Chowdhury and Noll 2007)	Mobile phone + Pwd	Access Control Policy	Distributed Computation	Limited Disclosure	Consistent Experience	–
	Security and Cloud Computing: ICIMI (Celesti et al. 2010)	SSO	–	Multiple IdPs	Limited Disclosure	Consistent Experience	–
	Strengthen Cloud Computing Security with FIM Using HIBC (Yan et al. 2009)	SSO	–	Hierarchical Storage	Limited Disclosure	Consistent Experience	–
Federated IDMS	Chord Based Identity Management for e-Healthcare Cloud Applications (Kim et al. 2010)	SSO	–	Distributed Computation	Encryption	Consistent Experience	–
	Security APIs for My Private Cloud: granting access to anyone (Chadwick and Casenove 2011)	SSO	Access Control Policy	Multiple IdPs	Limited Disclosure	Consistent Experience	Activity Monitoring

Table 5 Analysis of Deployment based IDMSs

Categories	Cloud IDMSs	Authn	Authz	Identity Federation	Privacy	User-Centricity	Audit & Logging
	An Identity-Based OTP Scheme with Anonymous Authentication (Luo et al. 2009)	Smart card + OTP	–	Smart Card	Pseudonyms	Self-Service	–
Anonymous IDMS	UIMM Based on Anonymous Credentials (Zhang and Chen 2010)	–	Access Control Policy	–	Pseudonyms + Data Disclosure Prefer.	–	–
	An Entity-centric Approach for Privacy & IDM in Coud Computing (Angin et al. 2010)	–	–	–	Limited Disclosure	–	History Maintenance
User-Centric IDMS	Enhancing Privacy and Dynamic Federation in IdM for Consumer Cloud Computing (Hoellrigl et al. 2010)	PKI	OAUTH	–	–	–	–
	User-Controlled Automated Identity Delegation (Sanchez et al. 2012)	User name + Pwd	OAUTH + Access Control Policy	Multiple IdPs	Data Disclosure Prefer.	–	–

2. ***Centralized IDMS***

   - ***An Identity-Centric Internet: Identity in the Cloud, IDaaS and other delights:*** Concept of Identity in Cloud Agents (IC-Agents) is presented in Ates et al. (2011), which is a logical identity proxy with Identity as a Service (IDaaS) component. The proposed concept of logical identity proxy (IC-Agent) ensures the segregation of user's PII while disseminating user's credentials across multiple SPs. Moreover, IC-Agents acquire *user consent* before disclosing their sensitive identity information mandatory for the verification process thus offering support to **Privacy** feature. Moreover, IC-Agents are capable of *monitoring and logging* all the operations of user (for instance authorizations and other service accesses), which may help to carry out the auditing processes as per SP's requirement, thus offering support to **Auditing and Logging** principle. Furthermore, IC-Agents provide interfaces called dashboards for user **Authentication** and are capable of functioning as a personal IdP to provide *SSO* facility that eliminates the need of repeated user logins for each independent resource access. IC-Agents is a **User-Centric** identity management concept where users are provided with full control over their identity information stored at IC-Agent host (Software and Hardware) by offering support to *self-service* mechanism. Proposed concept of IC-Agent is based on the *distributed storage architecture* for personal data storage where some data is stored by the IC-Agent, while other information is retrieved from multiple distinct IC-Agent sources thus follows the **Identity Federation** principle. IC-Agent applies the user defined *data disclosure policies* while forwarding user credentials to the SPs for secure authentication and **Authorization**.
   - ***Distributed Identity for Secure Service Interaction:*** Chowdhury and Noll (2007) present a Role based identity management architecture that utilizes users' mobile devices as unique attributes of their digital identity. Presented architecture ensures user's **Privacy** by following the *limited disclosure* principle, where each identifier discloses limited information to the SP while accessing multiple related services. System complements user privacy even more by accepting the segregation and distinction among the different identity contexts such as personal, social and corporate identity. **Authorization** is provided through role based access control model that implements *access control policies* considering user roles. The **Identity Federation** mechanism is ensured via *distributed storage and computation* mechanism, which distributes the identifying information across multiple locations, for instance partial user credentials are held in the network storage area, whereas the other half is stored at the user's mobile device (SIM card). Proposed system offers **User Centricity** since it offers *self-service* feature via my digital identity component that allows the user to alter, append or revoke his stored identity credentials. Moreover, network based identity data store (My digital identity) enables the user to define *data disclosure policies* that guarantees seamless and *consistent user experience* across multiple SPs. System provides **Authentication** via user's SIM card so belongs to the *"something you*

have" authentication mechanism; the system also proposes an extended SIM card architecture that holds multiple credentials for different authentication services. However, the system does not specify any **Auditing and Logging** functionality that is critical for maintaining and keeping the check and balance on CSCs as well as Cloud SPs for the security of IDMSs.

3. ***Federated IDMS***

   - ***Security and Cloud Computing-ICIMI:*** Inter-Cloud Identity Management Infrastructure (ICIMI) is described in Celesti et al. (2010), which is a federated IDMS. ICIMI uses IdP/SP model to provide **Authentication**, where Home Cloud performs one-time authentication to access the federated Foreign Cloud services that are in its shared trusted domain thus follows the *SSO* mechanism. This infrastructure offers support to **User-Centricity** by providing a solution that offers *consistent user experience* since it is independent of the underlying authentication mechanisms and incorporates standards such as SAML, which is capable of accepting assertions having different formats. Moreover, ICIMI is based on *distributed system architecture* comprising of multiple IdPs; therefore, the mechanism for storage and retrieval of user credentials follows the **Identity Federation** principle. In addition, this scheme offers **Privacy** via *limited disclosure principle* where Home Cloud hides the identity of CSCs by making the resource acquirement request on behalf of its consumers (users). However, the system incorporates no **Authorization** or access control mechanism that ensures legitimate access to Cloud services and resources. Secure logging to ensure **Auditing and Logging** is another significant feature of IDMSs but is not reflected in the design and architecture of the proposed infrastructure.
   - ***Strengthen Cloud Computing Security with FIM Using HIBC:*** A Federated Identity Management (FIM) system for Cloud along with Hierarchical Identity-Based-Cryptography (HIBC) is described in Yan et al. (2009), which allocates unique identities to users and servers in hierarchical fashion. This system provides **Authentication** via *SSO* mechanism that requires the user to use single identity credentials for accessing multiple services and resources that are provided by distinct SPs. HIBC offers support to **Identity Federation** principle by specifying the *hierarchical storage* approach for identity storage and maintenance. In addition, HIBC model ensures **User Centricity** by incorporating numerous authentication protocols supported by distinct Cloud SPs and offers a *consistent user experience* to the service consumers. The HIBC model ensures **Privacy** through *limited disclosure* principle where same exclusive identity information is disclosed in each resource access request. However, no access control policies are specified in the model for ensuring the legitimate access by authoritative users hence lacks **Authorization** capability. Similarly, proposed HIBC model lacks **Auditing and Logging** capability.
   - ***Chord Based Identity Management for e-Healthcare Cloud Applications:*** Kim et al. (2010) presents an algorithm, Chord for Cloud (C4C) so that the customers of one Cloud may use the services of the other Cloud environments, with single user identity. Proposed C4C algorithm offers the

Authentication via *SSO* principle where session values are extracted from the user's service request and are verified by the SP (node). After successful verification, user is given access to the requested service thus offering user with seamless service provisioning. This algorithm offers support to **User Centricity** by incorporating protocols such as SAML and OpenID that ensures *consistent user experience*. Moreover, C4C follows the **Identity Federation** principle by offering the *distributed storage and computation* architecture. User credentials are disseminated among multiple Cloud nodes for authentication purposes, whereas sensitive identity information at each processing node is secured via Intrusion Detection Systems. In addition, C4C is composed of multiple components out of which Session Manager (SM) module is responsible for maintaining the information about valid user sessions and after successful verification, this SM module initiates the **Authorization** process. SM module creates session for each authenticated user and specifies valid time slot for each created session. To ensure **Privacy**, user credentials are kept encrypted throughout the valid user sessions and revoked after the session is terminated thus following *limited disclosure* principle. The algorithm does not does not support **Auditing and Logging** feature.

– ***Security APIs for My Private Cloud:*** A conceptual model along with the security architecture is presented in Chadwick and Casenove (2011) that describes a set of three security APIs each designed for a specific functionality. Authentication API deals with the verification and identification of users and sends back user's identity credentials to the Cloud. **Authentication** is provided via *SSO*, where user logouts are restricted to the Cloud applications only and session with the authentication IdP/server remains active. **Authorization** API on the basis of user's identity attributes and *access control policies* decides what rights a particular Cloud user may possess and when to revoke those access rights. These security APIs enable the Cloud users to protect their **Privacy** in Cloud by following *limited disclosure* principle where users can specify their own privacy policies for the protection of their information from other users. This model enables the Cloud resource owners to monitor and log the activities of their potential users, in order to ensure that only the legitimate resource can be accessed by the user having valid delegated attributes thus offers support to **Auditing and Logging** principle. In order to provide support to **User-Centricity**, proposed model offers support to multiple IdP protocols such as SAML and OpenID and offers *consistent user experience*. Moreover, this model enables the user to relate his identity information across multiple SPs to raise his assurance level for authentication purposes thus offering support to **Identity Federation** feature as well.

Functional behavior class

1. ***Anonymous IDMS***

– ***An Identity-Based OTP Scheme with Anonymous Authentication:*** Identity based One-time Password (OTP) authentication scheme is presented in Luo et al. (2009) that operates on smart card based bilinear pairings. Proposed OTP scheme uses temporary identity information called *pseudonyms* to ensure anonymity and **Privacy** by service consumers. Scheme utilizes smart card that is based on bilinear pairings and generates *OTP* along with the temporary user identity to ensure anonymity and security in the **Authentication** process. **Identity Federation** feature is ensured through the combination of smart card and OTP where identity information is collected from *multiple sources* to generate user credentials that are finally used in the process of authentication or **Authorization**. OTP scheme offers support to **User-Centricity** by incorporating *self-service* feature through password change facility module thus enabling the user to change or update his password without any assistance from the server side. However, the proposed OTP scheme offers no **Auditing and Logging** that maintains the logs of all user activities.

- ***UIMM Based on Anonymous Credentials*** Zhang and Chen (2010) proposes a Universal Identity Management Model (UIMM) where users are allowed to prove the ownership of their identity credentials without having to communicate with the IdP. In the proposed model, **Privacy** is ensured via *pseudonym* based signature scheme that ensures the minimal and selective disclosure of user information. The **Authentication** module includes IdP that provides the *identity credentials* to users and is also responsible for performing their verification. UIMM also ensures **Authorized** access to multiple federated security realms. The identity meta-system component aims to provide **User-Centricity** via identity selector component that allows the user to choose the credentials for different resource accesses. On the contrary, UIMM offers no support for **Identity Federation**; even though it allows the user to access the information from anywhere, but that information is retrieved from a central location, which may result in single point of failure. This model does not offer support to **Auditing and Logging** principle.
- ***An Entity-centric Approach for Privacy and Identity Management in Cloud Computing*** Angin et al. (2010) present an entity-centric architecture for Identity Management in Cloud that uses Active Bundles (AB) scheme and ensures user's anonymity by applying privacy preserving policies. Presented scheme offers **Privacy** via *limited disclosure* principle, which implements information disclosure policy provided by AB scheme. Identity data module of the AB scheme comprises of the encrypted authentication information that the SP requires while performing user **Authentication**. AB scheme incorporates a *Disclosure history* module that offers support to future information disclosure decisions considering past user interactions and ensure **Logging and Auditing** feature. However, proposed entity-centric architecture lacks support for **User-Centricity** since, no support for Consistent user experience is provided by the architecture. All the identity information is stored in the AB; a centralized location, **Identity Federation** principle is not followed. **Authorization** module that can ensure legitimate resource access is not incorporated in this architecture.

2. ***User-Centric IDMS***

- ***Enhancing Privacy and Dynamic Federation in IdM for Consumer Cloud Computing:*** Sánchez et al. (2012) presents a dynamic privacy-enhanced federated identity management solution for on-demand resources provisioning in cloud computing scenarios. Proposed system provides **Privacy** management along with the complete control of users over their data through ECP and privacy engine modules. The system also offers **User Centricity** via user profile management module that allows the user to perform *self-service*. System incorporates **Authentication** module that supports various authentication mechanisms such as user *name/password* and *digital certificates*. User resource and service accesses are monitored via *activity monitoring* module, thus facilitating the **Auditing and Logging** feature. The **Authorization** module uses OAUTH protocol for the exchange of user's identity attributes in the process of authorization. Proposed system offers support to dynamic Cloud federation where Cloud trusts are established; however, it does not offer support to distributed storage mechanisms, hence, does not provide **Identity federation** features.

- ***User-Controlled Automated Identity Delegation:*** Hoellrigl et al. (2010) has described a user-controlled automated identity delegation solution that attempts to address the information consistency problem in a user-centric identity management system. Proposed system implements a User Access Management (UAM) module that enhances *OAUTH* protocol and offer support to **Authorization** via user defined *access control policies* and preferences. The system allows the user to *federate his identity* information across multiple IdPs thus offering support to **Identity Federation** feature. It offers **Authentication** by simple user *name and password* mechanism. It supports **User Centricity** by automating the process of disseminating the identity attributes in accordance with the user's *data disclosure preferences*. System fails to provide **Privacy**, since user information can be linked across various malicious SPs that may interact to correlate user identities. Similarly, this system does not monitor and log the user activities thus lack support to **Logging and Auditing** feature.

Discussion

Research findings & future research directions

We have performed an in-depth analysis of various Cloud based IDMSs in Section 'Results', which reveals that most of the systems do not offer support to all the essential features of Cloud IDMS and the ones that do, have their own weaknesses. None of the discussed techniques heuristically covers all the security features; moreover, they lack compliance to international standards which, understandably, undermines their credibility. Besides, identified features are worth considering, but are not mandatory as security and business requirements vary from one organization to the other. Organizations moving towards Cloud would benefit from this work by evaluating the security of the IDMSs in advance to avoid any adversity.

Furthermore, we have identified the current gaps, challenges and future key-trends related to Cloud identity management systems that must be considered by the Cloud identity service providers and subscribers. According to the most recent analysis, Identity with access management has been included among the top ten mega-trends for year 2014 by many well-known vendors and consultancy organizations including CA Technologies,

Gartner, Ping Identity and Oracle. Although the applications for managing identities vary from region to region, and even from business to business, they generally fall into one of the following areas.

1. ***Service automation:*** (Rose et al. 2011) By combining the individual's personal information into business processes organization shall be able to achieve efficient, automated and simplified transactions rather than the manual user input in activities such as user identification, authentication and authorization.
2. ***Increased user control:*** (Rose et al. 2011) Electronic identity enables the individuals to perform transactions on their own, without any assistance from the administrators or IT professionals. Self-service is a key example of user enablement. This aspect has recently achieved much attention from the IT industry and seems to continue its growth in new businesses as well.
3. ***Naming and identification of resources:*** Cloud providers offers wide range of services and resources including servers, storage and network to its subscribers. Therefore, to avoid any conflict and confusion, inter-Cloud IDMS must specify a mechanism for the naming and identification of Cloud services and resources.
4. ***Interoperability of Identity information:*** While performing user authentication and authorization, identity credentials could be expressed in a variety of ways for instance X.509 certificates, SAML assertions or WS-Federation security tokens. In inter-Cloud scenario, such representations result in syntactic and semantic issues hence requires interoperable identity information.
5. ***Identity life cycle Management:*** Throughout the life time of an entity, modifications in its attribute values, access permissions and service provisioning may occur. Consequently, all identity management systems are required to have updated and synchronized identity information to avoid any conflicts caused by the usage of old user data.
6. ***Single sign-on for interactions on the Inter-Cloud:*** Single sign-on must be implemented in a way that if users get successfully authenticated at one Cloud service provider, then throughput their valid session they must not be prompted again for authentication on any other (member) Cloud service provider.
7. ***Hardware components in cellular devices will become an increasingly important part of Identity management systems:*** (CA Technologies Predicts Key Trends for Identity and Access Management in 2014 (2014)) Due to the increase in the adoption of new mobile devices, advanced hardware protection technologies, such as ARM TrustZone, will become widely available. In addition to securing the hardware, users will be able to use their cell phones for authentication and identity proofing purposes while performing sensitive functions like financial transactions.
8. ***Lack of scalable identity proofing will continue to vex broader B2C/G2C deployments:*** As more and more users register for online services, validation of digital identities in an accurate and scalable manner is becoming a constant challenge (CA Technologies Predicts Key Trends for Identity and Access Management in 2014 (2014)). In addition to this, with the increase in identity proofing sources (such as new public and private records hosted by business or government organizations), the demand for scalable identity proofing will be further strengthened in 2014.

Consequently, it will require the IT industry to work in partnership and deliver scalable identity proofing mechanisms.

9. ***Risk-based authentication expands beyond Financial Services:*** The requirement for both strong authentication and consistent user experience will result in the widespread adoption of risk-based authentication systems. In such systems, the contextual information about service consumers, hardware devices and software applications is analyzed to determine the risk level for consumer's identity (CA Technologies Predicts Key Trends for Identity and Access Management in 2014 (2014)). Initially these services were adopted by the financial services only, however with the increase in requirement for improved security with improved user convenience; this technology will begin to be more widely adopted across other industries as well.

Above, we have provided a comprehensive list of state-of-the-art challenges and gaps that are needed to be addressed by the Cloud IDMS research community, and without which the very phenomenon of Cloud identity management is not fully achieved.

Conclusion

In this paper, we have presented research that is a step towards the development of an assessment criterion that can be used by other researchers and industry professionals to perform the evaluation of existing, as well as future Cloud based IDMSs. Since the domain of Cloud IDMS is still in its early stages, therefore, requires considerable attention from the research community and IT industry. We have contributed in this regard, by presenting a research work that holistically covers the domain of Cloud IDMSs. In particular, our contribution is multi-dimensional: first, we have presented a comprehensive list of security attacks that involve Cloud identity management systems and identity credentials as an attack tool or target, secondly we have identified the features that will act as countermeasure against the mentioned attacks. Thirdly, state-of-the-art mechanisms against each feature are filtered out with an objective to maximize their performance, security and interoperability. As another positive contribution, we have explicated the feature-mechanism relationship in the form of a well-organized taxonomy which is later used to analyze/evaluate various Cloud based IDMSs. Lastly, we have applied the previously discussed attacks against each Cloud IDMS to confirm their reliability and applicability in Cloud. Our findings from the analysis are presented in the subsequent tables which prove that Cloud IDMSs invariably have some pros and cons in their architecture and functionality. Furthermore, most of these Cloud IDMSs have limitations in terms of their reliability and applicability, as they are confined only to specific Cloud identity management scenarios. In conclusion, our work will augment the research which leads to the development and designing of a robust and holistically secure Cloud based IDMS covering all the identified features. Presented taxonomy will allow CSCs and CSPs to make a knowledgeable decision by selecting an appropriate IDMS that best satisfies their security and functional requirements.

Competing interests

The authors declare that they have no competing interests.

Authors' contributions

UH: has been lead author in drafting the manuscript, has conducted in-depth literature review, performed categorization, has done the detailed analysis of existing identity management solutions, and has revised the paper for important intellectual content. RM: has made key contributions to the paper conception and provided continuous support to the first author during all stages of the paper drafting,and gave final approval of the version to be published.

MAS: has significantly supported the first author during regular meetings, critically revised the paper for important intellectual content, provided new ideas, and gave final approval of the version to be published. MN: has significantly contributed to the research by introducing large-scale perspective for cloud IDMSs, critically revised the paper, and gave approval of the version to be published. All authors read and approved the final manuscript.

Acknowledgements
The authors would like to acknowledge the continual financial support provided by National University of Sciences and Technology (NUST) Pakistan in publishing related research papers. We would like to thank Ms. Yumna Ghazi for proofreading the paper.

Author details
[1] Department of Computing, School of Electrical Engineering & Computer Science, National University of Sciences & Technology (NUST), Sector H-12, Islamabad, Pakistan. [2] COMSATS Institute of IT, Islamabad, Pakistan.

References

Albeshri A, Caelli W: **Mutual protection in a cloud computing environment.** In *High Performance Computing and Communications (HPCC), 2010 12th IEEE International Conference on*. Piscataway, New Jersey, United States: IEEE; 2010:641-646.

Almorsy M, Grundy J, Müller I: **An analysis of the cloud computing security problem.** In *Proceedings of APSEC 2010 Cloud Workshop*. Sydney, Australia; 2010.

Alrodhan WA, Mitchell CJ: **Enhancing user authentication in claim-based identity management.** In *Collaborative Technologies and Systems (CTS), 2010 International Symposium on*. Piscataway, New Jersey, United States: IEEE; 2010:75–83.

Angin P, Bhargava B, Ranchal R, Singh N, Linderman M, Othmane LB, Lilien L: **An entity-centric approach for privacy and identity management in cloud computing.** In *Reliable Distributed Systems, 2010 29th IEEE Symposium on*. Piscataway, New Jersey, United States: IEEE; 2010:177–183.

Arias-Cabarcos P, Almenárez-Mendoza F, Marín-López A, Díaz-Sánchez D, Sánchez-Guerrero R: **A metric-based approach to assess risk for "on cloud" federated identity management.** *J Netw Syst Manag* 2012, **20:**513–533. Springer, 2012.

Ates M, Ravet S, Ahmat AM, Fayolle J: **An identity-centric internet: identity in the cloud, identity as a service and other delights.** In *2012 Seventh International Conference on Availability, Reliability and Security*. Piscataway, New Jersey, United States: IEEE; 2011:555–560.

Bertino E, Takahashi K: **IdentityManagement: Concepts, Technologies, and Systems:** ARTECH HOUSE, 16 Sussex Street, London SW1V 4RW UK; 2010.

Bhadauria R, Chaki R, Chaki N, Sanyal S: **A survey on security issues in cloud computing.** 2011. arXiv preprint arXiv:1109.5388.

Bhargav-Spantzel A, Camenisch J, Gross T, Sommer D: **User centricity: a taxonomy and open issues.** *J Comput Secur* 2007, **15:**493–527. IOS Press, 2007.

Brute Force Attack. *OWASP Testing Guide* 2013. [https://www.owasp.org/index.php/Brute_force_attack] [Online accessed September-2013].

CA Technologies Inc.: **CA Technologies Predicts Key Trends for Identity and Access Management in 2014.** 2014. [http://www.ca.com/us/news/press-releases/na/2014/ca-technologies-predicts-key-trends-for-identity-and-access-management-in-2014.aspx]. Online accessed April-2013.

Cameron K: *The laws of identity*: Microsoft Corp.; 2005.

Cao Y, Yang L: **A survey of identity management technology.** In *IEEE International Conference on Information Theory and Information Security (ICITIS)*: IEEE; 2010:287–293.

Celesti A, Tusa F, Villari M, Puliafito A: **Security and cloud computing: intercloud identity management infrastructure.** In *Enabling Technologies: Infrastructures for Collaborative Enterprises (WETICE), 2010 19th IEEE International Workshop on*. Piscataway, New Jersey, United States: IEEE; 2010:263–265.

Chadwick DW, Casenove M: **Security APIs for my private cloud-granting access to anyone, from anywhere at any time.** In *Cloud Computing Technology and Science (CloudCom), 2011 IEEE Third International Conference on*. Piscataway, New Jersey, United States: IEEE; 2011:792–798.

Chang C-C: **Some forgery attacks on a remote user authentication scheme using smart cards.** *Informatica* 2003, **14:**289–294. IOS Press, 2003.

Chen D, Zhao H: **Data security and privacy protection issues in cloud computing.** In *Computer Science and Electronics Engineering (ICCSEE), 2012 International Conference on. Volume 1*. Piscataway, New Jersey, United States: IEEE; 2012:647–651.

Chen J, Wu X, Zhang S, Zhang W, Niu Y: **A decentralized approach for implementing identity management in cloud computing.** In *Cloud and Green Computing (CGC), 2012 Second International Conference on*. Piscataway, New Jersey, United States: IEEE; 2012:770–776.

Chowdhury MM, Noll J: **Distributed identity for secure service interaction.** In *Wireless and Mobile Communications, 2007. ICWMC'07. Third International Conference on*. Piscataway, New Jersey, United States: IEEE; 2007:56.

Choudhury AJ, Kumar P, Sain M, Lim H, Jae-Lee H: **A strong user authentication framework for cloud computing.** In *Services Computing Conference (APSCC), 2011 IEEE Asia-Pacific*. Piscataway, New Jersey, United States: IEEE; 2011:110–115.

Cloud Security Alliance: **Cloud Security Alliance SecaaS Guidance, Category 1: Identity and Access Management, 2012 by Cloud Security Alliance.** [https://cloudsecurityalliance.org/download/secaas-category-1-identity-and-access-management-implementation-guidance/] [Online accessed: November 2013], 2011.

Conrado C, Kamperman F, Schrijen GJ, Jonker W: **Privacy in an identity-based DRM system.** In *the Proceedings of Database and Expert Systems Applications, 2003. Proceedings. 14th International Workshop on*. Piscataway, New Jersey, United States: IEEE; 2003:389–395.

Dhamija R, Dusseault L: **The seven flaws of identity management: usability and security challenges.** *IEEE Secur Privacy* 2008, **6:**24–29. IEEE, 2008.

Donevski A, Ristov S, Gusev M: **Security assessment of virtual machines in open source clouds.** In *Information & Communication Technology Electronics & Microelectronics (MIPRO), 2013 36th International Convention on*. Piscataway, New Jersey, United States: IEEE; 2013:1094–1099.

Eucalyptus Systems Inc.: **Eucalyptus Identity and Access Management (IAM).** 2012. [https://www.eucalyptus.com/docs/eucalyptus/4.0/security-guide/security_bp_access.html], [Online accessed August-2014].

Feng J, Chen Y, Ku WS, Liu P: **Analysis of integrity vulnerabilities and a non-repudiation protocol for cloud data storage platforms.** In *Parallel Processing Workshops (ICPPW), 2010 39th International Conference on*. Piscataway, New Jersey, United States: IEEE; 2010:251–258.

Ferdous MS, Poet R: **A comparative analysis of identity management systems.** In *High Performance Computing and Simulation (HPCS) 2012 International Conference on*. Piscataway, New Jersey, United States: IEEE; 2012:454–461.

Fox A, Griffith R, Joseph A, Katz R, Konwinski A, Lee G, Patterson D, Rabkin A, Stoica I: *Above the Clouds: a Berkeley View of Cloud Computing*. Berkeley: Dept. Electrical Eng. and Comput. Sciences, University of California; 2009. Rep. UCB/EECS, 28, 13.

Gopalakrishnan A: **Cloud computing identity management.** *SETLabs Briefings* 2009, **7:**45–54.

Gunjan K, Sahoo G, Tiwari RK: **Identity management in cloud computing–a review.** In *International Journal of Engineering Research and Technology (Vol. 1, No. 4 (June-2012))*. Kudasan, Gandhinagar, Gujarat, India: ESRSA Publications; 2012.

Halpert B: *Auditing Cloud Computing: a Security and Privacy Guide (Vol. 21)*. Hoboken, New Jersey, USA: John Wiley & Sons; 2011.

Hoellrigl T, Kühner H, Dinger J, Hartenstein H: **User-controlled automated identity delegation.** In *Network and Service Management (CNSM), 2010 International Conference on*. Piscataway, New Jersey, United States: IEEE; 2010:230-233.

Jansen WA: **Cloud hooks: Security and privacy issues in cloud computing.** In *System Sciences (HICSS), 2011 44th Hawaii International Conference on*. Piscataway, New Jersey, United States: IEEE; 2011:1–10.

Jansen W, Grance T: *Guidelines on Security and Privacy in Public Cloud Computing*: NIST special publication, 800, 144, NIST; 2011.

Jensen M, Schwenk J, Gruschka N, Iacono LL: **On technical security issues in cloud computing.** In *Cloud Computing, 2009. CLOUD'09. IEEE International Conference on*. Piscataway, New Jersey, United States: IEEE; 2009:109–116.

Jøsang A, Fabre J, Hay B, Dalziel J, Pope S: **Trust requirements in identity management.** In *Proceedings of the 2005 Australasian workshop on Grid computing and e-research-Volume 44*: Australian Computer Society, Inc.; 2005:99–108.

Kim IK, Pervez Z, Khattak AM, Lee S: **Chord based identity management for e-Healthcare cloud applications.** In *Applications and the Internet (SAINT), 2010 10th IEEE/IPSJ International Symposium on*. Piscataway, New Jersey, United States: IEEE; 2010:391–394.

Kumar R, Gupta N, Charu S, Jain K, Jangir SK: *Open Source Solution for Cloud Computing Platform Using OpenStack*; 2014.

Kumaraswamy S, Lakshminarayanan S, Reiter M, Stein J, Wilson Y: **Domain 12: Guidance for Identity & Access Management V2. 1.** *Cloud Security Alliance* 2010. [https://cloudsecurityalliance.org/guidance/csaguide-dom12-v2.10.pdf]

Lang U: **Openpmf scaas: authorization as a service for cloud & soa applications.** In *Cloud Computing Technology and Science (CloudCom) 2010 IEEE Second International Conference on*. Piscataway, New Jersey, United States: IEEE; 2010:634–643.

Leandro MA, Nascimento TJ, dos Santos DR, Westphall CM, Westphall CB: **Multi-tenancy authorization system with federated identity for cloud-based environments using shibboleth.** In *ICN 2012, The Eleventh International Conference on Networks*; 2012:88–93.

Leskinen J: **Evaluation criteria for future identity management.** In *Trust, Security and Privacy in Computing and Communications (TrustCom), 2012 IEEE 11th International Conference on*. Piscataway, New Jersey, United States: IEEE; 2012:801–806.

Li M, Yu S, Ren K, Lou W: **Securing personal health records in cloud computing: Patient-centric and fine-grained data access control in multi-owner settings.** In *Security and Privacy in Communication Networks*. Berlin, Heidelberg: Springer; 2010:89–106.

Luo S, Hu J, Chen Z: **An identity-based one-time password scheme with anonymous authentication.** In *Networks Security, Wireless Communications and Trusted Computing, 2009. NSWCTC'09. International Conference on. Volume 2*. Piscataway, New Jersey, United States: IEEE; 2009:864–867.

Mahmood Z: **"Data location and security issues in cloud computing".** In *Emerging Intelligent Data and Web Technologies (EIDWT), 2011 International Conference on*. Piscataway, New Jersey, United States: IEEE; 2011:49–54.

Mather T, Kumaraswamy S, Latif S: *Cloud security and privacy: an enterprise perspective on risks and compliance*. Sebastopol, CA, USA: O'Reilly Media, Inc.; 2009.

McCallister E: *Guide to Protecting the Confidentiality of Personally Identifiable Information*. Collingdale, PA, United States: Diane Publishing; 2010.

Maler E, Mishra P, Philpott R: **Assertions and Protocols for the OASIS Security Assertion Markup Language (SAML).** 2003. [https://www.oasis-open.org/committees/download.php/3406/oasis-sstc-saml-core-1.1.pdf]. Online accessed January-2013.

Meier JD, Farre C, Taylor J, Bansode P, Gregersen S, Sundararajan M, Boucher R: *Improving Web Services Security: Scenarios and Implementation Guidance for WCF*: Microsoft Developer Network; 2009.

Nabeel M, Bertino E, Kantarcioglu M, Thuraisingham B: **Towards privacy preserving access control in the cloud.** In *Collaborative Computing: Networking, Applications and Worksharing (CollaborateCom), 2011 7th International Conference on*. Piscataway, New Jersey, United States: IEEE; 2011:172–180.

O'Gorman L: **Comparing passwords, tokens, and biometrics for user authentication.** In *Proceedings of the IEEE. Volume 91(12)*. Piscataway, New Jersey, United States: IEEE; 2003:2021–2040.

Olden E: **Architecting a cloud-scale identity fabric.** *Computer* 2011, **44:**52–59. IEEE, 2011.

Pashalidis A, Mitchell CJ: **A taxonomy of single sign-on systems.** In *Information Security and Privacy*. Berlin, Heidelberg: Springer; 2003:249–264.

Pearson S, Benameur A: **Privacy, security and trust issues arising from cloud computing.** In *Cloud Computing Technology and Science (CloudCom) 2010 IEEE Second International Conference on*. Piscataway, New Jersey, United States: IEEE; 2010:693–702.

Ranchal R, Bhargava B, Othmane LB, Lilien L, Kim A, Kang M, Linderman M: **Protection of identity information in cloud computing without trusted third party.** In *Reliable Distributed Systems, 2010 29th IEEE Symposium on*. Piscataway, New Jersey, United States: IEEE; 2010:368–372.

Ratha NK, Connell JH, Bolle RM: **Enhancing security and privacy in biometrics-based authentication systems.** *IBM Syst J* 2001, **40:**614–634. IBM, 2001.

Rhoton J: **Discover OpenStack: the identity component keystone.** 2013. [http://www.ibm.com/developerworks/cloud/library/cl-openstack-keystone/index.html?ca=dat-]. Online accessed August 2014.

Rimal BP, Choi E, Lumb I: **A taxonomy and survey of cloud computing systems.** In *INC, IMS and IDC, 2009. NCM'09. Fifth International Joint Conference on*. Piscataway, New Jersey, United States: IEEE; 2009:44–51.

Rose J, Rehse O, Rober B: *The value of our digital identity*: The Boston Consulting Group; 2011. [http://www.libertyglobal.com/PDF/public-policy/The-Value-of-Our-Digital-Identity.pdf], Online accessed March-2013.

Sanchez R, Almenares F, Arias P, Diaz-Sanchez D, Marín A: **Enhancing privacy and dynamic federation in IdM for consumer cloud computing.** *IEEE Trans Consum Electron* 2012, **58:**95–103. IEEE, 2012.

Salsano S, Veltri L, Papalilo D: **SIP security issues: the SIP authentication procedure and its processing load.** *Network* 2002, **16:**38–44. IEEE, 2002.

Saripalli P, Walters B: **Quirc: A quantitative impact and risk assessment framework for cloud security.** In *Cloud Computing (CLOUD), 2010 IEEE 3rd International Conference on*. Piscataway, New Jersey, United States: IEEE; 2010:280–288.

Saroiu S, Wolman A: **Enabling new mobile applications with location proofs.** In *Proceedings of the 10th workshop on Mobile Computing Systems and Applications*: ACM; 2009:3.

Senk C, Dotzler F: **Biometric Authentication as a service for enterprise identity management deployment: a data protection perspective.** In *Availability, Reliability and Security (ARES), 2011 Sixth International Conference on*. Piscataway, New Jersey, United States: IEEE; 2011:43–50.

Shin D, Lopes R, Claycomb W: **Authenticated dictionary-based attribute sharing in federated identity management.** In *Information Technology: New Generations, 2009. ITNG'09. Sixth International Conference on*. Piscataway, New Jersey, United States: IEEE; 2009:504–509.

Slone S: *Identity management. A white paper*: The open group identity management work area; 2004.

Sodhi G: **User provisioning with SPML.** *Inf Secur Tech Rep* 2004, **9**(1):86–96.

Subashini S, Kavitha V: **A survey on security issues in service delivery models of cloud computing.** *J Netw Comput Appl* 2011, **34:**1–11. Elsevier.

Suriadi S, Foo E, Jøsang A: **A user-centric federated single sign-on system.** *J Netw Comput Appl* 2009, **32:**388–401. Elsevier, 2009.

Thompson DR, Chaudhry N, Thompson CW: **"RFID security threat model".** In *the proceedings of Conference on Applied Research in Information Technology*; 2006.

Van der Hof S: *Innovating Government: An Introduction to the Book (pp. 1-14)*. R.J. Schimmelpenninckklaan, JN Den Haag: TMC Asser Press; 2011.

Wang JJ, Mu S: **"Security issues and countermeasures in cloud computing".** In *Grey Systems and Intelligent Services (GSIS), 2011 IEEE International Conference on*. Piscataway, New Jersey, United States: IEEE; 2011:843–846.

Wang L, Wang L, Mambo M, Okamoto E: **New identity-based proxy re-encryption schemes to prevent collusion attacks.** In *Pairing-Based Cryptography-Pairing 2010*. Berlin, Heidelberg: Springer; 2010:327–346.

Windley PJ: *Digital Identity*. Sebastopol, CA, USA: O'Reilly Media, Inc.; 2005.

Yan L, Rong C, Zhao G: **Strengthen cloud computing security with federal identity management using hierarchical identity-based cryptography.** In *Cloud Computing*. Berlin, Heidelberg: Springer; 2009:167–177.

Yassin AA, Jin H, Ibrahim A, Qiang W, Zou D: **Efficient password-based two factors authentication in cloud computing.** *Int J Secur Appl* 2012, **6:**143–148.

You P, Peng Y, Liu W, Xue S: **Security issues and solutions in cloud computing.** In *Distributed Computing Systems Workshops (ICDCSW), 2012 32nd International Conference on*. Piscataway, New Jersey, United States: IEEE; 2012:573–577.

Youseff L, Butrico M, Da Silva D: **Toward a unified ontology of cloud computing.** In *Grid Computing Environments Workshop, 2008. GCE'08*. Piscataway, New Jersey, United States: IEEE; 2008:1–10.

Zhang Y, Chen JL: **Universal identity management model based on anonymous credentials.** In *Services Computing (SCC), 2010 IEEE International Conference on*. Piscataway, New Jersey, United States: IEEE; 2010:305–312.

Zissis D, Lekkas D: **Addressing cloud computing security issues.** *Future Generat Comput Syst* 2012, **28:**583–592. Elsevier, 2012.

Zhou Y, Feng D: **Side-channel attacks: ten years after its publication and the impacts on cryptographic module security testing.** In *IACR Cryptology ePrint Archive, Volume 2005*; 2005:388.

A generic and adaptive aggregation service for large-scale decentralized networks

Evangelos Pournaras[1*], Martijn Warnier[2] and Frances MT Brazier[2]

*Correspondence:
e.pournaras@tudelft.nl
[1]Faculty of Electrical Engineering, Mathematics and Computer Science, Delft University of Technology, Mekelweg 4, 2628 CD, Delft, Netherlands
Full list of author information is available at the end of the article

Abstract

Purpose: Aggregation functions are used in distributed environments to make system-wide information locally available in the nodes of a network. The computation of different aggregation functions, e.g., SUMMATION, AVERAGE, MAXIMUM etc., in large-scale distributed systems is challenging and crucial for a wide range of applications. This is especially the case when the input values of these functions dynamically change during system runtime. Related approaches of decentralized aggregation are function-dependent, interaction-dependent, assume static values or cannot always tolerate duplicates and continuously changing information.

Methods: This paper introduces DIAS, the Dynamic Intelligent Aggregation Service. DIAS is an agent-based middleware that addresses these issues with a holistic approach: an efficient availability of the distributed information in every node of the network that enables the simultaneous computation of almost any aggregation function. Such an abstraction initially requires a significant communication and storage cost and has a rather large overhead. These issues are resolved by introducing an implicit local representation and storage of the explicit distributed information: aggregation memberships in bloom filters.

Results: The performance impact of bloom filters in DIAS is critical for its applicability as it compensates and reduces the initial high communication and storage required for such an abstraction.

Conclusions: Experimental evaluation under various aggregation and resource-constrained settings shows that DIAS is an efficient and accurate decentralized aggregation service.

Keywords: Aggregation; Adaptation; Agent; Bloom filter; Consistency

Background

The increasing scale and decentralization of distributed systems and applications results in an information gap: Agents, with partial knowledge about a system, require the local availability of collective and summarized knowledge about the state of the whole system to perform decision-making, adapt execution of their tasks and meet global application objectives. Therefore, aggregation of information becomes a crucial requirement to acquire such collective and summarized knowledge for a wide range of distributed applications.

Centralized computation of aggregation functions is straightforward as the whole set of information is available in one location. However, centralized aggregation is not always an option for reasons that may concern scalability or privacy. This paper focuses on the problem of decentralized aggregation of information distributed across the nodes of a network. Aggregation functions such as SUMMATION, AVERAGE, MAXIMUM, etc. are locally computed by each node of the network. The input of these functions can be arithmetic values collected from each node of the network as well. Communication, storage and processing costs are fundamental issues that challenge the design of a generic service for decentralized aggregation.

Related aggregation methodologies are function-dependent, interaction-dependent, assume static values or cannot always tolerate duplicates and continuously changing information (Ahmed et al. 2006; Haridasan and van Renesse 2008; Jelasity et al. 2005; Kashyap et al. 2006; Kempe et al. 2003; Nath et al. 2008). In contrast, this paper introduces a generic, agent-based and middleware for dynamic decentralized aggregation, DIAS, the *Dynamic Intelligent Aggregation Service*. DIAS is based on a holistic approach: availability of distributed information in every node of the network that enables simultaneous computation of almost any aggregation function. DIAS is based on the concept of *aggregation membership* to make this holistic approach possible. Aggregation memberships are aggregation information derived and abstracted from the explicit aggregation values. For example, an agent has memberships of other agents whose information is aggregated. Complementarily, an aggregate of an agent has memberships of aggregated information in other agents. This paper shows that such implicit information can be locally and efficiently stored in probabilistic data structures, the bloom filters (Bloom 1970).

A known problem of bloom filters is that of false positives (Bloom 1970). A false positive incorrectly denotes that some information is stored in a bloom filter when it is actually not. DIAS is able to detect inconsistencies such as duplicate and outdated information under the effect of false positives in bloom filters. This paper shows how detection is possible by *mutually checking the memberships* between the remote agents of DIAS without introducing additional communication. Experimental evaluation illustrates the efficiency and performance trade-offs of DIAS. High accuracy is achieved under a range of aggregation and resource-constrained settings.

This paper is outlined as follows: Section "Problem description" illustrates the problem description and related work. Section "System overview" outlines the architecture of DIAS. Section "Modeling of dynamics" introduces the model of dynamic aggregation in DIAS. Section "Dissemination and collection" illustrates the information dissemination and collection in DIAS. Section "Consistent aggregation sessions" shows the concept of aggregation membership and Section "Computation of aggregates" outlines how they are used to accurately compute aggregation functions. Section "Realization based on bloom filters" follows with a bloom filter realization of the aggregation memberships. Section "Experimental evaluation" evaluates the performance of DIAS. Section "Discussion and future work" discusses the approach of DIAS and outlines future work. Finally, Section "Conclusions" concludes this paper.

Problem description

Assume an overlay network of nodes, all having an *aggregation value* about the state of a (application) parameter. In this paper, an aggregation value is represented by a numerical (real) value. *Aggregation* is defined in this paper as the computation of aggregation functions (aggregates), e.g., SUMMATION, by all of the nodes of an overlay network with input the total aggregation values in this overlay network. Aggregation is *decentralized* if it can be performed without using any centralized computational entity for this purpose. Most decentralized aggregation systems have the following features:

- *Function-dependence*: Distributed applications may require the computation of a wide range of aggregation functions. AVERAGE, SUMMATION, MAXIMUM and MINIMUM are common numerical aggregation functions. Textual and rule aggregation are more complex. Aggregation functions share different mathematical properties (Calvo et al. 2002) and, therefore, their computational requirements may vary significantly. Due to this reason, different aggregation methodologies have been developed for specific aggregation functions or classes of aggregation functions. For example, gossip based aggregation (Jelasity et al. 2005) calculates the AVERAGE function as an iterative variance reduction algorithm over the values of nodes in an overlay network. Nonetheless, the COUNT operator that estimates the number of participating nodes cannot be calculated without additional protocol complexity to effectively apply the 'inverse birthday paradox' (Jelasity et al. 2005). The SUMMATION operator is derived by the product estimation of AVERAGE and COUNT and therefore, two instances of gossiping protocols are required. Similar issues are raised (Kempe et al. 2003) together with inaccuracy issues when there are failures in the network.
- *Interaction-dependence*: Most aggregation methodologies are designed in line with the properties, strengths and constraints of the network interaction mechanism that supports them, i.e., gossiping or routing over tree topologies. Replacing the interaction mechanism of an aggregation methodology with a different one makes this methodology inaccurate, cost-ineffective and actually infeasible. The interaction-independence of aggregation methodologies that this paper focuses on concerns the actual option to use a single aggregation mechanism over different interaction mechanisms. However, this abstraction cannot satisfy that the performance of aggregation is comparable between different interaction mechanisms. The variance reduction algorithm applied in gossip-based aggregation (Jelasity et al. 2005) requires gossiping communication between peers in a network. Information diffusion based on which distributed aggregation is performed also depends on a similar gossiping protocol (Kempe et al. 2003). Aggregation over structured topologies, such as trees, relies on multicasting. For example, tree aggregation requires unique paths between nodes in an overlay network to avoid double-counting. This requirement is not satisfied in unstructured (random) overlay networks maintained by gossiping protocols.
- *Static aggregation values*: Aggregation values may change and be derived from a continuous or discrete domain of values. Speed of change matters. Distributed aggregation schemes may be infeasible if aggregation values are highly dynamic. Investigating the degree of tolerable changes in the aggregation values of nodes is crucial for realizing a dynamic aggregation system. Adapting the aggregates with the new aggregation values is potentially a better solution than performing an expensive re-computation.
- *Inaccuracies*: Inaccuracies are estimations of aggregates with significant deviations from the actual aggregates. Two types of inaccuracies are studied:

(i) double-counting and (ii) outdated aggregation values, i.e., values that have changed during runtime. In duplicate-sensitive aggregation functions, such as SUMMATION, summing aggregation values twice results in an inaccurate aggregate. The same holds if aggregation values of nodes in an overlay network change dynamically during system runtime. Aggregates require adaptation to converge to their most recent actual values. Other inaccuracies related to network uncertainties, fault-intolerance etc. are not the focus of this paper and are usually related to the adopted interaction mechanism (Kennedy et al. 2009).

The above features appear to a certain degree in most of the existing aggregation approaches (Ahmed et al. 2006; Haridasan and van Renesse 2008; Jelasity et al. 2005; Kempe et al. 2003; Kashyap et al. 2006; Nath et al. 2008) and are mentioned in the related surveys (Chitnis et al. 2008; Kennedy et al. 2009). These features are actually the limitations of these systems in the sense that they are not generic and adaptive enough to perform aggregation under different network conditions and application requirements. Section "Comparison with related work" discusses and compares these approaches and their limitations in detail. Appendix A summarizes the related aggregation mechanisms discussed in this paper. Motivated by these issues, this paper focuses on the problem of designing a service for dynamic, accurate and decentralized aggregation decoupled from a specific interaction mechanism and aggregation function.

System overview

This paper introduces DIAS, the Dynamic Intelligent Aggregation Service. DIAS is a middleware service that computes aggregation functions from an input set of continuously changing aggregation values distributed in every node of a network. Figure 1 illustrates the three-level architecture of DIAS.

Each level is built by an *aggregator* and *disseminator* agent. These two agents, within a node, provide aggregation values to the agents of other nodes and consume aggregation

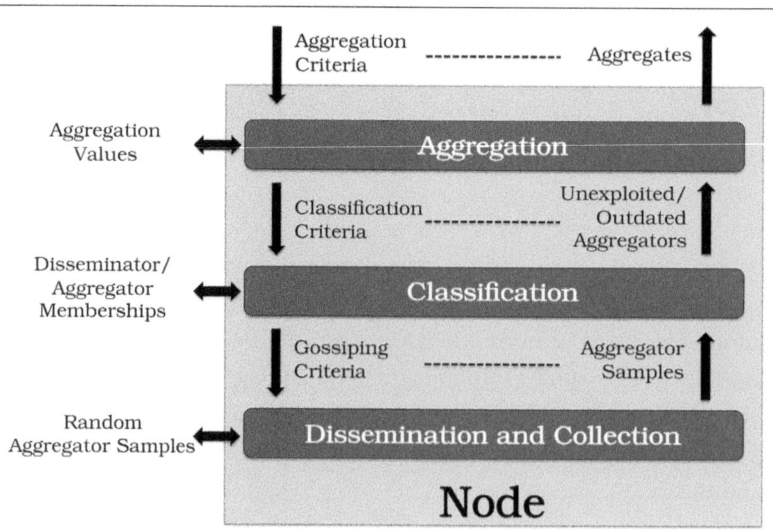

Figure 1 The DIAS middleware architecture.

values from them. However, note that, in practice, applications may not require all agents to disseminate and aggregate values. Section "Discussion and future work" discusses this issue in more detail.

The bottom level of DIAS is responsible for a gossip-based (Jelasity et al. 2007) dissemination and collection of *aggregator* samples. *Disseminators* gossip location information of agents to which the aggregation values need to be sent. Gossiping can be continuously parameterized by gossiping criteria provided by the middle level.

The discovered *aggregator* samples are provided to the middle level in which they are classified. Each *disseminator* classifies the received *aggregators* into three possible classes: (i) exploited, (ii) unexploited and (iii) outdated. These classes indicate if the aggregation value of a *disseminator* has been aggregated before by the classified *aggregators*, if it has not been aggregated or if an earlier (outdated) aggregation value has been aggregated that has changed. Classification is performed based on historical aggregation information generated during runtime[a]. The middle level provides to the top one contact information of possible *aggregators* to which aggregation values can be aggregated. DIAS is able to tune discovery of new aggregation values instead of updating the existing aggregated values and the other way around. These are the *adaptation strategies* of DIAS and are configured by the classification criteria provided by the top level.

Finally, the top level interacts with the remote *aggregators* to exchange aggregation values. These overlay interactions have two possible semantics: exploitation of a new aggregation value or update of aggregates with the most recent aggregation value. A number of aggregates are computed and delivered to the applications as defined by the aggregation criteria.

DIAS addresses the limitations illustrated in Section "Problem description" at a cost of higher communication overhead compared to related methodologies that specialize in specific aggregations functions or interaction mechanisms (Jelasity et al. 2005; Nath et al. 2008). As most of these limitations are related to a lack of abstraction, modularity and customization of aggregation mechanisms, DIAS is designed to split the complexity of dynamic decentralized aggregation into three organizational levels.

Memberships of DIAS are the means to detect inaccuracies such as double-counting and outdated aggregation values. However, a decentralized system cannot explicitly store memberships of all aggregation values locally in each node. This approach is neither scalable, efficient nor decentralized. To overcome this challenge, the probabilistic data structures of *bloom filters* (Bloom 1970) are used in DIAS for management of memberships. Bloom filters provide tremendous space savings at a cost of false positive memberships. DIAS, however, is able to detect false positive inconsistencies and, therefore, maintain a high accuracy level in the computed aggregates without introducing additional communication cost.

Modeling of dynamics

This section introduces a model for aggregation of states. A *state* represents a (aggregation) value of an application parameter at a specific point in time. The state of an application parameter changes during runtime. Decentralized aggregation computes aggregation functions that receive as input the states of different nodes for the same application parameter.

Assume that each of the n nodes of DIAS contains an *aggregator* A_i and a *disseminator* D_i with a *selected state* s_i' that is the one to be aggregated by all nodes. During each runtime iteration, selected state s_i' can be equal to one and only one state from a finite number v of locally unique *possible states* $s_i' = s_i^0|s_i^1| \ldots |s_i^{v-1}$. For example, in a movie recommender system, movies are ranked with one to five stars. The number of stars are the possible states and an actual ranking of a movie is the selected state. Although the possible states in each node are unique, two possible states between different nodes may have the same value. As the selected state changes, an earlier selected state is indicated as $\hat{s}_i$.

The system goal is the aggregation $f(s_0', s_1', \ldots, s_{n-1}')$ of all of the selected states in the overlay network during an aggregation phase. An *aggregation phase* is defined as the time period in which the selected states may change but the set of possible states remains the same. During an aggregation phase, the aggregates change continuously as a result of changes in the local selected states. Aggregation does not converge to a single value but rather to a distribution of aggregates over time. Section "Discussion and future work" discusses the applicability of this model in distributed applications.

Dissemination and collection

Decentralized aggregation requires the means to access all of the locations of *aggregators* that acquire the selected states of *disseminators*. Dissemination and collection of *aggregator* samples via gossiping provide lookup in a distributed environment. An *aggregator* sample contains the network identifier of this *aggregator*, e.g., IP address and port number. Each agent of the bottom level maintains its random view that is a list of size r with random *aggregator* samples that are continuously updated via the gossiping protocol of the peer sampling service (Jelasity et al. 2007).

Gossiping provides a highly connected and dynamic overlay network for aggregation. Furthermore, continuous update of the random view enables the discovery of changing aggregation values. The bottom level can be realized with different mechanisms beyond gossiping, e.g., flooding (Jiang et al. 2003), random walks (Gkantsidis et al. 2006) and DHTs (Yuh-Jzer et al. 2005). However, these mechanisms require high customization and DHTs require a topological maintenance. Their utilization becomes more complex within a generic decentralized aggregation service.

Consistent aggregation sessions

The middle level of DIAS provides *aggregators* to the top level that guarantee consistent aggregation sessions. An (unidirectional) *aggregation session* concerns (re)computation of the aggregates by an *aggregator* A_j after the receipt of a selected state from a remote *disseminator* D_i. If (re)computation occurs in both *aggregators* of nodes i and j, this aggregation session is *bidirectional*. An aggregation session is *consistent* if the input selected state of performed (re)computation by an *aggregator* A_j is not (i) a duplicate or (ii) an outdated selected state that has now changed. A consistent aggregation session between an *aggregator* A_j and a *disseminator* D_i is mutually satisfied if and only if the following conditions hold:

- The *disseminator* D_i disseminates for first time (i) its selected state, or (ii) its updated selected state to the *aggregator* A_j.

- The *aggregator* A_j aggregates for first time (i) the selected state, or (ii) the updated selected state of the *disseminator* D_i.

An inconsistent aggregation session usually results in inaccurate aggregates. Note that double-counting does not always result in inaccuracies as some aggregation functions are insensitive to duplicates, i.e., MAXIMUM or MINIMUM. However, duplicates cause additional communication and processing overhead in nodes. For this reason, this paper treats inconsistent aggregation sessions as subject of prevention.

Selecting *aggregators* that result in consistent aggregation sessions requires some form of history information about the past aggregation sessions performed. This section introduces the concept of aggregation memberships and their use to classify *aggregators* in the outdated, exploited and unexploited classes. Beyond consistency, this classification provides the option to perform the update of aggregates in favor of (i) changing (outdated) aggregation values or (ii) unexploited aggregation values. These two options distinguish the two adaptation strategies of DIAS.

Note that classification is used as the means to guarantee consistent aggregation sessions that enable a more generic design for aggregation in order to overcome the limitations illustrated in Section "Problem description".

Aggregation memberships

If an arbitrary aggregation value is selected from the network during an aggregation phase, this aggregation value has a probability of membership in the computed aggregates. Aggregation membership $M_{group}(member)$ of a certain 'member' to a certain 'group' is either positive or negative. This concept can be applied to the aggregation dynamics illustrated in Section "Modeling of dynamics". Each agent of the middle level in a node i stores unique identifiers of possible states $\mathsf{S}_i^0, \ldots, \mathsf{S}_i^{\nu-1}$ corresponding to the actual possible states $s_i^0, \ldots, s_i^{\nu-1}$. Respectively, S_i' and $\hat{\mathsf{S}}_i$ refer to the unique identifiers of the selected s_i' and outdated $\hat{s}_i$ state in node i. The middle level stores a representation of the local states, their unique identifiers, and the top level stores the actual states, e.g., numerical or other type. The middle level also uses the local unique network identifier of the node to map the local *aggregator* A_i and *disseminator* D_i. Therefore, $\mathsf{A}_i = \mathsf{D}_i$. The following four aggregation memberships are defined in a unidirectional aggregation session between an *aggregator* A_j and a *disseminator* D_i in two nodes i and j:

Membership 1 ($M_{\mathsf{D}_i}(\mathsf{A}_j)$). *An aggregator in a disseminator.*

A *disseminator* D_i stores the identifier of an *aggregator* A_j to which it has disseminated its selected state at least once during an aggregation phase.

Membership 2 ($M_{\mathsf{S}_i^u}(\mathsf{A}_j)$). *An aggregator in a possible state.*

A *disseminator* D_i stores the identifier of an *aggregator* A_j for each possible state identified as S_i^u aggregated by this *aggregator*.

Membership 3 ($M_{\mathsf{A}_j}(\mathsf{D}_i)$). *A disseminator in an aggregator.*

An *aggregator* A_j stores the identifier of a *disseminator* D_i from which it has aggregated its selected state at least once during an aggregation phase.

Membership 4 ($M_{\mathsf{A}_j}(\mathsf{S}_i')$). *A selected state in an aggregate.*

An *aggregator* A_j stores the identifier of a selected state S'_i aggregated from a *disseminator* D_i.

Figure 2 illustrates the aggregation memberships of DIAS stored in the middle level. Aggregation memberships can be used as follows: Assume an aggregation session between *disseminator* D_i that sends its selected state S'_i and *aggregator* A_j that receives this state. *Disseminator* D_i knows if *aggregator* A_j aggregates its selected state S'_i for first time by checking aggregation membership $M_{D_i}(A_j)$. Furthermore, D_i knows if A_j has aggregated a different possible state earlier by checking all aggregation memberships $M_{S_i^u}(A_j)$. Respectively, *aggregator* A_j knows if it has aggregated a selected state from *disseminator* D_i by checking aggregation membership $M_{A_j}(D_i)$. Moreover, A_j knows if the specific selected state S'_i has been aggregated earlier by checking aggregation membership $M_{A_j}(S'_i)$. Therefore, both (i) duplicate and (ii) outdated selected states can be detected between an *aggregator* and a *disseminator* and the consistency of an aggregation session is satisfied.

Aggregation memberships represent two mutual conditions resulting in information redundancy: Both *aggregators* and *disseminators* store membership information about their in-between aggregation. Section "Realization based on bloom filters" shows how this redundancy is exploited in an efficient model realization of aggregation memberships based on bloom filters.

Classification

Classification performed in the middle level is based on an *aggregation pool* containing three *aggregation views*. These views are queues of a limited size in which *aggregators* are classified. Three aggregation views are defined in the aggregation pool: (i) exploited, (ii) unexploited and (iii) outdated. The exploited *aggregators* of a *disseminator* D_i are the ones that have aggregated its earliest selected state s'_i. The unexploited *aggregators* of a *disseminator* D_i are the ones with which a consistent aggregation session has not been established. Finally, the outdated *aggregators* of a *disseminator* D_i are the ones that have

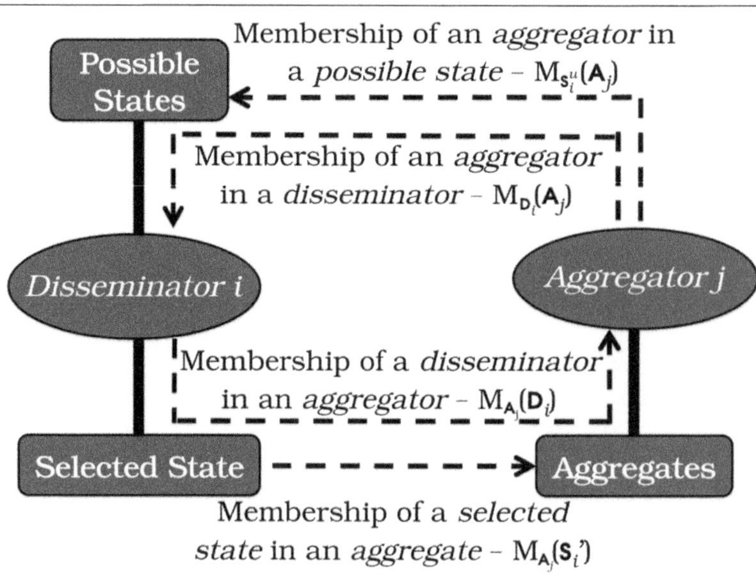

Figure 2 The aggregation memberships of DIAS.

aggregated a selected state of this *disseminator* earlier but since then this selected state has changed. Aggregation views are used as a buffer and have a limited size to allow scalability and decentralization.

Algorithm 1 illustrates the classification of an *aggregator* A_j in the aggregation pool based on the aggregation memberships $M_{D_i}(A_j)$ and $M_{S'_i}(A_j)$ of a *disseminator* D_i. When A_j is received by the bottom level, the middle level executes a membership query $M_{D_i}(A_j)$ that indicates if a consistent aggregation session has been performed between A_j and D_i. If membership is negative, *aggregator* A_j is classified as unexploited. Otherwise, if membership is positive, the next membership query $M_{S'_i}(A_j)$ is performed to indicate if *aggregator* A_j has computed in its aggregates the most recent selected state S'_i. If this membership is positive, *aggregator* A_j is exploited (duplicate aggregation value), otherwise, *aggregator* A_j has computed an earlier selected state of D_i and therefore A_j is classified as outdated.

Algorithm 1 Classification in the aggregation pool.

Require: *aggregator* A_j from bottom level

```
if M_{D_i}(A_j) : negative then
    A_j is unexploited
else if M_{S'_i}(A_j) : negative then
    A_j is outdated
else
    A_j is exploited
end if
```

If the selected state of *disseminator* D_i changes, the aggregation pool requires rearrangement. *Aggregators* contained in the exploited view before the change of the selected state move to the outdated view. In contrast, *aggregators* contained in the outdated view before the change of the selected state are queried again ($M_{S'_i}(A_j)$) and are classified as outdated or exploited. As a result of this querying, the aggregation pool remains consistent and adapts instantly after a change of the selected state.

Adaptation strategies

A consistent aggregation session is established with either an unexploited or an outdated *aggregator*. Priority is defined by the classification criteria received from the top level. These two options are the two adaptation strategies of DIAS and are referred to as EXPLOITATION and UPDATE respectively.

EXPLOITATION is a more relevant adaptation strategy if selected states do no change often and the aggregates still converge to their actual values, for example, at the beginning of aggregation or during network scaling with new nodes. In contrast, UPDATE is more relevant for steady size of networks and when aggregates have converged to the actual values. Changes of the selected states after convergence require adaptations of aggregates.

Selection of *aggregators* from the aggregation pool is conditional to the availability of *aggregators* in the class of preference for each adaptation strategy. This means that if EXPLOITATION is adopted but the unexploited view of the aggregation pool is empty, then outdated *aggregators* are selected corresponding to the selections of the UPDATE strategy.

The same holds if the UPDATE strategy is adopted and the view of outdated *aggregators* is empty: unexploited *aggregators* are selected. To this extent, the adaptation strategies of DIAS are dynamic.

Adoption of an adaptation strategy can be static, e.g., a system parameter contained in the classification criteria, or dynamic during system runtime. For example, the adopted adaptation strategy may change based on monitored parameters or based on a time period that aggregates do not change significantly.

Aggregation session

An aggregation session requires remote interactions between *aggregators* and *disseminators* to guarantee its consistency. Figure 3 illustrates these interactions between a node *i* and *j* during a bidirectional aggregation session. A unidirectional aggregation session is established with two messages depicted by the arrows (1) and (2). A bidirectional aggregation session includes a third message depicted by arrow (3).

The 'request' message, illustrated by arrow (1), initiates an aggregation session and contains the following information:

- Flag: This denotes a unidirectional 'uni' or bidirectional 'bi' aggregation session.
- Class: This denotes if the *aggregator* A_i, receiving this message, is classified by a *disseminator* D_j as unexploited or outdated.
- D_j: This is the identifier of the *disseminator* D_j that has performed the classification of the *aggregator* A_i.
- S'_j: This is the selected state identifier of D_j.
- S_j: This is the earlier selected state identifier of the *disseminator* D_j aggregated by A_i.

A 'response' message, illustrated by arrow (2) or (3), completes a unidirectional or bidirectional aggregation session and contains the following information:

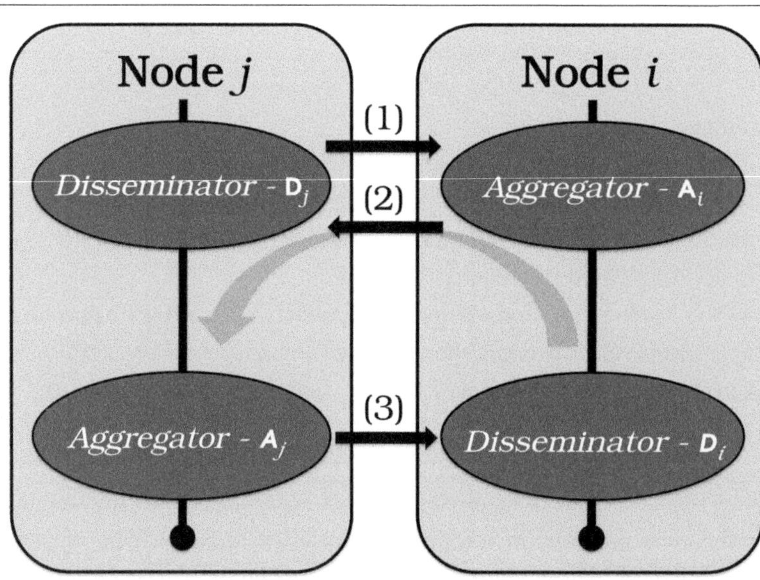

Figure 3 The remote interactions of DIAS.

- Flag: This denotes a unidirectional 'uni' or bidirectional 'bi' aggregation session. A third flag, the 'uni-bi', denotes the upgrade of a unidirectional aggregation session to a bidirectional one by including a 'request' message flagged as 'bi'.
- Class: This denotes if the *aggregator* A_j, sending this message, is classified by a *disseminator* D_i as unexploited or outdated.
- A_j: This is the identifier of the *aggregator* A_j.
- 'Request' message: This integrated message is optional. It upgrades the unidirectional aggregation session to a bidirectional one.

Note that the integrated 'request' message in the 'response' message provides one message fewer for a bidirectional session to complete.

Computation of aggregates

The top level is responsible for the computation of aggregates. An aggregate is continuously computed based on an aggregation function provided by the aggregation criteria. Aggregates are updated by sending the value of the selected state to *aggregators* provided by the middle level and classified as unexploited. If the provided *aggregators* are classified as outdated, the earlier selected state is sent as well.

The top level forms an overlay network between *aggregators* and *disseminators* linked with overlay links that have two possible semantic values: unexploited or outdated but not exploited. Therefore, the computed aggregation functions exclude overlay links from the top level that result in duplicates (exploited *aggregators*). The aggregation memberships, the classification, the selection of *aggregators* are all complexity hidden from the aggregation process of the top level. As explained in Section "Adaptation strategies", the adaptation strategies tune the aggregation process in favor of (i) updating aggregates with the most recent selected states (UPDATE) or (ii) discovering new selected states (EXPLOITATION). The top level has to only provide the classification criteria that trigger this optimization and inform about changes in the selected state.

Delivery of aggregates to applications may be performed periodically. Another option is a minimum deviation threshold over a certain time period that denotes convergence to the actual aggregate values. The aggregation criteria define these requirements.

Realization based on bloom filters

Explicit storage of aggregation memberships in every agent of the middle level is not a scalable, efficient and decentralized solution. Aggregation memberships can be a cost-effective and viable approach in large-scale decentralized environments by using an implicit storage mechanism: bloom filters (Bloom 1970).

A bloom filter is a probabilistic data structure for efficient membership storage and querying. A bloom filter is based on a number of k hash functions that hash an element in a limited binary space of 2^m size, where m is the size of the bit vector in which information is stored. More specifically, each hash function outputs a random index in this binary space.

A *simple bloom filter* supports insertions and membership queries. During an insertion, the bits that are indexed by the hash functions are set to 1. During membership queries, the membership of an element in the bloom filter is confirmed if all of the bits indexed by all of the hash functions are 1.

Counting bloom filters additionally support removal of memberships (Li et al. 2000). This is achieved by representing the storage space with integers, instead of single bits, that act as counters. Insertions increment the counters indexed by hash functions and removals decrement respectively. Data overflow by consecutive insertions is prevented by choosing an adequate size of 3 - 4 bits for the integers. Therefore, a counting bloom filter is 3 - 4 times larger than a simple one.

Each of the memberships illustrated in Figure 2 is stored in a bloom filter. More specifically, a *disseminator* D_i has a simple bloom filter for storing $M_{\mathsf{D}_i}(\mathsf{A}_j)$ memberships and v counting bloom filters, one for each possible state, for storing $M_{\mathsf{S}_i^u}(\mathsf{A}_j)$ memberships. The counting bloom filters provide the flexibility to reflect the changes of the selected states. For example, in an aggregation session between a *disseminator* D_i and an outdated *aggregator* A_j, the membership $M_{\hat{\mathsf{S}}_i'}(\mathsf{A}_j)$ is removed from the counting bloom filter of the earlier selected state $\hat{\mathsf{S}}_i$ and the membership $M_{\mathsf{S}_i'}(\mathsf{A}_j)$ is added in the counting bloom filter of the most recent selected state S_i'. Complementarily, the *aggregator* A_j has a simple bloom filter for storing the $M_{\mathsf{A}_j}(\mathsf{D}_i)$ memberships and a counting bloom filter for storing the $M_{\mathsf{A}_j}(\mathsf{S}_i')$ memberships. This provides a consistent update of aggregates by replacing textsfoutdated selected states with the most recent ones.

The space saving achieved by bloom filters come at the cost of false positives. False positive membership indicates that a state or agent identifier is hashed in a bloom filter when it is actually not hashed. The probability of false positives depends on (i) the number of elements stored in the bloom filter, (ii) the number k of hash functions and (iii) the size 2^{m} of the storage space. The minimum number of bits in a simple bloom filter x that hashes n elements and results in a certain probability $P_{fp}(x)$ of false positives is computed as $2^{\mathsf{m}} = -n\frac{\ln P_{fp}(x)}{(\ln 2)^2}$ (Deke et al. 2010). False positives can cause inconsistent aggregation sessions (inaccurate aggregates) and additional communication overhead if they are not detected and eliminated.

The space savings computed for a bloom filter can be outlined as follows: Assume at least $128n$ bits stored in conventional data structures such as an array. The $128n$ bits are actually n number of agent or state memberships represented by global unique identifiers of 128 bits. A hash table requires even a higher storage space due to the additional storage of indexes that enhance searching operations. In contrast, assume a bloom filter x with a probability $P_{fp}(x) = 0.01$ of false positives that stores the same number n of memberships. The relation $2^{\mathsf{m}} = -n\frac{\ln P_{fp}(x)}{(\ln 2)^2}$ shows that, in this case, an array stores $128/9.6 \approx 13$ times the space of this bloom filter. For a bloom filter with $P_{fp}(x) = 0.1$ and $P_{fp}(x) = 0.001$, its space storage is approximately 56 and 9 times lower respectively.

Note that false negatives in counting bloom filters may occur if an erroneous element removal is performed. This removal may result in a biased and inconsistent probabilistic data structure. For example, if a removed element is not actually hashed, then its removal changes bits indicating memberships of other elements that are actually hashed (Deke et al. 2010). This paper assumes that false negatives cannot be generated in principle if and only if removals are not performed from counting bloom filters. Otherwise, Section "Second level check" illustrates how false negatives are prevented in DIAS if removals are performed.

The mutual membership check

DIAS deals with the problem of false positives in bloom filters by taking advantage of decentralized mutual membership checks between *disseminators* and *aggregators*. A *mutual memberships check*, denoted as '⋒' in this paper, is the process of querying two memberships in a *disseminator* and an *aggregator* that are assumed to either be both present or not. For example, the aggregation memberships $M_{\mathsf{D}_i}(\mathsf{A}_j)$ and $M_{\mathsf{A}_j}(\mathsf{D}_i)$ are mutual. During an aggregation phase, a *disseminator* stores memberships of *aggregator* identifiers and, respectively, these *aggregators* store memberships of the respective *disseminator* identifiers resulting in mutual aggregation memberships. $M_{\mathsf{S}'_i}(\mathsf{A}_j)$ and $M_{\mathsf{A}_j}(\mathsf{S}'_i)$ are also mutual memberships. Selected state S'_i of a *disseminator* D_i is associated with the $M_{\mathsf{S}'_i}(\mathsf{A}_j)$ membership of an *aggregator* A_j. Respectively, *aggregator* A_j stores the $M_{\mathsf{A}_j}(\mathsf{S}'_i)$ membership of the selected state identifier S'_i.

Mutual membership checks provide detection of false positives in the bloom filters of DIAS. Only if multiple false positives occur between $M_{\mathsf{D}_i}(\mathsf{A}_j)$-$M_{\mathsf{A}_j}(\mathsf{D}_i)$ and $M_{\mathsf{S}'_i}(\mathsf{A}_j)$-$M_{\mathsf{A}_j}(\mathsf{S}'_i)$ in a single aggregation session, then an inconsistent aggregation session may come as a result of these false positives.

Assume two arbitrary memberships $M_x(a)$ and $M_y(b)$ based on the unique identifiers of two members a and b in the groups x and y respectively. Assume also that these two memberships are mutual, meaning that they should be both positive or negative such as $M_x(a) \Cap M_y(b)$: *positive* | $M_x(a) \Cap M_y(b)$: *negative*. $M_x(a)$ and $M_y(b)$ are stored in two simple bloom filters with false positive probabilities $P_{fp}(x)$ and $P_{fp}(y)$ respectively. The possible outcomes of the mutual membership check are the following:

Check 1. ***if*** $M_x(a)$: *positive* ***and*** $M_y(b)$: *positive* ***then*** $M_x(a) \Cap M_y(b)$: *positive*

$M_x(a)$ and $M_y(b)$ memberships are confirmed with a probability of $1 - P_{fp}(x)P_{fp}(y)$. This confirmation is false if and only if both bloom filters generate a false positive that is the product $P_{fp}(x)P_{fp}(y)$ of their false positive probabilities.

Check 2. ***if*** $M_x(a)$: *positive* ***and*** $M_y(b)$: negative, ***or***, $M_x(a)$: negative ***and*** $M_y(b)$: *positive* ***then*** $M_x(a) \Cap M_y(b)$: *negative*

$M_x(a)$ and $M_y(b)$ memberships are not confirmed with a probability of 1. In this case, one of the bloom filters generates a false positive.

Check 3. ***if*** $M_x(a)$: *negative* ***and*** $M_y(b)$: *negative* ***then*** $M_x(a) \Cap M_y(b)$: *negative*

$M_x(a)$ and $M_y(b)$ memberships are not confirmed with a probability of 1.

Mutual membership checks provide (i) a decrease in the probability that an inconsistent aggregation session occurs (Check 1) and (ii) detection of false positives (Check 2). This section introduces a consistency mechanism of aggregation sessions for accurate aggregates. This mechanism is based on two nested mutual membership checks between the bloom filters of an *aggregator* A_j and a *disseminator* D_i that define the four possible *outcomes* of an aggregation session:

- Exploitation: *Aggregator* A_j and *disseminator* D_i are involved for a first time in a consistent aggregation session as defined in Section "Consistent aggregation sessions". A selected state has not been aggregated before and the aggregates are updated with new information. The $M_{D_i}(A_j)$, $M_{A_j}(D_i)$, $M_{S'_i}(A_j)$ and $M_{A_j}(S'_i)$ memberships are added in the respective bloom filters.

- Update: *Aggregator* A_j and *disseminator* D_i have been involved before in a consistent aggregation session, however, this time the selected state has changed. The *aggregator* A_j updates its aggregates with the new selected state. The $M_{\hat{S}_i}(A_j)$ membership is replaced by the $M_{S'_i}(A_j)$ membership and $M_{A_j}(\hat{S}_i)$ is replaced by $M_{A_j}(S'_i)$.
- Duplicate: *Aggregator* A_j and *disseminator* D_i have been involved before in an aggregation session with the same selected state. Aggregation is not performed.
- Inconsistency: *Aggregator* A_j and the *disseminator* D_i are involved for a first time in a consistent aggregation session but the mutual membership check cannot confirm this. Alternatively, *aggregator* A_j and *disseminator* D_i have been involved before in an aggregation session with a different selected state. However, the consistency check cannot identify the textsfoutdated selected state to replace. These uncertainties are treated as an inconsistency and are a result of multiple false positives in the bloom filters.

The two nested mutual membership checks illustrated in Section "First level check" and "Second level check" show how an aggregation session reaches each of the above possible outcomes. The results of the memberships are exchanged in the messages defined in Section "Aggregation session".

First level check

This mutual membership check identifies if a consistent aggregation session has not been performed between an *aggregator* A_j and a *disseminator* D_i. *Disseminator* D_i queries the $M_{D_i}(A_j)$ membership of the A_j identifier in its bloom filter. Complementarily, *aggregator* A_j queries $M_{A_j}(D_i)$ membership. The $M_{D_i}(A_j)$ and $M_{A_j}(D_i)$ memberships are mutual as they are either both added in the bloom filters or not. Therefore, a mutual membership check provides the following benefits at the first level of the nested mutual membership check: (i) A decrease in the probability of an inconsistent aggregation session that requires two false positives generated by the two bloom filters. (ii) Detection of a false positive in either the $M_{D_i}(A_j)$ or $M_{A_j}(D_i)$ membership. Algorithm 2 illustrates the first level of the nested mutual membership check.

Algorithm 2 The first level check.

Require: access to *disseminator* D_i and *aggregator* A_j

```
if M_{D_i}(A_j) ⋒ M_{A_j}(D_i) : negative then
    outcome=exploitation
else
    go to Algorithm 3
end if
```

Ensure: outcome

This mutual membership check detects an exploitation outcome in an aggregation session if and only if $M_{D_i}(A_j) \Cap M_{A_j}(D_i)$: *negative*. This outcome is generated if at least one of the $M_{D_i}(A_j)$ and $M_{A_j}(D_i)$ memberships, in case of a single false positive, or both memberships, in case of no false positives, cannot be confirmed. On this first level, the exploitation outcome is reached with an absolute certainty. However, two simultaneous false positives in the $M_{\mathsf{D}_i}(\mathsf{A}_j)$ and $M_{\mathsf{A}_j}(\mathsf{D}_i)$ memberships are possible. Therefore, further examination is required on a second level of a mutual membership check to detect multiple false positives and lower the uncertainties of the outcomes.

Second level check

The second level of the mutual membership check detects if there is an textsfoutdated selected state $\hat{S}_i$ aggregated from a *disseminator* D_i that differs from its new selected state S'_i. The detection is performed by querying every $M_{\mathsf{S}_i^u}(\mathsf{A}_j)$ bloom filter membership of the respective possible state $\mathsf{S}_i^u \in \left\{\mathsf{S}_i^0, \ldots, \mathsf{S}_i^{v-1}\right\}$. $M_{\mathsf{A}_j}(\mathsf{S}_i^u)$ membership is also queried for every possible state S_i^u. The number o of positive mutual memberships $M_{\mathsf{S}_i^u}(\mathsf{A}_j) \Cap M_{\mathsf{A}_j}(\mathsf{S}_i^u)$ define the outcome of an aggregation session as illustrated in Algorithm 3.

Algorithm 3 The second level check.

Require: access to *disseminator* D_i and *aggregator* A_j

1: $o = 0$, $\hat{\mathsf{S}}_i = \mathsf{S}'_i$
2: **for** $u = 0$ to $v - 1$ **do**
3:     **if** $M_{\mathsf{S}_i^u}(\mathsf{A}_j) \Cap M_{\mathsf{A}_j}(\mathsf{S}_i^u)$: *positive* **then**
4:         $\hat{\mathsf{S}}_i = \mathsf{S}_i^u$
5:         $o = o + 1$;
6:     **end if**
7: **end for**
8: **if** $o = 0$ **then**
9:     outcome=exploitation
10: **else if** $o = 1$ **then**
11:     **if** $\hat{\mathsf{S}}_i \neq \mathsf{S}'_i$ **then**
12:         outcome=update
13:     **else**
14:         outcome=duplicate
15:     **end if**
16: **else** $\{o > 1\}$
17:     outcome=inconsistency
18: **end if**

Ensure: outcome

If there are no positive mutual memberships detected ($o = 0$ in line 8 and 9 of Algorithm 3), there is no positive $M_{\mathsf{A}_j}(\mathsf{S}_i^u)$ membership (no selected state aggregated before from D_i) and/or there is no positive $M_{\mathsf{S}_i^u}(\mathsf{A}_j)$ membership in any bloom filter of the possible states. This condition conflicts with the positive result of the mutual membership check $M_{\mathsf{D}_i}(\mathsf{A}_j) \Cap M_{\mathsf{A}_j}(\mathsf{D}_i)$ in the first level. Both $M_{\mathsf{D}_i}(\mathsf{A}_j)$ and $M_{\mathsf{A}_j}(\mathsf{D}_i)$ memberships are false positives. The outcome in this case is an exploitation.

If there is one positive mutual membership detected ($o = 1$ in lines 10-15), the system can derive the textsfoutdated selected state $\hat{\mathsf{S}}_i$. The outcome is either a duplicate, if the textsfoutdated selected state $\hat{\mathsf{S}}_i$ is the same with the new selected state S'_i, or an update in the opposite case. The uncertainty of this outcome is minimized by the nested mutual membership checks.

Finally, if more than one positive mutual membership is detected ($o > 1$ in lines 16-18), multiple false positives occur that cannot be identified. These false positives concern $M_{\mathsf{S}_i^u}(\mathsf{A}_j)$ and $M_{\mathsf{A}_j}(\mathsf{S}_i^u)$, or $M_{\mathsf{D}_i}(\mathsf{A}_j)$ and $M_{\mathsf{A}_j}(\mathsf{D}_i)$ in the first level of the nested mutual membership check. The outcome is an inconsistency and therefore, any aggregation at this point may result in inaccuracies of the aggregates.

The 'safer' approach to handle inconsistencies is to ignore these aggregation sessions and not perform any aggregation that may result in inaccurate aggregates. However, not only the aggregates can be influenced in this case. Recall from the beginning of this section that removal of a membership from a counting bloom filter that is actually not present introduces false negatives (Deke et al. 2010). Therefore, the following aggregation sessions are prone to inaccuracies as the assumption of no false negatives does not hold anymore. By skipping inconsistent aggregation sessions, DIAS makes sure that the condition of no false negatives in counting bloom filters is not violated.

Experimental evaluation

DIAS is implemented and evaluated in Protopeer (Galuba et al. 2009), a prototyping toolkit for distributed systems. The experimental settings illustrated in this section are summarized in Appendix B. A network of $n = 1500$ nodes runs DIAS for $t(\mathsf{DIAS}) = 800$ epochs. The agents of each node act both as *aggregators* and *disseminators*. Each epoch lasts for $T(\mathsf{DIAS}) = 1000$ ms that is the default parameter value in Protopeer. In practice, the selection of this parameter is performed based on factors such as the available bandwidth in the network. The system initially bootstraps a ring topology. The bootstrapping period is $t'(\mathsf{DIAS}) = 6$ epochs and the size of the ring view is $|\mathbf{v}(ring)| = 5$ for each node.

A simulated application of dynamically changing states is bootstrapped in $t'(application) = 15$ epochs. Each application instance in each node generates $\nu = 5$ numerical possible states during each aggregation phase. The possible states are selected randomly from the range $[0, 1)$ defined by five different beta distributions, one for each possible state. Appendix C illustrates these beta distributions. The selected state changes cyclically as $s_i' = s_i^0, s_i^1, \ldots, s_i^{\nu-1}, s_i^0$, etc. Two factors trigger these changes: (i) time and (ii) the parameter itself that the possible states represent. These factors are modeled based on two probabilities: (i) the probability $P_c(time)$ of changing a selected state every period $T(application)$ and (ii) the probability $P_c(parameter)$ of change in a specific type of application parameter. The probability $P_c(s_i')$ of a node i to change its selected state is $P_c(s_i') = P_c(time)P_c(parameter)$ assuming that the two probabilities $P_c(time)$ and $P_c(parameter)$ are independent.

Two types of changes in the selected states are examined: synchronous and asynchronous. In *synchronous changes*, the selected states of all nodes in the network change simultaneously. Synchronous changes are modeled as $P_c(time) = 1$ and $P_c(parameter) = 1$ for $T(application) = 200$ epochs. In contrast, *asynchronous changes* occur arbitrary over time. A dynamic setting of asynchronous changes is modeled as $P_c(time) = 0.4$ and $T(application) = 0.7$ for $T(application) = 10$ epochs. In practice, the changes in the selected states depend on the dynamics of the application.

The execution period of the top level is synchronized with the one of the middle level as $T(top) = T(middle) = 1000$ ms. The AVERAGE, SUMMATION and MAXIMUM aggregation functions are computed. The messages exchanged by the middle and top level are integrated. This minimizes the number of exchanged messages $\lambda(sessions)$ to the three ones illustrated in Section "Aggregation session". The integrated messages additionally contain the actual states for the computation of the aggregation functions. The aggregates are provided to the application after every computation.

The middle level is periodically executed at $T(middle) = 1000$ ms during which $z = 10$ bidirectional aggregation sessions are initiated at maximum. The size of the aggregation pool is selected to $q = 3 * 15 = 45$ with each of the unexploited, exploited and outdated containing 15 *aggregators* at maximum. The aggregation pool is filled by classifying $e = 15$ random *aggregator* samples collected from the bottom level in each execution period. Static adoptions of the EXPLOITATION and UPDATE strategies are evaluated.

Aggregation memberships are realized in the bloom filters of the XSiena BloomFilter library (Jerzak and Fetzer 2008). Double hashing (Dillinger and Manolios 2004) is used for collision resolution in the hashed elements of bloom filters. The size 2^{m} of the bloom filters and the number of hash functions k are selected empirically using the testing tools of XSiena BloomFilter. The expected number of hashed elements during the performed experiments is equal to the network size n. This selection is performed manually during system parameterization or in an automated fashion. In the latter case, DIAS is initialized with a default size of bloom filters and computes the system size using the COUNT aggregation function.

Three schemes are adopted in DIAS: (i) m = 16, k = 24, (ii) m = 14, k = 24 and (iii) m = 14, k = 6. The first scheme, with $2^{16} = 65536$ bits $= 8.192$ KB, does not result in false positives during the performed library tests, whereas false positives appear in the other two schemes because of the fewer number of bits available for hashing: $2^{14} = 16384$ bits $= 2.048$ KB. The relation $2^m = -n\frac{\ln p}{(\ln 2)^2}$ verifies the probability of false positives. For $n = 1500$, the probability of false positives in the first scheme is $0.76 * 10^{-9}$, whereas, for the other two schemes is 0.005. The second scheme introduces higher randomness compared to the third one due to the higher number of hash functions. However, the second scheme causes a higher number of bit changes during insertions. This results in a higher number of potential collisions (Dillinger and Manolios 2004) that cause a higher number of false positives.

The bottom level is realized by the peer sampling service (Jelasity et al. 2007). The size of the random view is $r = 50$ and the execution period is $T(bottom) = T(\text{DIAS})/5 = 250$ ms. The values of the 'view selection', 'view propagation' and 'peer selection' policies (Jelasity et al. 2007) are selected to maximize the randomness and dissemination speed.

The efficiency of DIAS is related to how close the values of the computed aggregates are to the actual ones. This closeness is quantified by two evaluation metrics: (i) accuracy α and (ii) matching μ. *Accuracy* α is defined as $\alpha = 1 - \varepsilon/\varepsilon_{max}$ where ε is the *absolute error* and ε_{max} is the *maximum probable absolute error*. The absolute error is the absolute difference of the actual aggregate from the computed aggregate. The maximum probable absolute error is the maximum possible absolute difference that the actual aggregate and the computed aggregate can have. Note that the convergence of accuracy is particularly interesting for the evaluation of DIAS as it outlines its speed and adaptivity in the computed aggregates. *Matching* μ is based on the calculation of the correlation coefficient and indicates the closeness of the distribution of the computed aggregates to the distribution of the actual aggregates. This metric is especially useful for the evaluation of DIAS under asynchronous changes.

The source data from which accuracy is computed are illustrated in Appendix C. Accuracy and matching are studied in line with the communication cost of the aggregation sessions in terms of the number of messages $\lambda(sessions)$ exchanged. The communication cost of the bottom level is excluded from the illustrated results as it is constant (Jelasity et al. 2007). The results are interpreted based on the number of aggregation outcomes that aggregation sessions result in. Finally, the effect of (i) the size of aggregation pool, (ii) the size of aggregation classes, (iii) the number of *aggregator* samples, (iv) the number of aggregation sessions (v) and the periodical executions are factors that are experimentally evaluated by (Pournaras 2013).

Adaptation strategies

This section evaluates the efficiency of DIAS with and without adaptation strategies. For this reason, the bloom filter scheme of m = 16 and k = 24 is adopted that does not result in false positives. The case when DIAS does not employ adaptation strategies is referred to as the RANDOM strategy and concerns random *aggregator* samples without a classification in the aggregation pool.

Figures 4a-4c illustrate the accuracy convergence under synchronous changes. EXPLOITATION and UPDATE converge to the maximum accuracy $\alpha = 1$ and adapt the aggregates within 100 epochs. Matching μ is 0.79, 0.59 and 0.90 for AVERAGE, SUMMATION and MAXIMUM respectively. The distribution of the exploitation and update outcomes depicted in Figure 5a and 5b explains the convergence of accuracy. These outcomes represent $z = 10$ bidirectional aggregations sessions by $n = 1500$ *aggregators*: $10 * 2 * 1500 = 30000$ exploitation and update outcomes. Note that, under synchronous changes, EXPLOITATION and UPDATE have the same effect. The total number of aggregation sessions with an exploitation outcome are performed within the first $T(application) = 200$ epochs. The next aggregation sessions result in update outcomes.

RANDOM also achieves a high accuracy according to Figures 4a-4c, with 0.71, 0.33 and 0.90 matching μ for each aggregate respectively. However, RANDOM has a slower convergence of 150 additional epochs compared to EXPLOITATION and UPDATE. This is because of the number of duplicate outcomes that reaches 28000 during convergence as depicted in Figure 5c. EXPLOITATION and UPDATE do not cause duplicate outcomes as the exploited *aggregators* are not selected from the aggregation pool.

Figure 4d-4f illustrate the convergence of accuracy under asynchronous changes. Although $P_c(time)P_c(parameter)n = 0.4 * 0.7 * 1500 = 420$ selected states change on average every $T(application) = 10$ epochs, accuracy converges to the maximum. Matching μ between the actual and computed AVERAGE for EXPLOITATION and UPDATE is 0.57 and 0.70 respectively. RANDOM is not influenced significantly with a matching of $\mu = 0.66$ for AVERAGE. RANDOM reaches exploitation and update outcomes during the converge period in contrast to EXPLOITATION that mostly reaches exploitation outcomes in the first 100 epochs (Figure 5d) and update outcomes in the next epochs (Figure 5e). Similarly with the case of synchronous changes, RANDOM requires 150 additional epochs to converge compared to EXPLOITATION. A converged number of 10000 duplicate

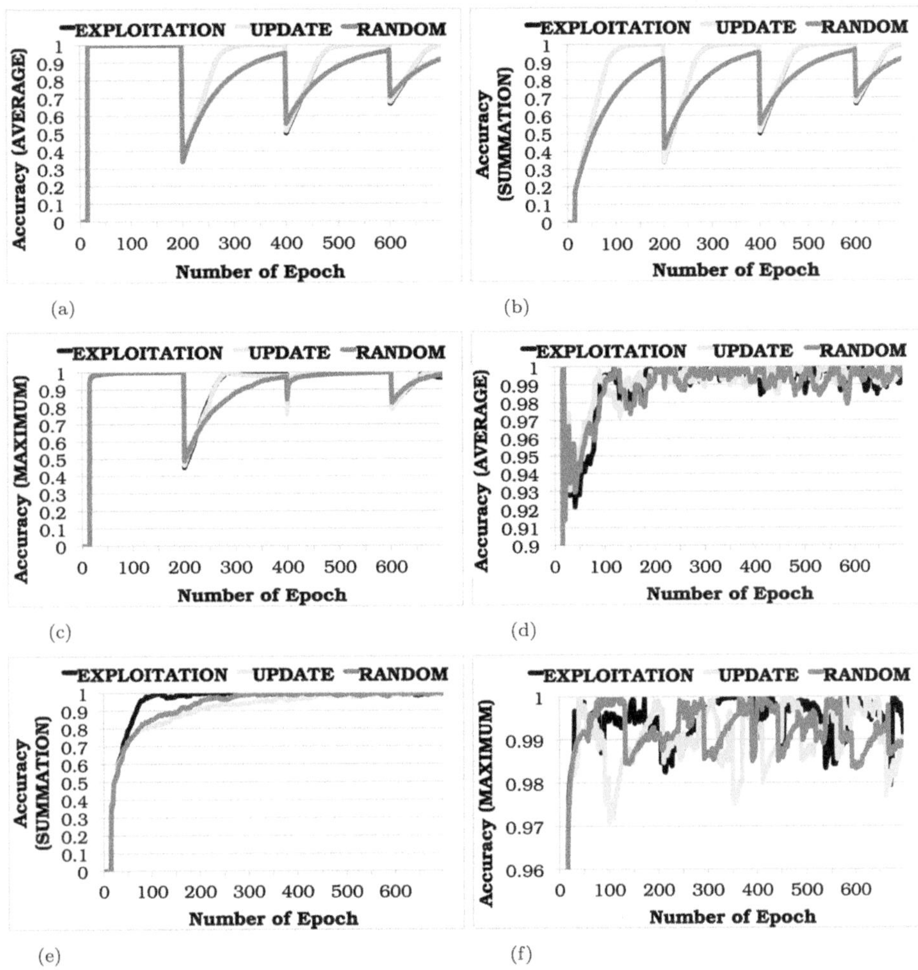

Figure 4 Accuracy of adaptation strategies. (a) AVERAGE under synchronous changes. **(b)** SUMMATION under synchronous changes. **(c)** MAXIMUM under synchronous changes. **(d)** AVERAGE under asynchronous changes. **(e)** SUMMATION under asynchronous changes. **(f)** MAXIMUM under asynchronous changes.

outcomes depicted in Figure 5f causes this delay. Matching μ in MAXIMUM is 0.67, 0.55 and 0.45 respectively for EXPLOITATION, UPDATE and RANDOM. SUMMATION is more challenging to compute. EXPLOITATION provides the fastest convergence within the first 100 epochs. RANDOM converges in approximately 250 epochs. UPDATE does not converge before the 400th epoch as it does not prefer *aggregators* from the unexploited view and is influenced by the changes of the selected states.

Figure 6 illustrates the messages $\lambda(sessions)$ sent during the aggregation sessions. Under synchronous changes, the distribution of the communication cost during runtime depicted in Figure 6a corresponds to the exploitation and update outcomes in Figure 6a and 6b respectively. EXPLOITATION and UPDATE minimize the messages exchanged to 0 when the aggregates converge to their actual values. This is not the case for RANDOM that continuously exchanges $\lambda(sessions) = nz3 = 1500 * 10 * 3 = 45000$ messages during runtime. These messages are generated by 1500 nodes that periodically establish 10 bidirectional aggregation sessions with 3 messages exchanged in each session. This is the how

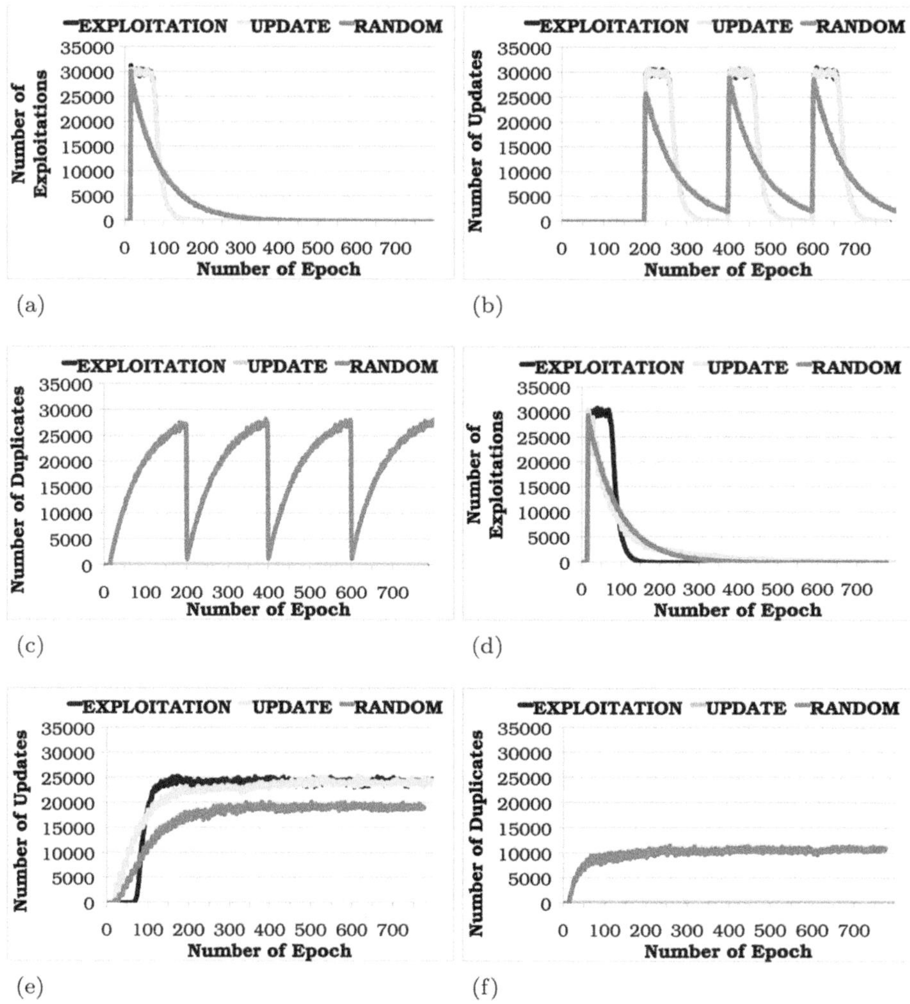

Figure 5 Outcomes of adaptation strategies. **(a)** Exploitation outcomes under synchronous changes. **(b)** Update outcomes under synchronous changes. **(c)** Duplicate outcomes under synchronous changes. **(d)** Exploitation outcomes under asynchronous changes. **(e)** Update outcomes under asynchronous changes. **(f)** Duplicate outcomes under asynchronous changes.

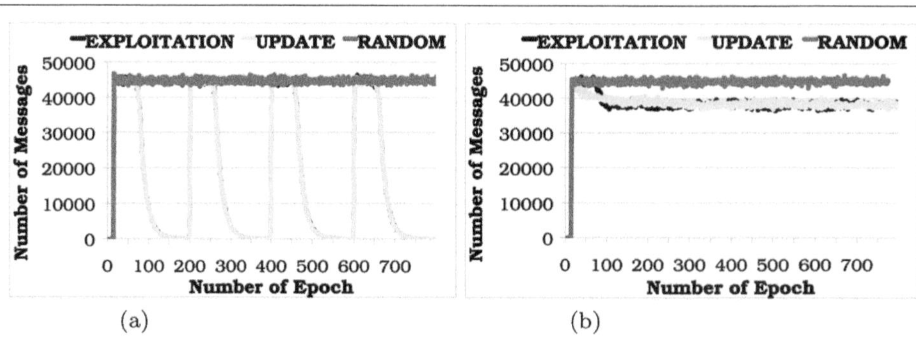

Figure 6 Communication cost of adaptation strategies. **(a)** Synchronous changes. **(b)** Asynchronous changes.

the communication cost is estimated for larger networks or a different frequency of aggregation sessions. Under asynchronous changes and during convergence in the first 100 epochs, EXPLOITATION and UPDATE exchange the maximum number of $38000 - 45000$ messages that converges to 38000 in the next epochs during which update outcomes are mainly reached.

This communication cost is significantly lower if the nodes of the network do no run both an *aggregator* and a *disseminator* agent. For example, if the network has 500 of its nodes with an *aggregator* and the rest 1000 nodes with a *disseminator*, the communication cost is computed in this case as $1000 * 10 * 2 = 20000$ messages that is significantly lower than the aforementioned upper communication cost.

Bloom filter aggregation memberships

This section investigates the impact of false positives in the accuracy α of aggregates and the communication cost. Specifically, the bloom filters scheme of $\mathsf{m} = 16$ and $\mathsf{k} = 24$ is compared with two other schemes prone to false positives according to the empirical investigations: (i) $\mathsf{m} = 14$, $\mathsf{k} = 24$ and (ii) $\mathsf{m} = 14$ and $\mathsf{k} = 6$.

Concerning the accuracy of the computed aggregates, no significant influence is observed in the two schemes prone to false positives. The matching α between for both (i) aggregation strategies and (ii) synchronous/asynchronous changes remains almost intact. For example, the bloom filter scheme with $\mathsf{m} = 14$ and $\mathsf{k} = 24$ results in a 0.01 lower matching of AVERAGE under synchronous changes compared to the one with $\mathsf{m} = 16$ and $\mathsf{k} = 24$.

Figure 7 compares the outcomes of aggregation sessions for EXPLOITATION in each bloom filter scheme. The results of UPDATE are similar and, therefore, are omitted. The exploitation outcomes are also omitted as they show similar results to those of Figure 7a and 7d. The bloom filter schemes with $\mathsf{m} = 14$, $\mathsf{k} = 24$ and $\mathsf{m} = 14$, $\mathsf{k} = 6$ miss update outcomes that actually result in inconsistency outcomes. This is shown in Figure 7a and 7b under synchronous changes and in Figure 7d and 7e under asynchronous changes. Figure 7c and 7f show that the scheme with $\mathsf{m} = 14$, $\mathsf{k} = 24$ generates 6 and 9 duplicate outcomes as a result of false positives during classification.

False positives result in a higher number of messages $\lambda(sessions)$. During classification, false positives result in inconsistency outcomes. An *aggregator* A_j is incorrectly classified in the outdated and exploited views if the memberships $M_{\mathsf{D}_i}(\mathsf{A}_j)$ and $M_{\mathsf{S}'_i}(\mathsf{A}_j)$ are false positive respectively. Under synchronous changes, incorrectly classified exploited *aggregators* move to the outdated view causing inconsistency outcomes. Figure 8 illustrates the number of exchanged messages during runtime for the three bloom filter schemes. EXPLOITATION is adopted. The results of UPDATE are similar and, therefore, are omitted.

Inconsistency outcomes raise the total number of messages exchanged by 15%. The same holds for asynchronous changes but the effect is much smaller as changes in the selected states occur more frequently. In this case, the increase is 2%.

Comparison with related work

Providing a fair quantitative comparison of DIAS with related mechanisms is challenging as DIAS is designed to be a more generic aggregation service and therefore, it serves a different purpose. Yet, this section illustrates a number of quantitative observations concerning the performance of DIAS in comparison with related methodologies.

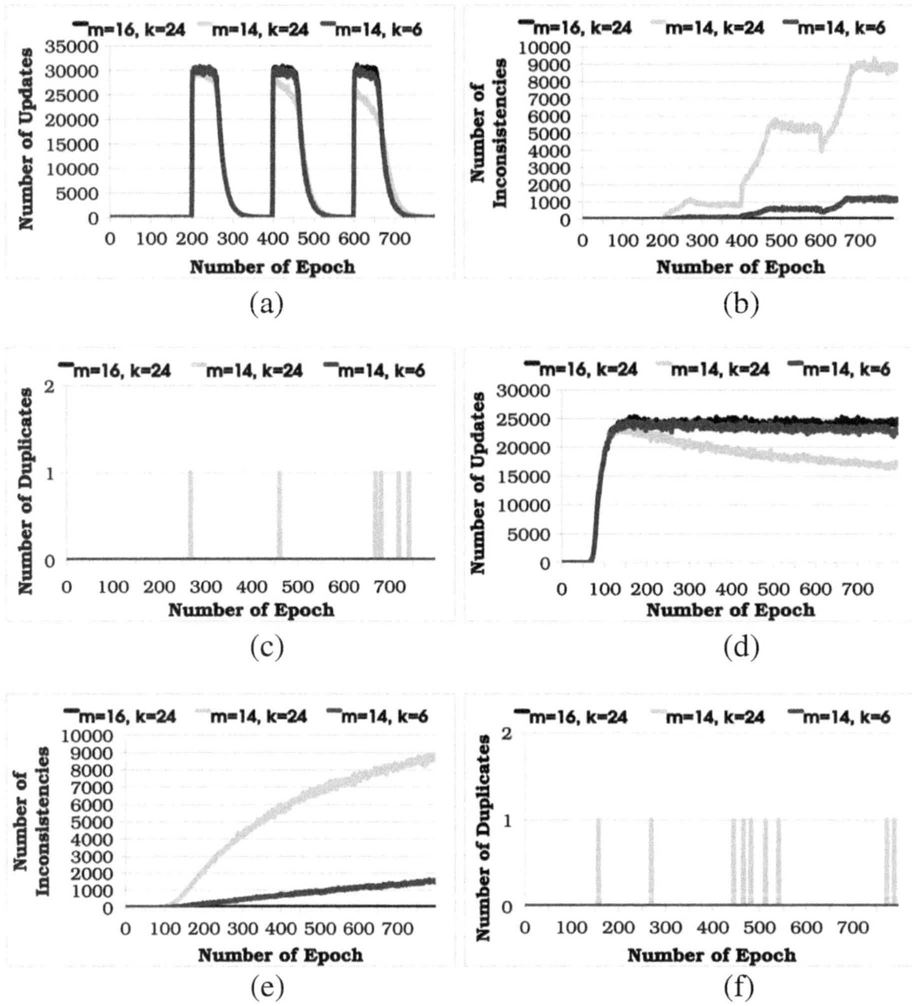

Figure 7 Outcomes of bloom filter schemes. (a) Update outcomes under synchronous changes. **(b)** Inconsistency outcomes under synchronous changes. **(c)** Duplicate outcomes under synchronous changes. **(d)** Update outcomes under asynchronous changes. **(e)** Inconsistency outcomes under asynchronous changes. **(f)** Duplicate outcomes under asynchronous changes.

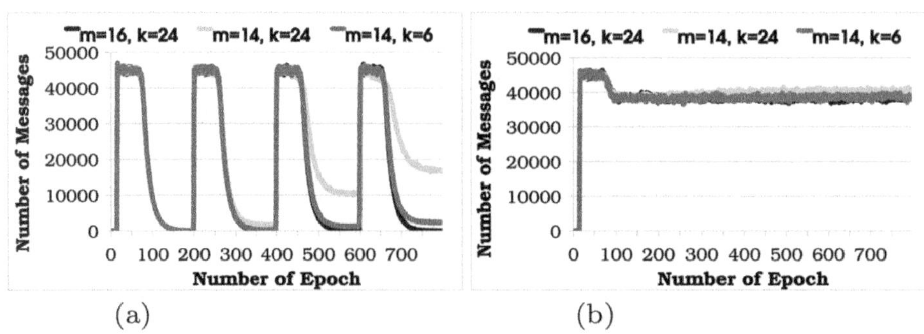

Figure 8 Communication cost of bloom filter schemes. (a) Synchronous changes. **(b)** Asynchronous changes.

For example, gossip-based variance reduction (Jelasity et al. 2005) computes AVERAGE approximately $4 - 5$ faster than DIAS under static aggregation values. This because the accuracy convergence of DIAS requires approximately 100 epochs, whereas the gossip-based variance reduction converges in $20 - 25$ (Jelasity et al. 2005) iterations. For synchronous changes, the performance of the two aggregation methodologies, i.e., number of messages and convergence speed, becomes comparable as the iterative variance reduction algorithm requires recomputation of the aggregates. This performance impact becomes more significant as the frequency of changes increases, for example, more than $4 - 5$ times faster convergence for DIAS. Furthermore, if changes become asynchronous, gossip-based aggregation (Jelasity et al. 2005) becomes infeasible. Recomputations of aggregates cannot be performed as they require some type of synchronization.

Finally, DIAS does not require any changes in its aggregation methodology if different aggregation functions need to be computed simultaneously. This is the most cost-effective use of DIAS that motivates its selection for aggregation over related methodologies.

Diffusion methodologies cannot be applied to a wide range of aggregation functions and are usually interaction-dependent. For example, MAXIMUM and MINIMUM require the communication cost of epidemics (Jelasity et al. 2005; Kashyap et al. 2006) that approaches the one of DIAS.

Other information diffusion and gossiping aggregation mechanisms (Haridasan and van Renesse 2008; Jelasity et al. 2005; Kennedy et al. 2009; Kempe et al. 2003; Nath et al. 2008) do not consider dynamic changes of the aggregation values and assume synchronized recomputations. Coordination of these recomputations in distributed environments in not straightforward. Synopsis diffusion mechanisms (Ahmed et al. 2006; Nath et al. 2008) incorporate incremental updates of aggregates if changes in the aggregation values occur. However, only a relatively low number of changes can be tolerated compared to DIAS. For example, DIAS tolerates in the illustrated experiments 33600 changes compared to 1000 changes (Ahmed et al. 2006). A high number of items in the bit vectors of synopsis diffusion causes significant inaccuracies. The false positives of DIAS do not influence the accuracy of aggregates as they can be detected and eliminated.

Robust tree overlays are a flexible methodology to compute a wide range of aggregates but require topology self-management (Pournaras et al. 2010) in decentralized environments. Communication and storage complexity can be higher than the aggregation itself. Performing a relevant evaluation and comparison of aggregation trees with other more dedicated to aggregation mechanisms, such as DIAS, requires a use-case context and a specific application scenario. If tree topologies are reused between different distributed applications, including aggregation, the allocated cost is shared between these applications something that makes the use of trees more effective Fei et al. 2001. The unique paths of tree topologies are not required in DIAS as unique aggregation values are identified by the classification in the middle level. Furthermore, tree aggregation suffers from an unequal load distribution in nodes and the impact of failures (Ogston and Jarvis 2010). The nodes close to the root receive a high number of forwarded messages from the bottom nodes. Similarly, the impact of a failure close to the leaves is small whereas a single failure close to the root partitions the overlay network. These issues do not concern DIAS as it does not depend on a specific interaction mechanism. Nonetheless, the realization of the bottom level by the peer sampling service (Jelasity et al. 2007) results in a uniform communication overhead between nodes.

Discussion and future work

The DIAS architecture provides three levels of abstraction and modularity. The top level does not have any knowledge about the underlying complexity of classification and aggregation memberships. A wide range of aggregation functions can be accurately computed as the middle level guarantees that *aggregator* samples are classified as unexploited or outdated. Similarly, the middle level receives *aggregator* samples discovered by the bottom level.

A key feature of DIAS is the predefined number of possibles states during an aggregation phase. A large number of applications are fundamentally based on this assumption and design. User ranking aggregation in recommender systems (Garcin et al. 2009), is based on a finite and often restricted number of options for a user to rank an element. In applications of demand-side energy management (James et al. 2006; Pournaras et al. 2010), aggregate information about a finite number of alternative demand options improve the stability of the Smart Power Grid.

Dissemination and collection of all aggregation values in every agent of the network requires a significant communication cost. One way to decrease this cost is to eliminate the number of *aggregators* and *disseminators* in a network. Section "Adaptation strategies" shows that the communication cost of DIAS is decreased more than half if the network is split into the 2/3 of the nodes running *disseminators* and 1/3 *aggregators*. It is not always necessary for each node to perform both aggregation and dissemination as various applications do not require this. This is especially the case if nodes have different roles in a network, e.g., consumers and producers in the Smart Power Grid.

DIAS is based on the exchange of *aggregator* samples instead of *disseminator* samples. In the current design of DIAS, aggregation values are disseminated to *aggregators* instead of the *aggregators* requesting the aggregation values. The $M_{A_i}\left(S'_j\right)$ membership of *aggregators* cannot be used during the classification process as the selected state S'_j is not known. This issue can be overcome by injecting the selected state in *disseminator* samples exchanged by the bottom level.

Experimental evaluation illustrates the high accuracy and matching achieved even in the case of false positives in bloom filters. Tolerance to false positives provides large data space savings. Accuracy is maintained even if the size of bloom filters decreases significantly, resulting in a high number of detected false positives. A future extension is the dynamic and automated allocation of larger space in the bloom filters based on accuracy requirements under false positives. Alternative approaches to bloom filters are also considered in future work, e.g., hash compaction (Dillinger and Manolios 2004).

The classification of *aggregator* samples in the aggregation pool proactively prevents duplicate outcomes that increase communication overhead. The mutual membership checks reactively detect duplicate outcomes not detected during classification due to false positives. Mutual membership checks guarantee highly accurate aggregates, especially in the case of duplicate-sensitive aggregation functions such as SUMMATION, without introducing additional communication cost. The performance of RANDOM shows the large communication cost that duplicate outcomes cause and the large savings achieved by EXPLOITATION and UPDATE. Other future work concerns the evaluation of DIAS and its applications in various network conditions, such as churn (Kennedy et al. 2009) and latency.

Conclusions

This paper concludes that DIAS is a generic and middleware service for dynamic decentralized aggregation in large-scale distributed networks. The aggregation approach of DIAS is holistic: a local and duplicate-free availability of the distributed aggregation values that enables the simultaneous computation of almost any aggregation function. Achieving this abstraction in a cost-effective manner and without depending on a specific interaction mechanism is a challenge that has not been addressed in related work. DIAS meets these requirements by introducing an implicit representation and storage of the explicit distributed aggregation values: aggregation memberships in bloom filters. Ultimately, the generic design and applicability of DIAS results in a higher communication overhead compared to methodologies based on information diffusion (Jelasity et al. 2005; Nath et al. 2008). This is the trade off end users of such aggregation systems have to deal with: more generic applicability versus higher communication overhead.

The experimental evaluation shows that DIAS achieves high accuracy under synchronous and asynchronous changes of the aggregation values. Even when using bloom filters with a high number of false positives, accuracy is maintained almost entirely due to the mutual membership checks. The classification of *aggregator* samples and their selection based on two adaptation strategies provide (i) the minimization of duplicates that increase inaccuracies and communication overhead and (ii) the intelligent adaptation of aggregation in different network conditions.

Appendix A: Overview of related work

Table 1 summarizes the related aggregation mechanisms discussed in this paper.

Table 1 An overview of related decentralized mechanisms to DIAS

	Aggregation function	Aggregation values	Interaction requirements	Storage requirements
DIAS	any	highly dynamic	dissemination and collection	bloom filters
(Ahmed et al. 2006)	SUMMATION[a], COUNT, AVERAGE, STANDARD, DEVIATION[b]	dynamic	flooding, gossiping or random walks	counting sketches
(Haridasan and van Renesse 2008)	distribution of aggregation values	static	gossiping	synopsis diffusion
(Jelasity et al. 2005)	AVERAGE, COUNT[c], SUMMATION[a]	static, recomputations	gossiping	hash maps for COUNT
(Kashyap et al. 2006)	algorithm variations for MINIMUM, MAXIMUM, SUMMATION, AVERAGE, RANK	static	group formation and gossiping	synopsis diffusion

Table 1 An overview of related decentralized mechanisms to DIAS *(Continued)*

(Kempe et al. 2003)	algorithm variations for SUMMATION, AVERAGE and quantiles	static	gossiping	synopsis diffusion
(Nath et al. 2008)	SUMMATION, COUNT	static	ring/tree topologies, flooding	synopsis diffusion
(Ogston and Jarvis 2010)	SUMMATION[d] queries	dynamic	tree topology	parent and children

[a]It is derived by the AVERAGE and COUNT aggregates.
[b]It is derived by the SUMMATION and its squares.
[c]It is computed using the 'inverse birthday paradox' as explained in Section "Problem description".
[d]Others aggregates could be potentially computed.

Appendix B: Experimental settings

Table 2 summarizes the selected experimental settings. Note that multiple values for a single parameter denote the tested variations of this parameter in some of the illustrated experiments. The values depicted with bold are the default ones.

Table 2 The experimental settings for the evaluation of DIAS

	Parameter	Value
Protopeer	n	1500
	t(DIAS)	800
	T(DIAS)	1000
	t'(DIAS)	6
	$\lvert\mathbf{v}(ring)\rvert$	5
Application	$t'(application)$	15
	v	5
	type of states	numerical
	input domain of states	[0,1)
	generation of possible states	beta distribution
	distribution for s_i^0	alpha=5, beta=25
	distribution for s_i^1	alpha=25, beta=5
	distribution for s_i^2	alpha=10, beta=5
	distribution for s_i^3	alpha=5, beta=10
	distribution for s_i^4	alpha=5, beta=5
	selection of a possible state	cyclical
	$T(application)$	10 (asynchronous), 200 (synchronous)
	$P_c(time)$, $P_c(parameter)$	(1.0, 1.0), (0.4, 0.7)
Top Level	$T(top)$	1000
	$f()$	AVERAGE, SUMMATION, MAXIMUM
Middle Level	$T(middle)$	1000
	z	10
	q	45
	e	15
	adaptation strategy adoption	static
	hashing scheme	double hashing
	m, k	**(16, 24)**, (14, 24), (14, 6)

Table 2 The experimental settings for the evaluation of DIAS *(Continued)*

	r	50
	T(*bottom*)	250
Bottom Level	view selection policy	swapper
	view propagation policy	push-pull
	peer selection	policy random

The bold values are the default ones in the performed experiments.

Appendix C: Source experimental data

Figure 9 shows the five beta distributions[b] used for the parameterization of the possible states in the experimental evaluation of DIAS. In each aggregation epoch, a random value from a given beta distribution is assigned to a possible state.

Figures 10, 11, 12 and 13 illustrate the source data of the experimental results based on which the accuracy and matching are computed.

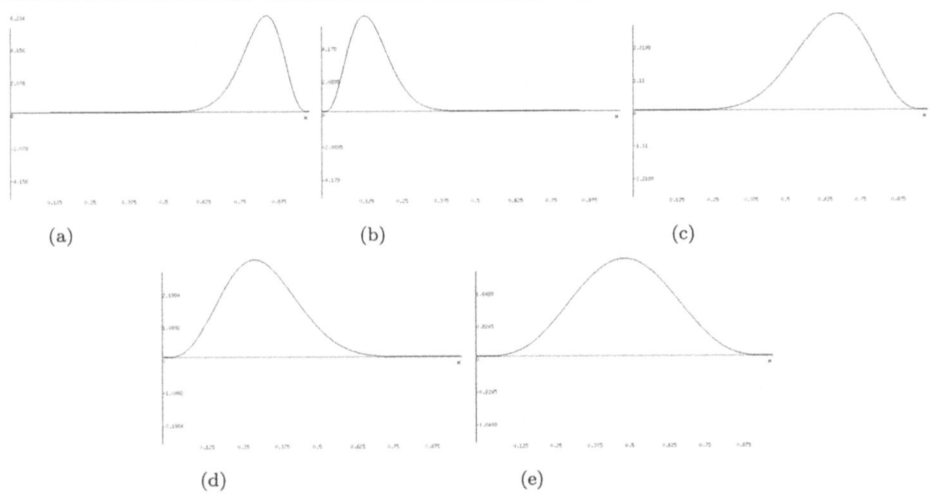

Figure 9 Five beta distribution used as the input domain of the possible states. (a) *alpha* = 25, *beta* = 5. **(b)** *alpha* = 5, *beta* = 25. **(c)** *alpha* = 10, *beta* = 5. **(d)** *alpha* = 5, *beta* = 10. **(e)** *alpha* = 5, *beta* = 5.

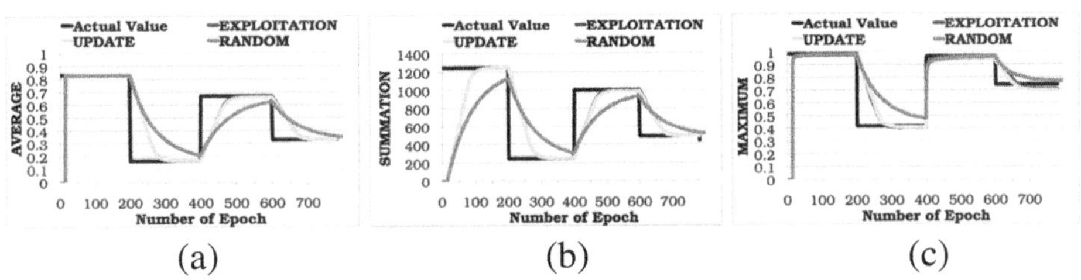

Figure 10 Aggregates of adaptation strategies under synchronous changes. (a) AVERAGE. **(b)** SUMMATION. **(c)** MAXIMUM.

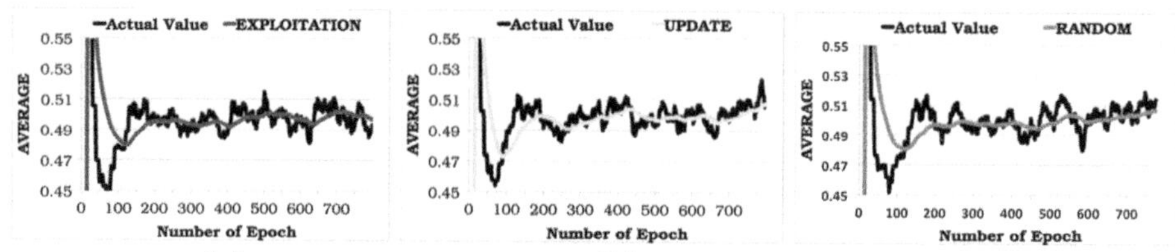

(a) (b) (c)

Figure 11 AVERAGE of the adaptation strategies under asynchronous changes. (a) EXPLOITATION. **(b)** UPDATE. **(c)** RANDOM.

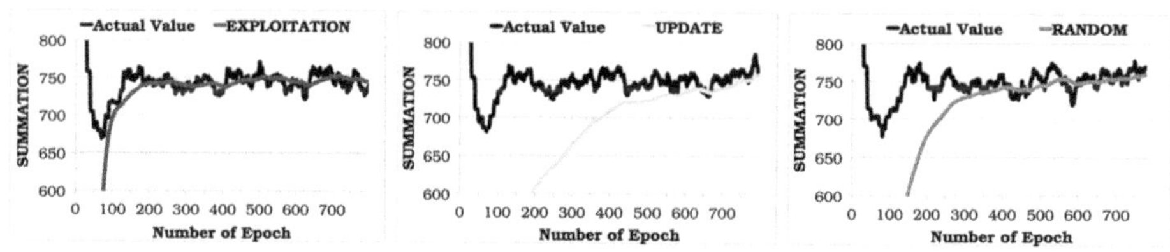

(a) (b) (c)

Figure 12 SUMMATION of adaptation strategies under asynchronous changes. (a) EXPLOITATION. **(b)** UPDATE. **(c)** RANDOM.

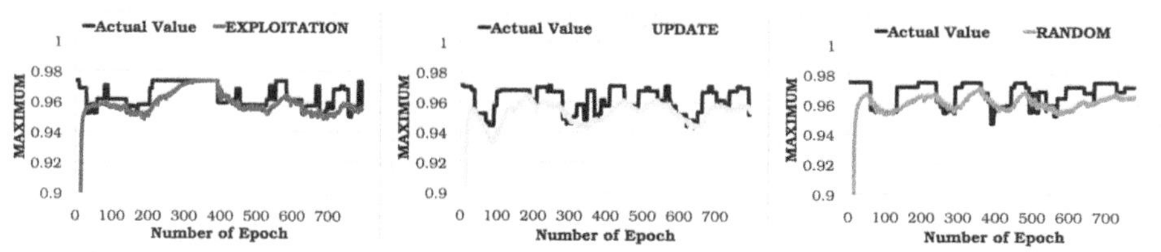

(a) (b) (c)

Figure 13 MAXIMUM of adaptation strategies under asynchronous changes. (a) EXPLOITATION. **(b)** UPDATE. **(c)** RANDOM.

Endnotes

[a]During system bootstrapping, there is no need for available historical information to distinguish between different classes as each aggregation value is by default unexploited.

[b]Generated by the Wessa online statistics software, available at: http://www.wessa.net/ (Last accessed: January 2013).

Author details

[1]Faculty of Electrical Engineering, Mathematics and Computer Science, Delft University of Technology, Mekelweg 4, 2628 CD, Delft, Netherlands. [2]Faculty of Technology, Policy and Management, Delft University of technology, Jaffalaan 5, 2628 BX Delft, Netherlands.

References

Ahmed N, Hadaller N, Keshav S: **Incremental maintenance of global aggregates.** 2006. Tech. rep., University of Waterloo, Waterloo, Ontario.

Bloom BH: **Space/time trade-offs in hash coding with allowable errors.** *Commun ACM* 1970, **13**(7):422–426.

Calvo T, Kolesárová A, Komorníková M, Mesiar R: **Aggregation operators: properties, classes and construction methods.** In *Aggregation Operators: New Trends and Applications, Volume 97 of Studies in Fuzziness and Soft Computing*, Heidelberg, Germany: Physica-Verlag GmbH; 2002:3–104.

Chitnis L, Dobra A, Ranka S: **Aggregation methods for large-scale sensor networks.** *ACM Trans Sensor Netw* 2008, **4**(2):1–36.

Deke G, Yunhao L, Xiangyang L, Panlong Y: **False negative problem of counting bloom filter.** *IEEE Trans Knowl Data Eng* 2010, **22**(5):651–664.

Dillinger PC, Manolios P: **Bloom filters in probabilistic verification.** In *Proceedings of the 5th International Conference on Formal Methods in Computer-Aided Design, FMCAD 2004, Volume 3312 of Lecture Notes in Computer Science,* Heidelberg: Springer-Verlag, Berlin; 2004:367–381.

Fei A, Cui J, Gerla M, Faloutsos M: **Aggregated multicast with inter-group tree sharing.** In *Proceedings of the 3rd International Workshop on Networked Group Communication, NGC 2001, Volume 2233 of Lecture Notes in Computer Science,* Heidelberg: Springer-Verlag Berlin; 2001:172–188.

Galuba W, Aberer K, Despotovic Z, Kellerer W: **ProtoPeer: a P2P toolkit bridging the gap between simulation and live deployement.** In *Proceedings of the Second International Conference on Simulation Tools and Techniques, ICST 2009 Gent,* Belgium: ACM; 2009:1–9.

Garcin F, Faltings B, Jurca R, Joswig N: **Rating aggregation in collaborative filtering systems.** In *Proceedings of the 3rd ACM Conference on Recommender Systems, RecSys 2009,* New York, NY, USA: ACM Press; 2009:349–352.

Haridasan M, van Renesse R: **Gossip-based distribution estimation in peer-to-peer networks.** In *Proceedings of the 7th International Workshop on Peer-to-peer Systems, IPTPS 2008.* Berkeley, CA, USA: USENIX Association; 2008:13.

James G, Cohen D, Dodier R, Platt G, Palmer D: **A deployed multi-agent framework for distributed energy applications.** In *Proceedings of the 5th International Conference on Autonomous Agents and Multiagent Systems, AAMAS 2006,* New York, NY, USA: ACM Press; 2006:676–678.

Jelasity M, Montresor A, Babaoglu O: **Gossip-based aggregation in large dynamic networks.** *ACM Trans Comp Syst* 2005, **23**(3):219–252.

Jelasity M, Voulgaris S, Guerraoui R, Kermarrec AM, van Steen M: **Gossip-based peer sampling.** *ACM Trans Comp Syst* 2007, **25**(3).

Jerzak Z, Fetzer C: **Bloom filter based routing for content-based publish/subscribe.** In *Proceedings of the 2nd International Conference on Distributed Event-based Systems, DEBS 2008,* New York, NY, USA: ACM Press; 2008:71–81.

Jiang S, Guo L, Zhang X: **LightFlood: an efficient flooding scheme for file search in unstructured peer-to-peer systems.** In *Proceedings of the 2003 International Conference on Parallel Processing, ICPP 2003,* Los Alamitos, CA, USA: IEEE; 2003:627–635.

Gkantsidis C, Mihail M, Saberi S: **Random walks in peer-to-peer networks: algorithms and evaluation.** *Perform Eval* 2006, **63**(3):241–263.

Kashyap S, Deb S, Naidu KVM, Rastogi R, Srinivasan A: **Efficient gossip-based aggregate computation.** In *Proceedings of the 25th Symposium on Principles of Database Systems - PODS 2006,* New York, NY, USA: ACM Press; 2006:308–317.

Kennedy O, Koch C, Demers A: **Dynamic approaches to in-network aggregation.** In *Proceedings of the 25th International Conference on Data Engineering, ICDE 2009.* Los Alamitos, CA, USA: IEEE; 2009:1331–1334.

Kempe D, Dobra A, Gehrke J: **Gossip-based computation of aggregate information.** In *Proceedings of the 44th Annual IEEE Symposium on Foundations of Computer Science, FOCS 2003,* Washington, DC, USA: IEEE Computer Society; 2003:482–491.

Li F, Pei C, Jussara A, Andrei BZ: **Summary cache: a scalable wide-area web cache sharing protocol.** *IEEE/ACM Trans Netw* 2000, **8**(3):281–293.

Nath S, Gibbons PB, Seshan S, Anderson Z: **Synopsis diffusion for robust aggregation in sensor networks.** *ACM Trans Sensor Netw* 2008, **4**(2):1–40.

Ogston E, Jarvis SA: **Peer-to-peer aggregation techniques dissected.** *Int J Parallel, Emergent and Distributed Syst* 2010, **25:**51–71.

Pournaras E: **Multi-level reconfigurable self-organization in overlay services.** *PhD thesis.* Delft University of Technology, Netherlands 2013.

Pournaras E, Warnier M, Brazier FMT: **Adaptation strategies for self-management of tree overlay networks.** In *Proceedings of the 11th IEEE/ACM International Conference on Grid Computing, Grid 2010.* Los Alamitos, CA, USA: IEEE; 2010:401–409.

Pournaras E, Warnier M, Brazier FMT: **Local agent-based self-stabilisation in global resource utilisation.** *Int J Auton Comput* 2010, **1**(4):350–373.

Yuh-Jzer J, Chien-Tse F, Li-Wei Y: **Keyword search in DHT-based peer-to-peer networks.** In *Proceedings of the 25th IEEE International Conference on Distributed Computing Systems, ICDCS 2005.* Los Alamitos, CA, USA: IEEE; 2005:339–348.

7

Generalized Thompson sampling for sequential decision-making and causal inference

Pedro A Ortega[1*] and Daniel A Braun[2*]

*Correspondence:
ope@seas.upenn.edu;
daniel.braun@tuebingen.mpg.de
[1]GRASP Laboratory, Electrical and Systems Engineering Department, University of Pennsylvania, Philadelphia, PA 19104, USA
[2]Max Planck Institute for Biological Cybernetics and Max Planck Institute for Intelligent Systems, Speemanstrasse 38, Tübingen 72076, Germany

Abstract

Purpose: Sampling an action according to the probability that the action is believed to be the optimal one is sometimes called Thompson sampling.

Methods: Although mostly applied to bandit problems, Thompson sampling can also be used to solve sequential adaptive control problems, when the optimal policy is known for each possible environment. The predictive distribution over actions can then be constructed by a Bayesian superposition of the policies weighted by their posterior probability of being optimal.

Results: Here we discuss two important features of this approach. First, we show in how far such generalized Thompson sampling can be regarded as an optimal strategy under limited information processing capabilities that constrain the sampling complexity of the decision-making process. Second, we show how such Thompson sampling can be extended to solve causal inference problems when interacting with an environment in a sequential fashion.

Conclusion: In summary, our results suggest that Thompson sampling might not merely be a useful heuristic, but a principled method to address problems of adaptive sequential decision-making and causal inference.

Keywords: Thompson sampling; Adaptive control; Bounded rationality; Decision-making; Causal inference

Background

In a research paper from 1933, Thompson studied the problem of finding out which one of two drugs was better when testing them on a patient population under the constraint that as few people as possible should be subjected to the inferior drug during the course of testing (Thompson 1933). Given a current (Bayesian) probability estimate P of one treatment being better than the other, he suggested that it might be a good idea to adjust the proportions of future test subjects that take the two drugs to the respective probabilities P and $1 - P$. This way one would not run into the danger of *permanently* cutting off all future test subjects from a potentially superior treatment that so far seems inferior due to statistical fluctuations, while only *temporarily* risking exposure to a potentially inferior drug for a smaller proportion of the population. Randomizing actions based on the probability that this action is believed to be optimal when faced with an unknown environment is now sometimes called *Thompson sampling.*

Today, Thompson's problem is generally thought of as a *bandit problem* that consists in determining which lever to pull at which point in time when facing a set of one-armed slot machines, each one having an unknown distribution over a reward variable (Russell and Norvig 1995; Sutton and Barto 1998). In the case of known prior probabilities and geometrically discounted future rewards, Gittins (1979) provides an optimal policy for the bandit problem that maximizes the expected future cumulative discounted reward. In contrast, Thompson sampling is usually considered as a heuristic approach to solve bandit problems (Wyatt 1997; Granmo 2008, 2010; Asmuth et al. 2009; Graepel et al. 2010; Scott 2010; May and Leslie 2011; Chapelle and Li 2011; Agrawal and Goyal 2011; Granmo and Glimsdal 2013; May et al. 2012; Kaufmann et al. 2012; Russo and Roy 2013; Korda et al. 2013; Bubeck and Liu 2013). However, the basic idea of Thompson sampling—that is, sampling actions from a mixture distribution of policies according to their probability of being optimal–can also be applied to solve more general problems in sequential adaptive control (Dearden et al. 1998; Strens 2000; Ortega and Braun 2010a, 2010b, 2012a; Braun and Ortega 2010; Osband and Russo 2013; Cao and Ray 2012; Tziortziotis et al. 2013a, 2013; Dimitrakakis 2013; Dimitrakakis and Tziortziotis 2013; Mellor and Shapiro 2013).

In this paper, our aim is to discuss some of the basic properties of Thompson sampling as a modeling approach. In particular, we want to argue that

1. Thompson sampling can be considered as an application of Bayes' rule for acting where actions are treated as causally intervened random variables within the framework of statistical causality.
2. Thompson sampling can be considered as a form of optimal adaptive control under bounded rationality where limited information processing capabilities are modeled by entropic search costs.
3. Thompson sampling provides a natural strategy for causal induction when interacting with an environment with unknown causal structure.

Although the third section contains an algorithmic extension to previous work (Ortega and Braun 2010a; 2010b), it should be emphasized that the main contribution of the paper is not so much to present a novel algorithm, but to discuss basic properties of Thompson sampling, in particular how it relates to the information-theoretic bounded rationality model in (Ortega and Braun 2013), how this boundedness can be interpreted in terms of sampling complexity, and how this method can be applied to solve problems of causal inference.

The paper is structured as follows. In Section "Problem statement" we clarify the problem statement and recapitulate the main result of (Ortega and Braun 2010b). In Section "Decision-making with limited resources" we analyze the decision-making problem faced by agents that are unable to compute the single best policy. In Section "Causal induction" we investigate how this approach can be applied to adaptive agents that need to discover the causal structure of their environment. Finally, we discuss the significance of these results in Section "Discussion".

Methods

Problem statement

In an adaptive control problem a decision-maker faces an environment Q_θ drawn from a set of potential environments $\mathcal{Q} = \{Q_\theta | \theta \in \Theta\}$. In general θ could be a continuous variable, but we restrict our exposition to the discrete case. Each environment Q_θ can be characterized by a set of conditional distributions $Q(o_t|\theta, a_{\leq t}, o_{<t})$ that indicate the probability of observing o_t given past observations $o_{<t} = o_1 \dots o_{t-1}$ and past actions $a_{\leq t} = a_1 \dots a_t$. This class of environments is very general, and it encompasses multi-armed bandits, (partially observable) Markov decision processes and others –compare Chapter 3 (Legg 2008). To allow for self-optimizing agents, the environment is typically assumed to be ergodic, so agents can recover from their mistakes –compare Section 3.5 (Legg 2008). The decision-maker has perfectly fitting prediction models $P(o_t|\theta, a_{\leq t}, o_{<t}) = Q(o_t|\theta, a_{\leq t}, o_{<t})$, but is uncertain about θ. The uncertainty about θ can be represented by a prior distribution $P(\theta)$. The interaction proceeds as follows. First an environment θ is sampled from $P(\theta)$. The agent picks an action a_0 and receives an observation o_0, to which the agent responds with a_1 and receives observation o_1 etc. The agent's policy can be described by a set of conditional distributions $P(a_t|o_{<t}, a_{<t})$.

Problem statement: decision-theory

In order to solve the problem within the framework of maximum expected utility theory, one requires

- a prior $P(\theta)$ over possible environments Q_θ
- a class of prediction models $P(o_t|\theta, a_{\leq t}, o_{<t})$
- a utility function $U(o_{\leq T}, a_{\leq T})$.

Then one can reduce the problem of the unknown environment to a problem with known environment. Such a "known" environment can be created from the set of possible environments by marginalizing over the parameter of the possible environments, thus, obtaining the Bayesian mixture distribution, where

$$P(o_t|o_{<t}, a_{\leq t}) = \sum_\theta P(\theta|o_{<t}, a_{\leq t}) P(o_t|\theta, o_{<t}, a_{\leq t}). \tag{1}$$

The adaptive control problem is then solved by finding the optimal policy $p(a_t|a_{<t}, o_{<t}) = \delta(a_t - a_t^*)$, where

$$\begin{aligned} a_t^* = \arg\max_{a_t} \sum_{o_t} P(o_t|o_{<t}, a_{\leq t}) \max_{a_{t+1}} \sum_{o_{t+1}} P(o_{t+1}|o_{\leq t}, a_{\leq t+1}) \cdots \\ \cdots \max_{a_T} \sum_{o_T} P(o_T|o_{<T}, a_{\leq T}) U(o_{\leq T}, a_{\leq T}) \end{aligned} \tag{2}$$

maximizes the expected utility under the mixture distribution (Hutter 2004). Equations (1) and (2) define a Bayesian adaptive control problem (Martin 1967; Duff 2002). This problem formulation becomes quickly intractable, as the number of reachable information states grows exponentially in the time horizon (Duff 2002).

Problem statement: probability theory & statistical causality

Ignoring the notion of utility for a moment and treating actions purely as (causally intervened) random variables (Pearl 2000), one could think of another kind of adaptive control problem that is defined entirely in probabilistic and causal terms. This requires the following ingredients

- a prior $P(\theta)$ over possible environments Q_θ
- a class of prediction models $P(o_t|\theta, a_{\leq t}, o_{<t})$
- a class of policy models $P(a_t|\theta, a_{<t}, o_{<t})$

such that for every possible environment indexed by θ there is a perfectly fitting predictor $P(o_t|\theta, a_{\leq t}, o_{<t}) = Q_\theta(o_t|a_{\leq t}, o_{<t})$ and a desirable custom-built[a] policy $P(a_t|\theta, a_{<t}, o_{<t})$. The problem statement is: What is the next action a_t given the uncertainty over θ and given the history of previous actions $\hat{a}_{<t}$ and previous observations $o_{<t}$? As for any random variable, answering this question for the random variable a_t simply requires computing the predictive distribution conditioned on the past, that is

$$P(a_t|\hat{a}_{<t}, o_{<t}) = \sum_\theta P(a_t|\theta, \hat{a}_{<t}, o_{<t}) P(\theta|\hat{a}_{<t}, o_{<t}). \quad (3)$$

As there can be only one action at any one time, single actions can be obtained as samples from $P(a_t|\hat{a}_{<t}, o_{<t})$. Importantly, sampling from $P(a_t|\hat{a}_{<t}, o_{<t})$ is equivalent to first sampling a random belief θ from the posterior $P(\theta|\hat{a}_{<t}, o_{<t})$ and then sampling an action from $P(a_t|\theta, \hat{a}_{<t}, o_{<t})$. This componentwise sampling from a mixture distribution is known as *hierarchical sampling* and corresponds here to a generalized Thompson sampling procedure, where first a random belief is sampled and then the associated policy with respect to this belief is executed. If we assume now that each of the custom-built policies is optimal in their respective environments, we effectively select an action according to the probability that it is the optimal action, because we first sample the environment θ according to its posterior probability of being the true environment and then we perform the policy that is optimal in that environment. The question is how this problem formulation can be reconciled with a decision-theoretic problem statement that involves utilities. This is the topic of Section "Decision-making with limited resources".

Statistical causality

While both actions and observations are treated as random variables in (3), there is an important difference between actions and observations. Observations are produced by the environment and can be used to update the agent's state of knowledge about the environment. In contrast, actions are set by the agent itself and hence they do not provide information about the environment. This distinction becomes crucial when conditioning on the history of actions and observations. The theory that deals with the distinction between exogenous and endogenous information is *statistical causality* (Pearl 2000; Glymour et al. 2000).

What is a causal intervention?

A typical example for causal intervention is the manipulation of a barometer (Pearl 2000). In a barometer the atmospheric pressure changes the height of the mercury: if it rises, we expect good weather; and if it drops rapidly, we expect rain. A simple Bayesian model captures this relation:

$$P(w|b) = \frac{P(b|w)P(w)}{P(b)},$$

where w and b are the weather and the Barometer variables respectively, $P(w)$ is the prior probability of the weather (e.g. good or bad) and $P(b|w)$ is the likelihood of the barometer change given the weather. The posterior $P(w|b)$ allows us to infer the weather from the barometer reading.

Now, imagine you decide to change the level of the mercury yourself, say (using a bit of imagination) by means of a pressurizing device. Now, *you* set the value of the random variable—and intuition tells us that we cannot predict the weather anymore from the barometer reading. Apparently, our previous Bayesian model is useless now. This shouldn't come as a surprise, as our intervention effectively changed the relation between the barometer and the weather.

Mathematically, we can model the causal relationship between different random variables by a particular factorization of the joint probability distribution into conditional probabilities reflecting generative mechanisms, *where causes are in the conditional and effects in the argument.* In our example the weather causes the barometer to rise and fall and not the other way around, that is in causal terms we have $P(b, w) = P(b|w)P(w)$ and not $P(b, w) = P(w|b)P(b)$, even though in purely probabilistic terms the two factorizations are of course equivalent. The causal factorization becomes important when modifying the causal relationships by intervention, as in our example. Manipulating the barometer setting directly severs the causal between weather and barometer. Consequently, we have to modify our joint probability distribution from $P(b, w)$ to

$$P(\hat{b}, w) = \delta(b)P(w),$$

that is, where $P(b|w)$, viewed as a generative mechanism, has been replaced by $\delta(b)$, thereby rendering the two random variables independent. The hat-notation (due to (Pearl 2000)) is just a shorthand referring to this particular transformation of the probability distribution. When we evaluate the posterior under the intervention $\hat{b}$, we get

$$P(w|\hat{b}) = \frac{P(w, \hat{b})}{\sum_w P(w, \hat{b})} = \frac{1 \cdot P(w)}{1} = P(w).$$

In other words, we don't gain knowledge about the weather—as expected. Notice that intervening the alternative factorization, $P(\hat{b}, w) = \delta(b)P(w|b)$, would give a different result that is inconsistent with our causal story: we have assumed that the mercury level of the barometer depends functionally on the weather, and not the other way around. The reason for this special treatment of actions is that when we set the value of a random variable ourselves, we change Nature's probability law.

Causal interventions in Thompson sampling

To calculate the effect of an intervention, the causal model has to be known, that is the unique factorization of the joint distribution into conditional probabilities matching the causal dependencies over the random variables. In our case, the causal dependencies are straightforward: first, the environment secretly chooses a true parameter $\theta^* \in \Theta$, and then the interactions $a_1, o_1, a_2, o_2, \ldots$ follow chronologically. This causal model allows us to study the effect of interventions in the problem statement given in (3). We start by re-expressing the posterior $P(\theta|\hat{a}_{<t}, o_{<t})$ as

$$\begin{aligned}
&P(\theta|\hat{a}_{<t}, o_{<t}) \\
&\overset{(1.)}{=} \frac{P(\theta, \hat{a}_{<t}, o_{<t})}{\sum_{\theta'} P(\theta', \hat{a}_{<t}, o_{<t})} \\
&\overset{(2.)}{=} \frac{P(\theta)\prod_{k=1}^{t} P(\hat{a}_k|\theta, \hat{a}_{<k}, o_{<k})P(o_k|\theta, \hat{a}_{\leq k}, o_{<k})}{\sum_{\theta'} P(\theta')\prod_{k=1}^{t} P(\hat{a}_k|\theta', \hat{a}_{<k}, o_{<k})P(o_k|\theta', \hat{a}_{\leq k}, o_{<k})} \\
&\overset{(3.)}{=} \frac{P(\theta)\prod_{k=1}^{t} P(\hat{a}_k|\theta, a_{<k}, o_{<k})P(o_k|\theta, a_{\leq k}, o_{<k})}{\sum_{\theta'} P(\theta')\prod_{k=1}^{t} P(\hat{a}_k|\theta', a_{<k}, o_{<k})P(o_k|\theta', a_{\leq k}, o_{<k})} \\
&\overset{(4.)}{=} \frac{P(\theta)\prod_{k=1}^{t} P(o_k|\theta, a_{\leq k}, o_{<k})}{\sum_{\theta'} P(\theta')\prod_{k=1}^{t} P(o_k|\theta', a_{\leq k}, o_{<k})},
\end{aligned} \tag{4}$$

where we first expand the probabilities in terms of the joint distribution, second rewrite the joint distribution as the causal factorization, third remove the intervention tags from the intervened random variables that are in the probability conditions (Pearl's second rule of do-calculus (Pearl 2000)), and fourth replace each conditional probability having an intervened variable in the argument by a delta function over its chosen value—compare Chapter 4.2 in (Pearl 2000).

These equations show that beliefs are updated only using past observations, and that past actions provide no further evidence. Intuitively, the reason for this is that the agent can be surprised about his past observations and learn from them, but he cannot be surprised about his own actions chosen by himself in the past. Also, due to Pearl's second law of do-calculus (Pearl 2000) we have

$$P(a_t|\theta, \hat{a}_{<t}, o_{<t}) = P(a_t|\theta, a_{<t}, o_{<t}). \tag{5}$$

Using (3), (4) and (5), we arrive at the desired result, that is the predictive distribution for $a_t \in \mathcal{A}$ given by

$$P(a_t|\hat{a}_{<t}, o_{<t}) = \sum_{\theta} P(a_t|\theta, a_{<t}, o_{<t})P(\theta|\hat{a}_{<t}, o_{<t}), \tag{6}$$

obtained only by applying probability theory and causal calculus.

Results

Decision-making with limited resources

The difference between the random belief model supposed by Thompson sampling and the standard maximum expected utility model for decision-making is highlighted by contrasting two simple decision scenarios depicted in Figure 1. The goal is to predict the outcome when throwing one of two possible biased coins. A rational decision maker places bets (shown inside speech bubbles) such that his subjective expected utility is maximized. These subjective beliefs are delimited within dotted boxes. A Thompson sampling agent first samples a random belief and then chooses the best prediction with respect to this belief.

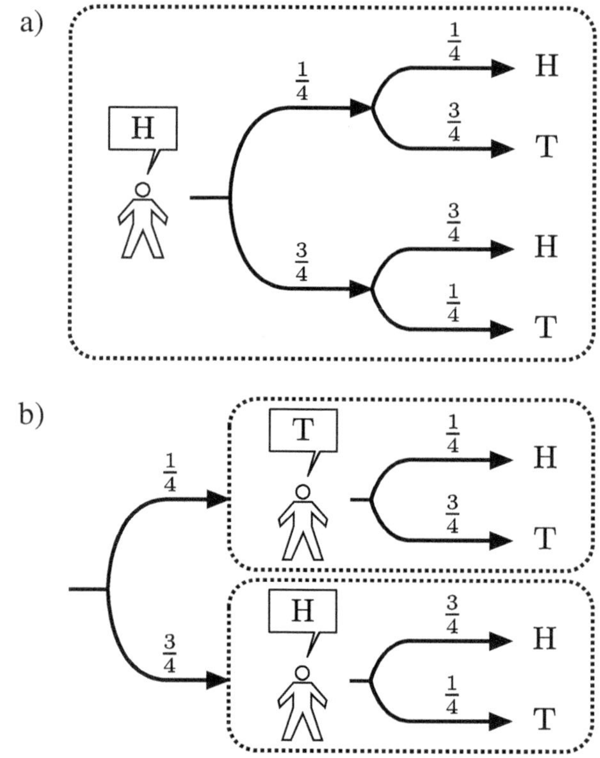

Figure 1 Comparison between (a) a fully rational decision maker and (b) a decision maker with random beliefs. There are two biased coins. The probability of throwing the first coin is $\frac{1}{4}$, the probability of throwing the second coin is $\frac{3}{4}$. The first coin has a probability of $\frac{1}{4}$ for the outcome Head and the second coin has a probability of $\frac{3}{4}$ for Head. The decision-maker has to predict the most likely outcome. A correct guess is rewarded with \$1, an incorrect guess yields \$0.

The difference between the two becomes clear by inspecting the expected utility in each case: they are

$$\max_{P'} \sum_{\theta} P(\theta) \left\{ \sum_{o} P'(a|\theta) P(o|\theta, a) U(o) \right\}, \qquad \text{(a)}$$

$$\text{and} \quad \sum_{\theta} P(\theta) \max_{P'} \left\{ \sum_{o} P'(a|\theta) P(o|\theta, a) U(o) \right\} \qquad \text{(b)}$$

respectively, where the labels (a) and (b) correspond to the labels in Figure 1. Here it is clearly seen that the difference between the two lies in the order in which we apply the expectation (over the environment parameter) and the maximization operator. It should also be noted that the expected utility of (a) is an upper bound on the expected utility of (b). Yet, both cases can constitute optimal decisions depending on constraints. In (a), the decision-maker picks his action taking into account the uncertainty over the bias, while in (b), the decision-maker picks his action only after his beliefs over the coin bias are instantiated—that is, *he is optimal w.r.t. his random beliefs.* Here we consider how this optimality w.r.t. random beliefs can be considered as a form of optimal decision-making under information processing constraints.

Modeling bounded rational decision-making

Here we consider a particular information-theoretic model of bounded rational decision-making that formalizes limited information processing resources by a variational principle that trades off expected utility gains (or losses) and entropic information costs (Ortega 2011a; Ortega and Braun 2011, 2012a, 2013). Information processing costs are usually ignored in the study of perfectly rational decision-makers. Given a choice set $\mathcal{X}$ with choices $x \in \mathcal{X}$ and utilities $U(x)$, a perfectly rational decision-maker would always choose the best option $x^* = \arg\max_x U(x)$—presupposing there is a unique maximum. In general, a bounded rational decision-maker is unable to pick out the best option with certainty, and his choice can be described by a probability distribution $P(x)$ reflecting uncertainty. Improving the choice strategy $P(x)$ can be understood as a costly search process.

Let us assume the initial strategy of the decision-maker can be described by a probability distribution $P_0(x)$. The search process for the optimum transforms this initial choice into a final choice $P(x)$. In case of the perfectly rational decision-maker the final choice is $P(x) = \delta_{x,x^*}$. In the general case of the bounded rational decision-maker the search is costly and he will not be able to afford such a stark reduction in uncertainty. Assuming that search costs are real-valued, additive and higher for rare events (Ortega and Braun 2010c), it can be shown that the cost of the search is determined by the information distance D_{KL} between P_0 and P, that is $D_{KL} = \sum_x P(x) \log \frac{P(x)}{P_0(x)}$. Both Bayesian search (Jaynes 1985) and Koopman's random search (Stone 1998) are compatible with these assumptions, as well as energetic costs that would have to be paid by a Maxwellian demon for reducing uncertainty in statistical physical systems (Ortega and Braun 2013). How this information-theoretic model of search costs relates to computational resources such as space and time complexity is still an open problem (Vitanyi 2005).

Simple decisions

The decision process is modeled as a transformation of a prior choice probability P_0 into a posterior choice probability P by taking into account the utility gains (or losses) and the transformation costs arising from information processing, such that

$$P = \arg\max_{\tilde{P}} \left\{ \sum_x \tilde{P}(x)U(x) - \frac{1}{\alpha} \sum_x \tilde{P}(x) \log \frac{\tilde{P}(x)}{P_0(x)} \right\}, \tag{7}$$

where the $x \in \mathcal{X}$ are the possible outcomes, $P_0(x)$ are their prior probablities, $U(x)$ are their utilities. The *inverse temperature* $\alpha \geq 0$ can be regarded as a rationality parameter that translates the cost of information processing measured in units of information into units of utility. If the limits in information processing capabilities are given as a constraint $D_{KL}(P||P_0) \leq K$ with some positive constant K, then α is determined as a Lagrange multiplier. The maximizing distribution $\tilde{P} = P$ is the *equilibrium distribution*

$$P(x) = \frac{1}{Z_\alpha} P_0(x) e^{\alpha U(x)}, \qquad \text{where} \qquad Z_\alpha = \sum_x P_0(x) e^{\alpha U(x)}, \tag{8}$$

and represents the choice probabilities after deliberation—see Theorem 1.1.3 (Keller 1998) and (Ortega and Braun 2013) for a proof. The *value* V of the choice set $\mathcal{X}$ under choice probabilities P can be determined from the same variational principle

$$V[P] = \max_{\tilde{P}} \left\{ \sum_x \tilde{P}(x)U(x) - \frac{1}{\alpha} \sum_x \tilde{P}(x) \log \frac{\tilde{P}(x)}{P_0(x)} \right\}$$
$$= \frac{1}{\alpha} \log \left(\sum_x P_0(x) e^{\alpha U(x)} \right) = \frac{1}{\alpha} \log Z_\alpha. \tag{9}$$

For the two different limits of α, the value and the equilibrium distribution take the asymptotic forms

$$\alpha \to +\infty \qquad \tfrac{1}{\alpha} \log Z_\alpha = \max_x U(x) \qquad P(x) = \mathcal{U}_{\max}(x) \qquad \text{(perfectly rational)}$$

$$\alpha \to 0 \qquad \tfrac{1}{\alpha} \log Z_\alpha = \sum_x P_0(x)U(x) \qquad P(x) = P_0(x) \qquad \text{(irrational)}$$

where $\mathcal{U}_{\max}$ is the uniform distribution over the maximizing subset $\mathcal{X}_{\max} := \{x \in \mathcal{X} : U(x) = \max_{x'} U(x')\}$. It can be seen that a perfectly rational agent with $\alpha \to \infty$ is able to pick out the optimal action—which is a deterministic policy in the case of a single optimum—, whereas finitely rational agents have stochastic policies with non-zero probability of picking a sub-optimal action.

The model of bounded rational decision-making also lends itself to an interpretation in terms of sampling complexity. If we use a rejection sampling scheme to obtain samples from $p(x)$ by first sampling from $p_0(x)$, we can ask how many samples we will need on average from p_0 to obtain one sample from p. In this scheme, we produce a sample $x \sim p_0(x)$ and then decide whether to accept or reject the sample based on the criterion

$$u \leq \frac{e^{\alpha U(x)}}{e^{\alpha T}}, \tag{10}$$

where u is drawn from the uniform $\mathcal{U}[0; 1]$ and T is the acceptance target value with $T \geq \max_x U(x)$. The equality holds for the most efficient sampler, but requires knowledge of the maximum. With this sampling scheme, the accepted samples will be distributed according to Equation (8). The average number of samples needed from p_0 to produce one sample of p is then

$$\overline{\sharp Samples} = \frac{1}{\sum_x p_0(x) \frac{e^{\alpha U(x)}}{e^{\alpha T}}} = \frac{e^{\alpha T}}{Z_\alpha}. \tag{11}$$

The important point about Equation (11) is that the average number of samples increases with increasing rationality parameter α. In fact, the average number of samples will grow exponentially for large α when $T > \max_x U(x)$, as

$$\frac{e^{\alpha T}}{Z_\alpha} \xrightarrow{\alpha \to \infty} \frac{e^{\alpha(T - U(x^*))}}{P_0(x^*)},$$

where $x^* = \arg\max U(x)$. It can also be straightforwardly seen that

$$\overline{\sharp Samples} \geq e^{D_{KL}(p||p_0)} = \frac{e^{\alpha \sum_x p(x)U(x)}}{Z_\alpha} \tag{12}$$

because $\sum_x p(x)U(x) \leq T$, that is the exponential of the Kullback-Leibler divergence provides a lower bound on the average number of samples.

Decisions in the presence of latent variables

To model a Thompson sampling agent, we need at least a two-step decision with a variable x that has to be chosen by the agent, and a variable θ that is chosen by the environment. In the example described in Figure 1, the variable x is the agent's prediction for the outcome of a coin toss, the variable θ indicates nature's choice which one of the two coins is tossed. The agent's prediction can take on the values $x = H$ and $x = T$ corresponding to the outcomes Head and Tail. The variable θ takes on the two values $\theta = \frac{1}{4}$ and $\theta = \frac{3}{4}$ corresponding to the biases of the two coins. The prior probability over θ is $p_0\left(\theta = \frac{1}{4}\right) = \frac{1}{4}$ and $p_0\left(\theta = \frac{3}{4}\right) = \frac{3}{4}$. The expected rewards for all combinations of x and θ are then $U\left(x = H, \theta = \frac{1}{4}\right) = \frac{1}{4}$, $U\left(x = T, \theta = \frac{1}{4}\right) = \frac{3}{4}$, $U\left(x = H, \theta = \frac{3}{4}\right) = \frac{3}{4}$ and $U\left(x = T, \theta = \frac{3}{4}\right) = \frac{1}{4}$.

In the case of two-step decisions, the variational problem can in general be formulated as a nested expression (Ortega and Braun 2011, 2012a, 2013)

$$\arg\max_{\tilde{p}(x,\theta)} \sum_x \tilde{p}(x) \left[U(x) - \frac{1}{\alpha} \log \frac{\tilde{p}(x)}{p_0(x)} + \sum_\theta \tilde{p}(\theta|x) \left[U(x,\theta) - \frac{1}{\beta} \log \frac{\tilde{p}(\theta|x)}{p_0(\theta|x)} \right] \right]. \quad (13)$$

with the two different rationality parameters α and β for the two different variables x and θ. Limited information processing resources with respect to these variables can also be thought of as different degrees of control. For example, if α assumed a large value, the decision-maker could basically hand-pick a particular x, or if θ was determined by a coin toss that the agent cannot influence, we could model this by setting β to zero. The utility can in general depend on both action and observation variables. However, since the action by itself does not yield a reward in our case, we have $U(x) \equiv 0$. Moreover, we see that in our case, nature's probability of flipping either coin does not actually depend on the agent's prediction, so we can replace the conditional probabilities $p(\theta|x)$ by $p(\theta)$. We have then an inner variational problem:

$$\arg\max_{\tilde{p}(\theta)} \sum_\theta \tilde{p}(\theta) \left[-\frac{1}{\beta} \log \frac{\tilde{p}(\theta)}{p_0(\theta)} + U(x,\theta) \right] \quad (14)$$

with the solution

$$p(\theta) = \frac{1}{Z_\beta(x)} p_0(\theta) \exp\left(\beta U(x,\theta)\right) \quad (15)$$

and the normalization constant $Z_\beta(x) = \sum_\theta p_0(\theta) \exp\left(\beta U(x,\theta)\right)$ and an outer variational problem

$$\arg\max_{\tilde{p}(x)} \sum_x \tilde{p}(x) \left[-\frac{1}{\alpha} \log \frac{\tilde{p}(x)}{p_0(x)} + \frac{1}{\beta} \log Z_\beta(x) \right] \quad (16)$$

with the solution

$$p(x) = \frac{1}{Z_{\alpha\beta}} p_0(x) \exp\left(\frac{\alpha}{\beta} \log Z_\beta(x)\right) \quad (17)$$

and the normalization constant $Z_{\alpha\beta} = \sum_x p_0(x) \exp\left(\frac{\alpha}{\beta} \log Z_\beta(x)\right)$. From Equation (17) we can derive both the perfectly rational decision-maker and the Thompson sampling agent. To simplify, we assume in the following that the agent has no prior preference for x, that is $p_0(x) = \mathcal{U}(x)$.

The perfectly rational decision-maker is obtained in the limit $\alpha \to \infty$ and $\beta \to 0$. If we first take the limit $\lim_{\beta \to 0} \frac{1}{\beta} \log Z_\beta(x) = \sum_\theta p_0(\theta) U(x, \theta)$, a decision-maker with rationality α chooses x with probability

$$p(x) = \frac{e^{\alpha \sum_\theta p_0(\theta) U(x,\theta)}}{Z_{\alpha\beta}}. \tag{18}$$

The perfectly rational expected utility maximizer as depicted in Figure 1a is then obtained from Equation (18) by taking the limit $\alpha \to \infty$.

In contrast, the Thompson sampling agent is obtained when $\beta = \alpha$. In this case, the choice probability for x is given by

$$p(x) = \sum_\theta p_{0(\theta)} \frac{e^{\alpha U(x,\theta)}}{Z_\alpha(x)}. \tag{19}$$

The resulting agent is a probabilistic superposition of agents that act optimally for any given θ as depicted in Figure 1b. It can be seen that in Equation (19) and in Equation (18) the order of the expectation operation and the (soft-)maximization operation are reversed.

Again we can interpret this formalism in terms of sampling complexity. Here we should accept a sample $x \sim p_0(x)$ if it fulfils the criterion

$$u \leq \frac{e^{\alpha \frac{1}{\beta} \log Z_\beta(x)}}{e^{\alpha T}} = \left[\frac{Z_\beta(x)}{e^{\beta T}}\right]^{\frac{\alpha}{\beta}}, \tag{20}$$

where $u \sim \mathcal{U}[0;1]$ and $T \geq \max_x \max_\theta U(x, \theta)$. From Equation (11) we know that the ratio $Z_\beta(x)/e^{\beta T}$ is the acceptance probability of $\theta \sim p_0(\theta)$. In order to accept one sample from x, we thus need to accept $\frac{\alpha}{\beta}$ consecutive samples of θ, with acceptance criterion

$$u \leq \frac{e^{\beta U(x,\theta)}}{e^{\beta T}} \tag{21}$$

with $u \sim \mathcal{U}[0;1]$ and T as set above. Since $\alpha \gg \beta$ we can assume $\alpha \approx N\beta$ with $N \in \mathbb{N}$, and we can see easily that the perfectly rational agent will require infinitely many θ samples ($\alpha \to \infty$ and $\beta \to 0$) to obtain one sample of x, whereas the Thompson sampling agent will only require a single sample ($\alpha = \beta$). The Thompson sampling agent is therefore the agent that can solve the optimization problem of Equation (16) for a given α with the least amount of samples. This can also be seen from Equation (18), when doing the Monte Carlo approximation $\sum_\theta p_0(\theta) U(x, \theta) \approx \frac{1}{N} \sum_i U(x, \theta_i)$ by drawing N samples $\theta_i \sim p_0(\theta_i)$. For infinitely many samples, the average approximates the expectation, for a single sample we can rewrite Equation (18) into Equation (19). This sampling procedure also allows estimating the upper and lower bounds of the optimal utility (Tziortziotis et al. 2013). Of course, the Thompson sampling agent will not achieve the same expected utility as the perfectly rational agent. But both agents can be considered optimal under particular information processing constraints.

Causal induction

A generalized Thompson-sampling agent can be thought of as a probabilistic superposition of models θ, where each model θ is characterized by a likelihood model $P(o_t|\theta, a_{\le t}, o_{<t})$ and a policy model $P(a_t|\theta, a_{<t}, o_{<t})$. In previous applications we assumed that all models θ have the same causal structure, i.e. considering multivariate random variables a_t and o_t, we assumed that the same variables a_t are intervened for all θ and the same causal model is used to predict the consequences of these interventions on the observational variables o_t. However, this need not be the case. In principle, different models θ could represent different causal structures and suggest intervention of different variables. Such a setup can be used for causal induction as illustrated in the following example.

> Imagine we are working on a medical treatment that involves two gene sites X and Y, each of which can be active or inactive. We encode the 'on' and 'off' states of X as $X = x$ and $X = \neg x$ and similarly $Y = y$ and $Y = \neg y$ to denote the 'on' and 'off' states of Y. Assume we are unsure about the causal mechanism between the two variables, that is we are unsure whether the activity of X influences the activity of Y or the other way around. Formally, we are interested in the explanatory power of two competing causal hypotheses: either 'X causes Y' ($\Theta = \theta$) or 'Y causes X' ($\Theta = \neg\theta$). Assume our goal is to have Y in an active state, but that it is much cheaper and easier to manipulate X instead of Y. This leaves us with the following policies. If X causes Y we prefer to manipulate X, because it is cheap and easy. If Y causes X we have no other choice, but to directly manipulate Y. When manipulating either gene, we can be 100% sure that the new state of the gene is set by us, but we only have a 50% chance that the state will be 'on'. Assume not manipulating either variable is not an option, because then both gene sites stay inactive. The question is how should we act if we do not know the causal dependency?

One of the main methods to deal with problems of causal inference is the framework of causal graphical models (Pearl 2000). Given a graph that represents a causal structure, we can intervene this graph and ask questions about the probabilities of the variables in the graph. However, in causal induction we would like to discover the causal structure itself, that is we would like to do inference over a multitude of graphs representing different causal structures (Heckerman et al. 1999). If one would like to represent the problem of causal discovery graphically, the main challenge is that the model Θ is a random variable that controls the causal structure itself. However, as argued in (Ortega 2011), this difficulty can be overcome by using a probability tree to model the causal structure over the random events. Probability trees can encode alternative causal realizations, and in particular alternative causal hypotheses (Shafer 1996). For instance, Figure 2a encodes the probabilities and functional dependencies among the random random variables of the previous problem.

In a probability tree, each (internal) node is a causal mechanism; hence a path from the root node to one of the leaves corresponds to a particular sequential real-

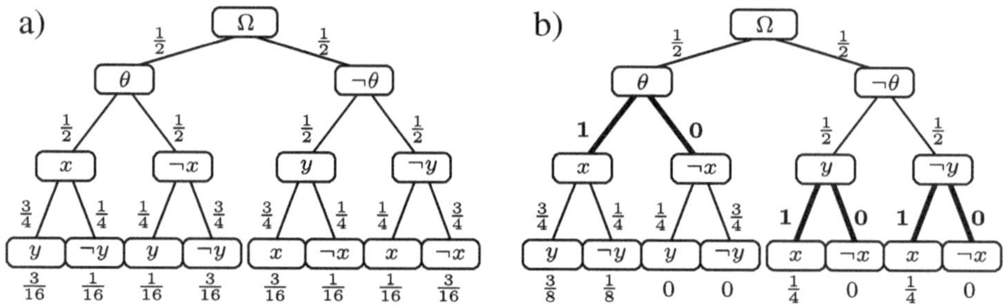

Figure 2 Panel a: An exemplary probability tree to represent the agent's prediction model about its environment. The probability of a realization (written under the leaves) is calculated by multiplying the probabilities starting from the root until a leave is reached. The immediate children of any note resolve the value of a single random variable. Note that the two hypotheses are statistically indistinguishable. Panel **b**: The probability tree resulting from **(a)** after intervening $\hat{x}$.

ization of causal mechanisms. The logic underlying the structure of this tree is as follows:

1. *Causal precedence:* A node causally precedes its descendants. For instance, the root node corresponding to the sure event Ω causally precedes all other nodes.
2. *Resolution of variables:* Each node resolves the value of a random variable. For instance, given the node corresponding to $\Theta = \theta$ and $X = \neg x$, either $Y = y$ will happen with probability $P(y|\theta, \neg x) = \frac{1}{4}$ or $Y = \neg y$ with probability $P(\neg y|\theta, \neg x) = \frac{3}{4}$.
3. *Heterogeneous order:* The resolution order of random variables can vary across different branches. For instance, X precedes Y under $\Theta = \theta$, but Y precedes X under $\Theta = \neg\theta$. This is precisely how we model competing causal hypotheses.

While the probability tree represents the agent's subjective model explaining the order in which the random values are resolved, it does not necessarily correspond to the temporal order in which the events are revealed to us. So for instance, under hypothesis $\Theta = \theta$, the value of the variable Y might be revealed before X, even though X causally precedes Y; and the causal hypothesis Θ, which precedes both X and Y, is never observed.

Consider a Thompson sampling agent that uses the beliefs outlined in Figure 2 that runs a single experiment. The agent does so by first manipulating X and observing Y:

1. *Manipulating X:* First, the agent instantiates his random beliefs by sampling the value of Θ from the prior:

$$P(\Theta = \theta) = P(\Theta = \neg\theta) = \frac{1}{2}.$$

Assume that the result is θ. Treating θ as if it was the true parameter, he proceeds to sample the action from $P(X|\theta)$ given by

$$P(X = x|\theta) = P(X = \neg x|\theta) = \frac{1}{2},$$

as indicated in the left branch of the probability tree. Assume that outcome is x, and this is the action that the agent executes. Because of this, the agent has to update its beliefs first by intervening the probability tree for $\hat{x}$ and second by conditioning on x. The intervention $\hat{x}$ is carried out by replacing all the nodes in the tree that resolve the value of X with new nodes assigning probability one to x and zero to $\neg x$. Figure 2b illustrates the result of this intervention. The posterior is then given by

$$P(\theta|\hat{x}) = \frac{P(\hat{x}|\theta)P(\theta)}{P(\hat{x}|\theta)P(\theta) + P(\hat{x}|\neg\theta)P(\neg\theta)} = \frac{1 \cdot \frac{1}{2}}{1 \cdot \frac{1}{2} + 1 \cdot \frac{1}{2}} = \frac{1}{2}.$$

In other words, the agent has switched on X, and has so far learned nothing from this fact.

2. *Observing Y:* Now, the agent observes the activity of Y, and assume that it is active, i.e. $Y = y$. Then, the posterior beliefs of the agent are given as

$$\begin{aligned} P(\theta|\hat{x}, y) &= \frac{P(y|\theta, \hat{x})P(\hat{x}|\theta)P(\theta)}{P(y|\theta, \hat{x})P(\hat{x}|\theta)P(\theta) + P(\hat{x}|\neg\theta, y)P(y|\neg\theta)P(\neg\theta)} \\ &= \frac{\frac{3}{4} \cdot 1 \cdot \frac{1}{2}}{\frac{3}{4} \cdot 1 \cdot \frac{1}{2} + 1 \cdot \frac{1}{2} \cdot \frac{1}{2}} = \frac{3}{5}. \end{aligned}$$

Since $P(\theta) < P(\theta|\hat{x}, y)$, the agent has gathered evidence favoring the hypothesis "X causes Y". This was only possible because the intervention introduced a statistical asymmetry among the two hypotheses that did not exist in the beginning. In comparison, if the action is not treated as an intervention, then the posterior is

$$\begin{aligned} P(\theta|x, y) &= \frac{P(y|\theta, x)P(x|\theta)P(\theta)}{P(y|\theta, x)P(x|\theta)P(\theta) + P(x|\neg\theta, y)P(y|\neg\theta)P(\neg\theta)} \\ &= \frac{\frac{3}{4} \cdot \frac{1}{2} \cdot \frac{1}{2}}{\frac{3}{4} \cdot \frac{1}{2} \cdot \frac{1}{2} + \frac{3}{4} \cdot \frac{1}{2} \cdot \frac{1}{2}} = \frac{1}{2} = P(\theta), \end{aligned}$$

that is, the agent doesn't learn anything just from observing. This also highlights the importance of interventions (Box 1966).

Naturally, multiple interventions and observations can be executed in consecution. In this case Thompson sampling is used in each time step to decide which policy model to use, which implies the decision which variables to intervene. Then, after the intervention, all variables are revealed simultaneously at every time step of the inference process. The update of the observational probabilities is done the same way as in the one step case, taking into account which variables were intervened. A simulation of the repeated Thompson sampling process for causal induction of our example system is shown in Figure 3. This very simple example contains the principles of causal induction using Thompson sampling. Of course, more complex causal structures require richer model classes as is customary in Bayesian modeling. But importantly, the essence of causal induction is already contained in our simple illustration.

Discussion

The main contribution of the present paper is to show in how far generalized Thompson sampling can be regarded as an optimal solution method for adaptive decision-making in the presence of information-processing constraints and how this framework can be extended to solve problems of causal induction. We previously proposed Equation (3) as a *Bayesian rule for acting* in (Ortega and Braun 2010a, 2010b) that optimally solves the adaptive coding problem for actions and observations. In practice, it is implemented by

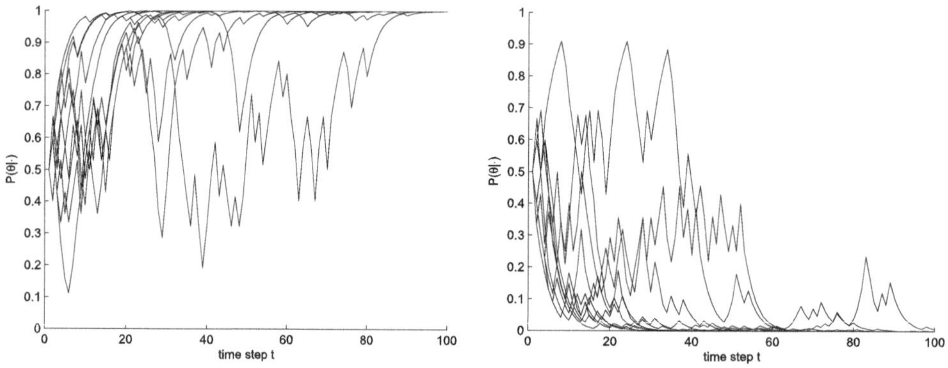

Figure 3 Thompson sampling for causal induction. (Left) Posterior distribution $P(\theta|\cdot)$ for 10 runs when the true system is given by $\Theta = \theta$. (Right) Posterior distribution $P(\theta|\cdot)$ for 10 runs when the true system is given by $\Theta = \neg\theta$. In both cases the agent is able to identify the causal structure of the environment with high confidence when $P(\theta|\cdots)$ is close to one or zero respectively.

sampling an environment parameter $\hat{\theta}_t$ for each time step from the posterior distribution $P(\theta|\hat{a}_{<t}, o_{<t})$, and then treating it as if it was the true parameter—that is, issuing the action a_t from $P(a_t|\hat{\theta}_t, a_{<t}, o_{<t})$. This action-sampling method where beliefs are randomly instantiated was first proposed as a heuristic in (Thompson 1933) and is now known as *Thompson sampling*. Importantly, this method can be generalized and applied to solve general sequential adaptive decision-making problems.

So far Thompson sampling has been mainly applied to multi-armed bandit problems. Multi-armed bandits can be represented by a parameter θ that summarizes the statistical properties of the reward obtained for each lever. Reward distributions range from Bernoulli to Gaussian (with unknown mean and variance), and they can also depend on the particular context or state (Graepel et al. 2010; May and Leslie 2011; Granmo 2010; Scott 2010). In particular, the work of (May and Leslie 2011) and the work of (Granmo 2010) prove asymptotic convergence of Thompson sampling. The performance of bandit algorithms has also been studied in terms of the rate of growth of the regret (Lai and Robbins 1995), and recent bandit algorithms have been shown to match this lower bound (Cappé et al. 2013), including Thompson sampling algorithms for Bernoulli bandits (Kaufmann et al. 2012). Also, the work of (Chapelle and Li 2011) presents empirical results that show Thompson sampling is highly competitive, matching or outperforming popular methods such as UCB (Lai and Robbins 1995; Auer et al. 2002).

Another class of problems, where Thompson sampling has been applied in the past, are Markov decision processes (MDPs). MDPs can be represented by parameterizing the dynamics and reward distribution (model-based) (Strens 2000) or by directly parameterizing the *Q-table* (model-free) (Dearden et al. 1998; Ortega and Braun 2010a). The first approach samples a full description of an MDP, solves it for the optimal policy, and then issues the optimal action. This is repeated in each time step. The second approach avoids the computational overhead of solving for the optimal policy in each time step by directly doing inference on the Q-tables. Actions are chosen by picking the one having the highest Q-value for the current state. The same ideas can also be applied to solve adaptive control problems with linear system equations, quadratic cost functions and Gaussian noise (Braun and Ortega 2010).

Optimality

While maximum expected utility is formally appealing as a principle for the construction of adaptive agents, its strict application is in practice often problematic. This is mainly due to two reasons:

1. *Computational complexity.* The computations required to find the optimal solution (for instance, the computational complexity of solving the Bellman optimality equations) are prohibitive in general and scale exponentially with the length of the horizon. The problem is tractable only in very special cases under assumptions that reduce the effective size of the problem.
2. *Causal precedence of policy choice.* The choice of the policy has to be made before the interaction with the environment starts. That is, an agent has to have a unique optimal policy before it has even interacted once with the environment. An optimal policy constructed by the maximum expected utility principle is therefore a very risky bet, as a lot of resources have to be spent before any evidence exists that the underlying model or prior is adequate.

Because of these two reasons, it is practically often impossible to apply the maximum expected utility principle. This has led to the development of theories of bounded rational decision-making that take the information processing limitations of decision-makers into account. The modern study of bounded rationality was famously broached by Simon (1956, 1972, 1984) and has since been extensively investigated in psychology (Gigerenzer and Selten 2001; Camerer 2003), cognitive science (Howes et al. 2009; Janssen et al. 2011; Lewis et al.), economics (Aumann 1997; Rubinstein 1998; Kahneman 2003), game theory (McKelvey and Palfrey 1995, 1998; Wolpert 2004), political science (Jones 2003), industrial organization (Spiegler 2011), computer science and artificial intelligence research (Lipman 1995; Russell 1995; Russell and Subramanian 1995). Different conceptions of bounded rationality are divided as to whether bounded rational behavior is thought to be fundamentally non-optimizing or whether it can be expressed as a (constrained) optimization problem and as to whether it involves any kind of meta-reasoning (Klein 2001). While the variational formulation in the free energy can also be thought of as a constrained optimization problem, this optimization is only implicit in an agent that runs an anytime algorithm to obtain samples that directly optimize the original (unconstrained) utility function. The average number of samples that can be afforded is determined by an inverse temperature parameter, such that the search for the optimum is aborted after some time, thereby generating some kind of *satisficing* solution. The free energy formulation of bounded rationality also allows reinterpreting a wider research program that has investigated relative entropy as a particular cost function for control (Kappen 2005; Todorov 2006, 2009; Theodorou et al. 2010; Peters et al. 2010; Braun and Ortega 2011; Kappen et al. 2012) and has inspired the formulation of optimal control problems as inference problems (Tishby and Polani 2011; Kappen et al. 2012; Rawlik et al. 2012). In Section "Decision-making with limited resources" we have argued that Thompson sampling can be regarded as an instantiation of free energy optimizing bounded rationality requiring the minimal amount of samples of the latent variable θ in the decision-making process determining the next action. An agent that follows such a Thompson sampling strategy randomly samples beliefs θ and acts optimally with respect to these random beliefs. In contrast, a perfectly rational agent optimizes his utility over the entire belief tree.

Policy Uncertainty. Given a problem specification in terms of the predictive model and the utility function, we can think about policy uncertainty in terms of policy search methods. The task of a policy search method is to calculate a policy that approximates the optimal policy. More specifically, let π be a parameter in a set Π indexing the set of candidate policies $P(a_t|\pi, a_{1:t-1}, o_{1:t-1})$ indexed by $\theta \in \Theta$. *Then, in the most general case, a policy search method returns a probability distribution $P(\pi)$ over Π representing the uncertainty over the optimal policy parameters.* If the algorithm solves the maximum expected utility problem, then the support of this distribution will exclusively cover the set of optimal policies $\Pi^* \subset \Pi$. Otherwise there remains uncertainty over the optimal policy parameters. However, many policy search methods do not explicitly deal with the uncertainty over the policy parameters. Some methods only return a point estimate $\hat{\pi} \in \Pi$. For instance, reinforcement learning algorithms (Sutton and Barto 1998) start from a randomly initialized point estimate $\hat{\pi}_0$ of the optimal policy and then generate refined point estimates $\hat{\pi}_1, \hat{\pi}_2, \hat{\pi}_3, \ldots$ in each time step $t = 1, 2, 3, \ldots$ using the data provided by experience. In order to converge to the optimal policy, these algorithms have to deal with the exploration-exploitation trade-off. This means that the agents cannot just greedily act according to these point estimates; instead, they have to produce explorative actions as well, that is, actions that deviate from the current estimate of the optimal policy—for instance producing optimistic actions based on UCB (Lai and Robbins 1995; Auer et al. 2002).

Crucially, when sampling actions from the predictive distribution, the policy index π is identical to the index θ that identifies a particular environment with the likelihood model $P(o_t|a_{1:t-1}, o_{1:t-1})$. By turning the reinforcement learning problem thus into an inference problem, the exploration-exploitation trade-off becomes a *bias-variance trade-off* (Geman et al. 1992) in policy space. This highlights the essence of the exploration-exploitation trade-off: any action issued by the agent has to respect the uncertainty over the policy parameter—otherwise they are biased. In particular, if the agent acts deterministically and greedily (i.e. it treats the estimate $\hat{\pi}$ as if it were the true policy parameter) then it is overfitting the experience and introducing a bias; likewise, an agent that follows a stochastic policy introduces variance and will not produce the highest possible reward compared to the case when the optimal policy is known. An excessively stochastic agent therefore underfits its experience.

The operational distinction of having policy uncertainty has important algorithmic consequences. When there is policy uncertainty, the belief of the decision-maker is itself a random variable. This means that the very policy is undefined until the random variable is resolved. Hence, the computation of the optimal policy can be delayed and determined dynamically. It is precisely this fact that is (implicitly) exploited in popular reinforcement learning algorithms, and explicitly in the algorithms based on random beliefs. This is in stark contrast to the case when there is no policy uncertainty, where the policy is pre-computed and static. Another example where random beliefs play a crucial role is in *games with incomplete information* (Osborne and Rubinstein 1999). Here, having incomplete information about the other player leads to a infinite hierarchy of meta-reasoning about the other player's strategy. To avoid this difficulty, Harsanyi introduced *Bayesian games* (Harsanyi 1967). In a Bayesian game, incomplete knowledge is modeled by randomly instantiating the player's types, after which they choose their strategies optimally—thus eliminating the need for recurrent reasoning about the other players' strategy. Similarly, a Thompson sampling agent randomly instantiates his belief at every point in time and acts optimally with respect to this belief. An important consequence of this is that agents have uncertainty about their policy.

Adaptive Coding. The adaptive control problem can also be construed as an adaptive coding problem both for actions and observations (Ortega and Braun 2010b, 2012b). The question then is: How can we construct a system P defined by $P(o_t|\hat{a}_{\le t}, o_{<t})$ and $P(a_t|\hat{a}_{<t}, o_{<t})$ such that its behavior is as close as possible to the custom-made system $P(o_t|\theta, \hat{a}_{\le t}, o_{<t})$ and $P(a_t|\theta, \hat{a}_{<t}, o_{<t})$ under any realization of Q_θ? Using the Kullback-Leibler divergence as a distance measure, we can formulate a variational problem in **Pr**, where **Pr** defines an input-output system trough a distribution over interaction sequences $a_1o_1a_2o_2\ldots$, such that

$$P := \arg\min_{\mathbf{Pr}} \left\{ \limsup_{t\to\infty} \sum_{\theta} P(\theta) \sum_{\tau=1}^{t} \left(D_m^{a_\tau}(\mathbf{Pr}) + D_m^{o_\tau}(\mathbf{Pr}) \right) \right\}$$

with

$$D_\theta^{a_t}(\mathbf{Pr}) = \sum_{\hat{a}_{<t}, o_{<t}} P(\hat{a}_{<t}, o_{<t}|\theta) \sum_{a_t} P(a_t|\theta, \hat{a}_{<t}, o_{<t}) \log \frac{P(a_t|\theta, \hat{a}_{<t}, o_{<t})}{\mathbf{Pr}(a_t|\hat{a}_{<t}, o_{<t})}$$

$$D_\theta^{o_t}(\mathbf{Pr}) = \sum_{a_{\le t}, o_{<t}} P(\hat{a}_{\le t}, o_{<t}|\theta) \sum_{o_t} P(o_t|\theta, \hat{a}_{\le t}, o_{<t}) \log \frac{P(o_t|\theta, \hat{a}_{\le t}, o_{<t})}{\mathbf{Pr}(o_t|\hat{a}_{\le t}, o_{<t})}.$$

In the case of observations, this is a well-known variational principle for Bayesian inference, as it describes a predictor that requires, on average, the least amount of extra bits to capture informational surprise stemming from the behavior of the environment. In the case of actions, the same principle can be harnessed to describe resourceful generation of actions in a way that requires random bits with minimum length on average, when trying to match the optimal policy most suitable for the unknown environment (MacKay 2003). When thinking about the adaptive control problem in this way, the aim of the adaptive agent is simply to avoid surprise. The fact that each custom-built policy $P(a_t|\theta, \hat{a}_{<t}, o_{<t})$ can be thought of as maximizing a utility in environment Q_θ is not crucial, as this policy could also be given by a teacher's demonstration in the absence of an explicitly stated utility function. The avoidance of surprise of adaptive systems has recently been discussed in the context of active inference and the free energy principle (Friston 2009, 2010).

Causality

In Section "Causal induction", we could demonstrate that generalized Thompson sampling can also be applied to the problem of causal induction, by designing policy and prediction models with different causal structures. This way generalized Thompson sampling can be used as a general method for causal induction that is Bayesian in nature. It is based on the idea of combining probability trees (Shafer 1996) with interventions (Pearl 2000) for predicting the behavior of a manipulated system with multiple causal hypotheses. Both the interventions and the constraints on the causal hypotheses introduce statistical asymmetries that permit the extraction of causal information. Unlike frameworks that aim to extract causal information from observational data alone (Shimizu et al. 2006; Griffiths and Tenenbaum 2009; Janzing and Schölkopf 2010), the proposed method is designed for agents that interact with their environment and use these interactions to discover causal relationships.

To construct the Bayes-causal solution (3), we needed to treat actions as interventions. This raises the question about why this distinction was not made for deriving classical expected utility solutions. Since,

$$P(a_t|a_{<t},o_{<t}) = \sum_\theta P(a_t|\theta,a_{<t},o_{<t})P(\theta|\hat{a}_{<t},o_{<t})$$
$$P(o_t|a_{\le t},o_{<t}) = \sum_\theta P(o_t|\theta,a_{\le t},o_{<t})P(\theta|\hat{a}_{\le t},o_{<t}),$$

determining the conditions boils down to analyzing when the equalities

$$\begin{aligned} P(\theta|a_{<t},o_{<t}) &= P(\theta|\hat{a}_{<t},o_{<t}) \\ P(\theta|\hat{a}_{\le t},o_{<t}) &= P(\theta|a_{\le t},o_{<t}) \end{aligned}$$

hold. Replacing both sides yields,

$$\frac{P(\theta)\prod_{k=1}^t P(a_k|\theta,a_{<k},o_{<k})P(o_k|\theta,a_{\le k},o_{<k})}{\sum_{\theta'} P(\theta')\prod_{k=1}^t P(a_k|\theta',a_{<k},o_{<k})P(o_k|\theta',a_{\le k},o_{<k})} = \frac{P(\theta)\prod_{k=1}^t P(o_k|\theta,a_{\le k},o_{<k})}{\sum_{\theta'} P(\theta')\prod_{k=1}^t P(o_k|\theta',a_{\le k},o_{<k})}$$

and we conclude that

$$P(a_k|\theta,a_{<k},o_{<k}) = \delta_{\bar{a}_k}(a_k),$$

i.e. the actions have to be issued deterministically (but possibly history-dependent) from a unique policy. Intuitively speaking, this is because the operations of intervening and conditioning coincide when the random variables are deterministic.

Convergence

There are important cases where random belief approaches can fail. Indeed, it is easy to devise experiments where having policy uncertainty converges exponentially slower (or does not converge at all) than the Bayes adaptive optimal policy. Consider, for example, two k-order Markov chains with only one observable state when applying k times the same action, but we do not know which action it is. For two possible actions and a uniform prior over the two possible environments the distribution over possible worlds stays uniform as long as no reward has been observed. Choosing actions randomly according to this distribution would require 2^k actions to accidentally choose a sequence of the same action of length k. Thus, the Bayes adaptive optimal policy converges in time k, while the agent with policy uncertainty needs exponentially longer. A simple way to remedy this problem is, of course, to sample random beliefs only every k time steps (Strens 2000). But this problem can be exacerbated in non-stationary environments. Take for instance, an increasing MDP with two actions and number of states $k = \lceil 10\sqrt{t}\rceil$, in which the optimal policy converges in 100 steps, while an agent with policy uncertainty would not converge at all in most realizations. Although (Ortega and Braun 2010b) prove asymptotic convergence for general environments fulfilling a restrictive form of ergodicity condition, this condition needs to be weakened for the convergence proof to be applicable to most real problems. But it is clear that a form of ergodicity is required for an agent with policy uncertainty to be able to learn to act optimally. Intuitively, this means that an agent can only learn if the environment has temporally stable statistical properties.

Conclusion

In this paper we have argued that Thompson sampling is a bounded rational strategy in decision-making that can be considered optimal under given information processing constraints. Thompson sampling agents have uncertainty over their policy, which is a natural phenomenon that arises whenever there are not enough computational resources to apply the maximum expected utility principle to single out a unique optimal policy. Having policy uncertainty effectively weakens the two assumptions of the maximum expected utility principle: the optimal policy can be chosen and refined during interactions, and the computational complexity is lower. We have shown that treating this uncertainty in a Bayesian way with actions as random variables that obey causal calculus naturally leads to Thompson sampling and its Bayesian generalization. This generalized Thompson sampling can be straightforwardly applied to the problem of causal induction. Maintaining and updating Bayesian probabilities is an optimally efficient way to deal with uncertainty—be it with respect to the policy or the environment (Ortega and Braun 2010a). As these random-belief approaches can be derived simply from probability theory and causal calculus we suggest that they cannot only be regarded as heuristic approximations to optimal decision-making, but as principled solution methods in their own right.

Endnote

[a]Each custom-built policy $P(a_t|\theta, a_{<t}, o_{<t})$ can be thought to maximize a utility function in its environment θ, but this is not essential—the policy could also just be given by a teacher's demonstration as in imitation learning (Schaal 1999).

Competing interests

The authors declare that they have no competing interests.

Authors' contributions

DAB and PAO conceived the analysis and wrote the paper.

Acknowledgements

This study was supported by the DFG, Emmy Noether grant BR4164/1-1.

References

Agrawal S, Goyal N: **Analysis of Thompson sampling for the multi-armed bandit problem.** In *JMLR: Workshop and Conference Proceedings vol 23 (2012) 39.1–39.26. 25th Annual Conference on Learning Theory*; 2011.

Asmuth J, Li L, Littman ML, Nouri A, Wingate D: **A Bayesian s+ in reinforcement learning.** In *Proceedings of the Twenty-Fifth Conference on Uncertainty in Artificial Intelligence*. UAI '09, Arlington, Virginia, United States: AUAI Press; 2009:19–26.

Auer P, Cesa-Bianchi N, Fisher P: **Finite-time analysis of the multiarmed bandit problem.** *Machine Learning* 2002, **47**:235–256.

Aumann RJ: **Rationality and bounded rationality.** *Games and Econ Behavior* 1997, **21**(1-2):2–14.

Box G: **Use and abuse of regression.** *Technometrics* 1966, **8**(4):625–629.

Braun DA, Ortega PA: **A minimum relative entropy principle for adaptive control in linear quadratic regulators.** In *The 7th Conference on Informatics in Control, Automation and Robotics, Volume 3*; 2010:103–108.

Braun DA, Ortega PA, Theodorou E, Schaal S: **Path integral control and bounded rationality.** In *IEEE Symposium on Adaptive Dynamic Programming and Reinforcement Learning*; 2011:202–209.

Bubeck S, Liu CY: **A note on the Bayesian regret of Thompson sampling with an arbitrary prior.** 2013. arXiv:1304.5758.

Camerer C: *Behavioral Game Theory: Experiments in Strategic Interaction.* Princeton: Princeton University Press; 2003.

Cao F, Ray S: **Bayesian hierarchical reinforcement learning.** In *Neural Information Processing Systems 25 (NIPS)*; 2012.

Cappé O, Garivier A, Maillard OA, Munos R, Stoltz G: **Kullback-Leibler upper confidence bounds for optimal sequential allocation.** *Ann Stat* 2013, **41**(3):1516–1541.

Chapelle O, Li L: **An empirical evaluation of Thompson sampling.** In *NIPS*; 2011:2249–2257.

Dearden R, Friedman N, Russell S: **Bayesian Q-learning.** In *AAAI '98/IAAI '98: Proceedings of the fifteenth national/tenth conference on Artificial intelligence/Innovative applications of artificial intelligence.* Menlo Park, CA, US: American Association for Artificial Intelligence; 1998:761–768.

Dimitrakakis C: **Monte-Carlo utility estimates for Bayesian reinforcement learning.** In *IEEE Conference on Decision and Control*; 2013.

Dimitrakakis C, Tziortziotis N: **ABC reinforcement learning.** In *Proceedings of The 30th International Conference on Machine Learning*; 2013:684–692.

Duff M: **Optimal learning: computational procedures for bayes-adaptive markov decision processes.** *PhD thesis* 2002. [Director-Andrew Barto].

Friston K: **The free-energy principle: a rough guide to the brain?** *Trends in Cognitive Science* 2009, **13:**293–301.

Friston K: **The free-energy principle: a unified brain theory?** *Nat Rev Neurosci* 2010, **11:**127–138.

Geman S, Bienenstock E, Doursat R: **Neural networks and the bias/variance dilemma.** *Neural Comput* 1992, **4:**1–58.

Gigerenzer G, Selten R: *Bounded Rationality: The Adaptive Toolbox.* Cambridge, MA: MIT Press; 2001.

Gittins J: **Bandit processes and dynamic allocation indices.** *J R Stat Soc Ser B , Methodological* 1979, **41:**148–177.

Glymour C, Spirtes P, Scheines R: *Causation, Prediction, and Search, 2nd edition.* Cambridge, Massachusetts, USA: MIT Press; 2000.

Graepel T, Quiñonero Candela J, Borchert T, Herbrich R: **Web-scale Bayesian click-through rate prediction for sponsored search advertising in Microsoft's Bing search engine.** In *Proceedings of the Twenty-Seventh International Conference on Machine Learning*; 2010:25–26.

Granmo OC: **A Bayesian learning automaton for solving two-armed bernoulli bandit problems.** In *Proceedings of the 2008 Seventh International Conference on Machine Learning and Applications*: ICMLA '08; 2008:23–30.

Granmo OC: **Solving two-armed Bernoulli bandit problems using a Bayesian learning automaton.** *Int J Intell Comput Cybernetics* 2010, **3**(2):207–234.

Granmo OC, Glimsdal S: **Accelerated Bayesian learning for decentralized two-armed bandit based decision making with applications to the Goore game.** *Applied intelligence* 2013, **38**(4):479–488.

Griffiths TL, Tenenbaum JB: **Theory-based causal induction.** *Psychological Rev* 2009, **116:**661–716.

Harsanyi J: **Games with incomplete information played by "Bayesian" players.** *Management Sci* 1967, **14**(3):159–182.

Heckerman D, Meek C, Cooper G: **A Bayesian approach to causal discovery.** *Computation, causation, and discovery* 1999, **19:**141–166.

Howes A, Lewis RL, Vera A: **Rational adaptation under task and processing constraints: implications for testing theories of cognition and action.** *Psychological Rev* 2009, **116**(4):717–751.

Hutter M: *Universal Artificial Intelligence: Sequential Decisions based on Algorithmic Probability.* Berlin: Springer; 2004.

Janssen CP, Brumby DP, Dowell J, Chater N, Howes A: **Identifying optimum performance trade-offs using a cognitively bounded rational analysis model of discretionary task interleaving.** *Topics in Cognitive Sci* 2011, **3:**123–139.

Janzing D, Schölkopf B: **Causal inference using the algorithmic Markov condition.** *IEEE Trans Inf Theor* 2010, **56**(10):5168–5194.

Jaynes E: **Entropy and search theory.** In *Maximum entropy and Bayesian methods in inverse problems.* Heidelberg: Springer-Verlag; 1985.

Jones BD: **Bounded rationality and political science: lessons from public administration and public policy.** *J Public Administration Res Theory* 2003, **13**(4):395–412.

Kahneman D: **Maps of bounded rationality: psychology for behavioral economics.** *Am Econ Rev* 2003, **93**(5):1449–1475.

Kappen H: **A linear theory for control of non-linear stochastic systems.** *Phys Rev Lett* 2005, **95:**200201.

Kappen H, Gómez V, Opper M: **Optimal control as a graphical model inference problem.** *Machine Learn* 2012, **1:**1–11.

Kaufmann E, Korda N, Munos R: **Thompson sampling: an asymptotically optimal finite-time analysis.** In *ALT, Volume 7568 of, Lecture Notes in Computer Science.* Edited by Bshouty NH, Stoltz G, Vayatis N, Zeugmann T. Heidelberg, Germany: Springer; 2012:199–213.

Keller G: *Equilibrium States in Ergodic Theory.* London Mathematical Society Student Texts: Cambridge Univeristy Press; 1998.

Klein G: **The fiction of optimization.** In *Bounded rationality: The adaptive toolbox.* Edited by Gigerenzer G, Selten R. Cambridge, Massachusetts, USA: MIT Press; 2001.

Korda N, Kaufmann E, Munos R: **Thompson sampling for 1-dimensional exponential family bandits.** In *Advances in Neural Information Processing Systems*; 2013:1448–1456.

Lai T, Robbins H: **Asymptotically efficient adaptive allocation rules.** *Adv Appl Math* 1995, **6:**4–22.

Legg S: **Machine super intelligence.** *PhD thesis,* Department of Informatics, University of Lugano 2008.

Lewis R, Howes A, Singh S: **Computational rationality: linking mechanism and behavior through bounded utility maximization.** *Topics in Cognitive Science* 2014, (in press).
Lipman B: **Information processing and bounded rationality: a survey.** *Canadian J Econ* 1995, **28:**42–67.
MacKay D: *Information Theory, Inference, and Learning Algorithms*. Cambridge, UK: Cambridge University Press; 2003.
Martin J: *Bayesian Decision Problems and Markov Chains*: Publications in Operations Research, Wiley; 1967.
May B, Leslie D: **Simulation studies in optimistic Bayesian sampling in contextual-bandit problems.** In *Technical Report 11:02*. Statistics Group, Department of Mathematics, Bristol, UK: University of Bristol; 2011.
May BC, Korda N, Lee A, Leslie DS: **Optimistic Bayesian sampling in contextual-bandit problems.** *J Mach Learn Res* 2012, **98888:**2069–2106.
Mckelvey R, Palfrey TR: **Quantal response equilibria for extensive form games.** *Experimental Econ* 1998, **1:**9–41.
McKelvey RD, Palfrey TR: **Quantal response equilibria for normal form games.** *Games and Econ Behavior* 1995, **10:**6–38.
Mellor J, Shapiro J: **Thompson sampling in switching environments with Bayesian online change point detection.** 2013. arXiv:1302.3721.
Ortega PA: **A unified framework for resource-bounded autonomous agents interacting with unknown environments.** *PhD thesis*, Department of Engineering, University of Cambridge, UK 2011a.
Ortega PA: **Bayesian causal induction.** In *NIPS Workshop on Philosophy and Machine Learning*, Granada, 2011.
Ortega PA, Braun DA: **A Bayesian rule for adaptive control based on causal interventions.** In *Proceedings of the third conference on general artificial intelligence*. Paris, France: Atlantis Press; 2010a.
Ortega PA, Braun DA: **A minimum relative entropy principle for learning and acting.** *J Artif Intell Res* 2010b, **38:**475–511.
Ortega PA, Braun DA: **A conversion between utility and information.** In *Proceedings of the Third Conference on Artificial General Intelligence*. Paris, France: Atlantis Press; 2010c:115–120.
Ortega PA, Braun DA: **Information, utility and bounded rationality.** In *Lecture notes on artificial intelligence, Volume 6830*. Heidelberg, Germany: Springer-Verlag; 2011:269–274.
Ortega PA, Braun DA: **Free energy and the generalized optimality equations for sequential decision making.** In *European Workshop for Reinforcement Learning*. Edinburgh, UK; 2012a.
Ortega PA, Braun DA: **Adaptive coding of actions and observations.** *NIPS Workshop on Information in Perception and Action* 2012b.
Ortega PA, Braun DA: **Thermodynamics as a theory of decision-making with information-processing costs.** *Proc R Soc A: Mathematical, Physical and Engineering Science* 2013, **469:**2153.
Osband I, Russo D, Roy BV: **(More) efficient reinforcement learning via posterior sampling.** In *Advances in Neural Information Processing Systems*; 2013:3003–3011.
Osborne MJ, Rubinstein A: *A Course in Game Theory*. Cambridge, Massachusetts, USA: MIT Press; 1999.
Pearl J: *Causality: Models, Reasoning, and Inference*. Cambridge, UK: Cambridge University Press; 2000.
Peters J, Mülling K, Altun Y: **Relative entropy policy search.** In *AAAI*; 2010.
Rawlik K, Toussaint M, Vijayakumar S: **On stochastic optimal control and reinforcement learning by approximate inference.** In *Proceedings of Robotics: Science and Systems*. Sydney, Australia; 2012.
Rubinstein A: *Modeling Bounded Rationality*. Cambridge, Massachusetts, USA: MIT Press; 1998.
Russell S: **Rationality and Intelligence.** In *Proceedings of the Fourteenth International Joint Conference on Artificial Intelligence*. Edited by Mellish C. Englewood Cliffs, New Jersey, USA: Prentice-Hall; 1995:950–957.
Russell S, Norvig P: *Artificial Intelligence: A Modern Approach, 1st edition*. Prentice-Hall: Englewood Cliffs, NJ; 1995.
Russell S, Subramanian D: **Provably bounded-optimal agents.** *J Artif Intell Res* 1995, **3:**575–609.
Russo D, Roy BV: **Learning to optimize via posterior sampling.** 2013. arXiv:abs/1301.2609.
Schaal S: **Is imitation learning the route to humanoid robots?** *Trends in cognitive sciences* 1999, **3**(6):233–242.
Scott S: **A modern Bayesian look at the multi-armed bandit.** *Applied Stochastic Models in Business and Industry* 2010, **26:**639–658.
Shafer G: *The Art of Causal Conjecture*. Cambridge, Massachusetts, USA: MIT Press; 1996.
Shimizu S, Hoyer PO, Hyvärinen A, Kerminen A: **A Linear Non-Gaussian Acyclic Model for Causal Discovery.** *J Mach Learn Res* 2006, **7:**2003–2030.
Simon HA: **Rational choice and the structure of the environment.** *Psychological Rev* 1956, **63**(2):129–138.
Simon HA: **Theories of bounded rationality.** In *Decision and Organization*. Edited by McGuire CB, Radner R. Amsterdam: North-Holland Publishing; 1972:161–176.
Simon H A: *Models of Bounded Rationality. Cambridge*. Cambridge, Massachusetts, USA: MIT Press; 1984.
Spiegler R: *Bounded Rationality and Industrial Organization*. Oxford: Oxford University Press; 2011.
Stone L: *Theory of Optimal Search*. New York: Academic Press; 1998.
Strens M: **A Bayesian framework for reinforcement learning.** In *Proceedings of the Seventeenth International Conference on Machine Learning*; 2000.
Sutton R, Barto A: *Reinforcement Learning: An Introduction*. Cambridge, MA: MIT Press; 1998.
Theodorou E, Buchli J, Schaal S: **A generalized path integral approach to reinforcement learning.** *J Mach Learn Res* 2010, **11:**3137–3181.
Thompson WR: **On the likelihood that one unknown probability exceeds another in view of the evidence of two samples.** *Biometrika* 1933, **25**(3/4):285–294.
Tishby N, Polani D: **Information theory of decisions and actions.** In *Perception-reason-action cycle: Models, algorithms and systems*. Edited by Vassilis T Hussain. Heidelberg: Springer-Verlag; 2011:601–636.
Todorov E: **Linearly solvable Markov decision problems.** In *Advances in Neural Information Processing Systems, Volume 19*; 2006:1369–1376.
Todorov E: **Efficient computation of optimal actions.** *Proceedings of the National Academy of Sciences USA* 2009, **106:**11478–11483.
Tziortziotis N, Dimitrakakis C, Blekas K: **Cover tree Bayesian reinforcement learning.** 2013a. arXiv: 1305.1809.
Tziortziotis N, Dimitrakakis C, Blekas K: **Linear Bayesian reinforcement learning.** In *Proceedings of the Twenty-Third international joint conference on Artificial Intelligence*: AAAI Press; 2013:1721–1728.

Vitanyi P: **Time, space, and energy in reversible computing.** In *Proceedings of the 2nd ACM conference on Computing frontiers*; 2005:435–444.

Wolpert DH: **Information theory - the bridge connecting bounded rational game theory and statistical physics.** In *Complex Engineering Systems*. New York, USA: Perseus Books; 2004.

Wyatt J: **Exploration and inference in learning from reinforcement.** *PhD thesis*, Department of Artificial Intelligence, University of Edinburgh 1997.

8

Clustering datasets by complex networks analysis

Giuliano Armano and Marco Alberto Javarone*

*Correspondence:
marco.javarone@diee.unica.it
DIEE - Department of Electrical and Electronic Engineering University of Cagliari Piazza d'Armi, Cagliari - 09123, Italy

Abstract

This paper proposes a method based on complex networks analysis, devised to perform clustering on multidimensional datasets. In particular, the method maps the elements of the dataset in hand to a weighted network according to the similarity that holds among data. Network weights are computed by transforming the Euclidean distances measured between data according to a Gaussian model. Notably, this model depends on a parameter that controls the shape of the actual functions. Running the Gaussian transformation with different values of the parameter allows to perform multiresolution analysis, which gives important information about the number of clusters expected to be optimal or suboptimal.

Solutions obtained running the proposed method on simple synthetic datasets allowed to identify a recurrent pattern, which has been found in more complex, synthetic and real, datasets.

Keywords: Clustering, Community detection, Complex networks, Multiresolution analysis

Background

Complex networks are used in different domains to model specific structures or behaviors 2010. Relevant examples are the Web, biological neural networks, and social networks 2002, 2004, 2003. Community detection is one of the most important processes in complex network analysis, aimed at identifying groups of highly mutually interconnected nodes, called communities 2004, in a relational space. From a complex network perspective, a community is identified after modeling any given dataset as graph. For instance, a social network inherently contains communities of people linked by some (typically binary) relations –e.g., friendship, sports, hobbies, movies, books, or religion. On the other hand, from a machine learning perspective, a community can be thought of as a cluster. In this case, elements of the domain are usually described by a set of features, or properties, which permit to assign each instance a point in a multidimensional space. The concept of similarity is prominent here, as clusters are typically identified by focusing on common properties (e.g., age, employment, health records).

The problem of clustering multidimensional datasets without a priori knowledge about them is still open in the machine learning community (see, for example, 2010, 2001, 1998). Although complex networks are apparently more suited to deal with relations rather than properties, nothing prevents from representing a dataset as complex network. In fact, the idea of viewing datasets as networks of data has already been developed in previous works. Just to cite few, Heimo et al. 2008 studied the problem of multiresolution module detection in dense weighted networks, using a weighted version of the q-state Potts method. Mucha et al. 2010 developed a generalized framework to study community structures of arbitrary multislice networks. Toivonen et al. 2012 used network methods in analyzing similarity data with the aim to study Finnish emotion concepts. Furthermore, a similar approach has been developed by Gudkov et al. 2008, who devised and implemented a method for detecting communities and hierarchical substructures in complex networks. The method represents nodes as point masses in an $N - 1$ dimensional space and uses a linear model to account for mutual interactions.

The motivation for representing a dataset as graph lies in the fact that very effective algorithms exist on the complex network side to perform community detection. Hence, these algorithms could be used to perform clustering once the given dataset has been given a graph-based representation. Following this insight, in this paper we propose a method for clustering multidimensional datasets in which they are first mapped to weighted networks and then community detection is enforced to identify relevant clusters. A Gaussian transformation is used to turn distances of the original (i.e. feature-based) space to link weights of the complex networks side. As the underlying Gaussian model is parametric, the possibility to run Gaussian transformations multiple times (while varying the parameter) is exploited to perform multiresolution analysis, aimed at identifying the optimal or suboptimal number of clusters.

The proposed method, called *DAN* (standing for Datasets as Networks), makes a step forward in the direction of investigating the possibility of using complex network analysis as a proper machine learning tool. The remainder of the paper is structured as follows: Section Methods describes how to model a dataset as complex network and gives details about multiresolution analysis. For the sake of readability, the section briefly recalls also some informative notion about the adopted community detection algorithm. Section Results and discussion illustrates the experiments and analyzes the corresponding results. The section recalls also some relevant notions of clustering, including two well-known algorithms, used therein for the sake of comparison. Conclusions (i.e. Section Conclusions) end the paper.

Methods

The first step of the *DAN* method consists of mapping the dataset in hand to a complex network. The easiest way to use a complex network for encoding a dataset is to let nodes denote the elements of the dataset and links denote their similarity. In particular, we assume that the weight of a link depends only on the distance among the involved elements. To put the model into practice, we defined a family of Gaussian functions –used for computing the weight between two elements.

Computing similarity among data

Let us briefly recall that a metric space is identified by a set $\mathcal{Z}$, together with a distance function $d : \mathcal{Z} \times \mathcal{Z} \rightarrow \mathbb{R}$, like Euclidean, Manhattan and Chebyshev distances. In *DAN*, the underlying assumption is that a sample s can be described by N features $f_1, f_2, \ldots, f_N$, encoded as real numbers. In other words, the sample can be represented as a vector in an N-dimensional metric space $\mathcal{S}$. Our goal is to generate a fully connected weighted network taking into account the distances that hold in $\mathcal{S}$. Conversely, the complex network space will be denoted as $\mathcal{N}$, with the underlying assumption that for each sample $s_i \in \mathcal{S}$ a corresponding $n_i \in \mathcal{N}$ exists and vice versa. This assumption makes easier to evaluate the proximity value L_{ij} between two $n_i, n_j \in \mathcal{N}$, according to the distance d_{ij} between the corresponding elements $s_i, s_j \in \mathcal{S}$.

Without loss of generality, let us assume that each feature in $\mathcal{S}$ is normalized in $[0, 1]$ and that a function $\psi : \mathbb{R} \rightarrow \mathbb{R}$ exists for computing the similarity among data in $\mathcal{N}$, starting from the value of the distance function in $\mathcal{S}$. In symbols:

$$L(n_i, n_j) = L_{ij} \triangleq \psi(d_{ij}) = \psi(d(s_i, s_j)) \tag{1}$$

Evaluating similarity for all pairs of samples in $\mathcal{N}$ (i.e., evaluating their weighted links) allows to generate a fully connected complex network. Moreover, recalling that $\mathcal{S}$ is normalized in $[0, 1]$, we expect $L_{ij} \approx 0$ when $d_{ij} \approx \sqrt{N}$, N being the number of features of the space $\mathcal{S}$. The value $\sqrt{N}$ comes from the following inequality, which holds for any pair of samples $s_i, s_j \in \mathcal{S}$ (represented by their vector representation in terms of the given set of features $\mathbf{r_i}, \mathbf{r_j}$):

$$d_{ij} = \sqrt{\sum_{k=1}^{N} (\mathbf{r_i}[k] - \mathbf{r_j}[k])^2} \leq \sqrt{N} \tag{2}$$

where $\mathbf{r_i}[k]$ denotes the k-th component of $\mathbf{r_i}$.

The adopted community detection algorithm

Community detection is the process of finding communities in a graph (the process is also called "graph partitioning"). From a computational perspective, this is not a simple task and many algorithms have been proposed, according to three main categories: divisive, agglomerative, and optimization algorithms. In our work, we used the Louvain method 2008, an optimization algorithm based on an objective function devised to estimate the quality of partitions. In particular, at each iteration, the Louvain Method tries to maximize the so-called *weighted-modularity*, defined as:

$$Q = \frac{1}{2m} \cdot \sum_{i,j} \left[A_{ij} - \frac{k_i k_j}{2m} \right] \cdot \delta(s_i, s_j) \tag{3}$$

where A_{ij} is the generic element of the adjacency matrix, k is the degree of a node, m is the total "weight" of the network, and $\delta(s_i, s_j)$ is the Kronecker Delta, used to assert whether a pair of samples belongs to the same community or not.

Multiresolution analysis

Let us recall that multiresolution analysis is performed with the goal of extracting relevant information, useful for identifying the optimal or suboptimal number of communities (hence, of clusters). To perform multiresolution analysis on the network space, a parametric family $\Psi(\lambda) : \mathbb{R} \rightarrow \mathbb{R}$ of functions is required, where λ is a parameter that controls the shape of each ψ function. After setting a value for λ, the corresponding ψ can be used to convert the distance computed for each pair of samples in the given dataset into a proximity value. In particular, the following parametric family of Gaussian functions has been experimented:

$$\Psi(\lambda; x) = e^{-\lambda x^2} \tag{4}$$

As a consequence, L_{ij}, i.e. the weight of the link between two nodes $n_i, n_j \in \mathcal{N}$, can be evaluated according to Equation 4 as follows:

$$L_{ij} \triangleq \psi(\lambda; d_{ij}) = e^{-\lambda d_{ij}^2} \tag{5}$$

where the λ parameter is used as a constant decay of the link.

Following the definition of $\Psi(\lambda; x)$ as $e^{-\lambda x^2}$, multiresolution analysis takes place varying the value of the λ parameter. The specific strategy adopted for varying λ is described in the experimental section. As for now, let us note that an exponential function with negative constant decay ensures that distant points in an Euclidean space are loosely coupled in the network space and vice versa. Moreover, this construction is useful only if $\Psi(\lambda; x)$ models local neighborhoods, which gives further support to the choice of Gaussian functions 2007.

Results and discussion

Experiments have been divided in three main groups: i) *preliminary tests*, aimed at running *DAN* on few and relatively simple synthetic datasets, ii) *proper tests*, aimed at running *DAN* on more complex datasets, and iii) *comparisons*, aimed at assessing the behavior of *DAN* with reference to *k−Means* and spectral clustering.

Almost all datasets used for experiments (except for Iris) are synthetic and have been generated according to the following algorithm:

Inputs: number of samples (n), dimension in the Euclidean space (N), number of clusters (k), and radius of a cluster (r)

1. For each cluster $j = 1, 2, \ldots, k$, choose a random position c_j in the normalized Euclidean space;
2. Equally subdivide samples among clusters and randomly spread them around each position c_j, with a distance from c_j in $[0, r]$.

Preliminary tests

A first group of 4 synthetic datasets, called *TS/1* (i.e., Testing Set 1) hereinafter, has been generated. Their main characteristics are summarized in Table 1. Figure 1 shows the datasets with 3 and 10 clusters, together with the optimal solutions achieved by *DAN*.

Table 1 Features of datasets used for preliminary tests (*TS/1*)

Group	*Dim*	N_s	N_c	μ_r	σ_r
	2D	1897	5	0.4	0.3
	3D	1683	3	0.09	0.04
	3D	1500	10	0.42	0.22
	4D	1680	6	0.62	0.45

Dim, N_s, and N_c denote the dimension of datasets, the number of samples, and the intrinsic number of clusters. Moreover, μ_r and σ_r denote the average radium and the variance of samples.

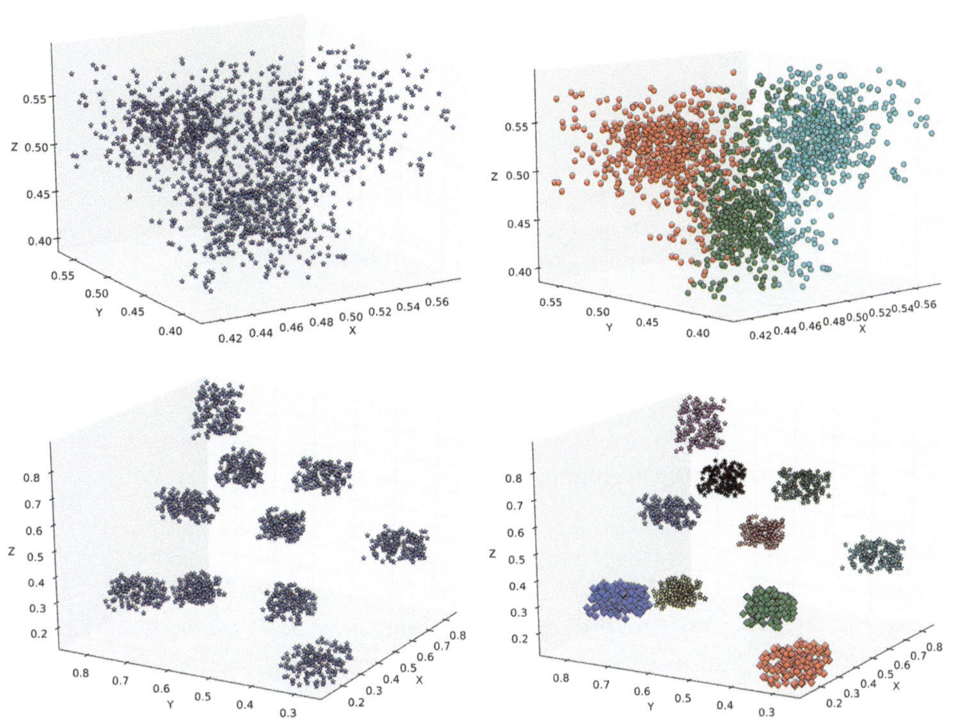

Figure 1 Second and third datasets of *TS/1*, together with the solutions achieved by *DAN* using $\log_{10}(\lambda) = 3$ (each cluster has been colored with a different color).

Multiresolution analysis has been performed varying the value of λ according to Equation 4. A logarithm scaling has been used for λ, as we experimentally found that small changes had a negligible impact on the corresponding algorithm for community detection. In particular, for each dataset, we calculated the adjacency matrix for all values of λ such that $\log_{10}(\lambda) = 0, 1, 2, 3, 4$. It is worth pointing out that the maximum value of $\log_{10}(\lambda)$ is expected to depend on the cardinality of the dataset in hand –the greater the cardinality, the greater the value of $\log_{10}(\lambda)$. However, for most datasets, a value of $\log_{10}(\lambda) = 4$, i.e., $\lambda = 10,000$, appears to be large enough to include all relevant information by means of multiresolution analysis. Table 2 shows the results of multiresolution analysis for preliminary tests.

As for the capability of identifying the optimal or suboptimal solutions[a] by means of multiresolution analysis, we observed the following pattern to occur: the optimal number of communities is robust with respect to the values of $\log_{10}(\lambda)$, as highlighted in Table 2.

Table 2 Results of multiresolution analysis achieved during preliminary tests

Group	N_c	**Number of Clusters**				
	5	2	3	**5**	**5**	**5**
	3	**3**	**3**	**3**	**3**	103
	10	2	3	**10**	**10**	151
	6	2	4	**6**	**6**	37
		0	1	2	3	4
				$\log_{10}(\lambda)$		

The number of communities is reported, calculated for $\log_{10}(\lambda) = 0, 1, 2, 3, 4$. Optimal values are highlighted in bold.

Table 3 Characteristics of datasets used for proper tests (*TS/2*), listed out according to the group they belong to

Group	*Dim*	N_s	N_c	μ_r	σ_r
	3D	350	5	0.35	0.19
	3D	2000	20	0.44	0.2
	3D	5000	30	0.51	0.24
	4D	535	4	0.64	0.46
	8D	1680	6	0.86	0.62
	12D	930	8	1.22	0.88
Iris	4D	150	3	0.49	0.26

Dim, N_s**, and** N_c **denote the dimension of datasets, the number of samples, and the intrinsic number of clusters. Moreover,** μ_r **and** σ_r **denote the average radium and the variance of samples.**

Our hypothesis was that this recurrent pattern could be considered as a decision rule for identifying the optimal number of communities (and hence of λ).

Proper Tests (*TS/2*)

We generated a second group of datasets, characterized by an increasing complexity with respect to *TS/1*. This second group of datasets is denoted as *TS/2* (i.e., Testing Set 2) hereinafter. We run *DAN* also on these new datasets, with the goal of verifying the validity of the pattern identified during preliminary tests. Moreover, we performed experiments using *Iris*, a well-known multivariate real dataset available at the UCI ML repository 2010. Iris contains 50 samples (described by 4 attributes) belonging to 3 species of Iris: setosa, virginica and versicolor. Table 3 summarizes the main characteristics of *TS/2* and *Iris*. The corresponding results, obtained with *DAN*, are shown in Table 4.

Looking at these results, we still observe the pattern identified by preliminary tests. Furthermore, one may note that a correlation often exists between the cardinality of the dataset in hand and the order of magnitude of its optimal λ (typically, the former and the latter have the same order of magnitude). It is also interesting to note that in some datasets of *TS/1* (i.e., 2nd, 3rd and 4th) and of *TS/2* (i.e., 4th, 5th and 6th) the optimal λ precedes a rapid increase in the number of communities. As a final note, we found no significant correlation between the optimal λ and the weighted-modularity parameter, notwithstanding the fact that this parameter is typically important to assess the performance of the adopted community detection algorithm.

Table 4 Results of multiresolution analysis on the selected datasets during proper tests, listed out according to the group they belong to

Group	N_c	**Pattern**	**Number of Clusters**				
	5	✓	3	**5**	**5**	8	84
	20	✓	3	4	16	**20**	**21**
	30	✓	4	5	21	**30**	**30**
	4	✓	2	**4**	**4**	105	181
	6	✓	2	4	**6**	**6**	1186
	8	✓	3	5	**8**	**8**	875
Iris	3	✓	**3**	**3**	10	82	147
			0	1	2	3	4
						$\log_{10}(\lambda)$	

The number of communities is reported, calculated for $\log_{10}(\lambda) = 0, 1, 2, 3, 4$**. Optimal values are reported in bold. The patterns observed on synthetic datasets (and reported in the table for the sake of completeness), allows to easily compute the expected optimal number of communities also for *Iris*.**

Comparison: *DAN* vs. *k*-Means and spectral clustering

Experimental results obtained with the proposed method have been compared with those obtained by running two clustering algorithms: the $k-Means$ and the spectral clustering. For the sake of readability, let us preliminarily spend few words on these algorithms.

K-*means*

As centroid-based clustering is one of the most acknowledged clustering strategies, the $k-Means$ algorithm (e.g., 1998), which belongs to this family, has been selected as one of the comparative tools. For the sake of completeness, let us briefly summarize it:

1. Randomly place k centroids in the given metric space;
2. Assign each sample to the closest centroid, thus identifying tentative clusters;
3. Compute the Center of Mass (CM) of each cluster;
4. IF CMs and centroids (nearly) coincide THEN STOP;
5. Let CMs become the new centroids;
6. REPEAT from STEP 2.

The evaluation function of $k - Means$, called *distortion* and usually denoted as J, is computed according to the formula:

$$J = \sum_{j=1}^{k} \sum_{i=1}^{n_j} \left| s_i^{(j)} - c_j \right|^2 \tag{6}$$

where n_j is the number of samples that belong to the j-th cluster, $s_i^{(j)}$ is the i-th sample belonging to j-th cluster, and c_j its centroid. Note that different outputs of the algorithm can be compared in terms of distortion only after fixing k –i.e., the number of clusters. In fact, comparisons performed over different values of k are not feasible, as the more k increases the lower the distortion is. For this reason, the use of $k-Means$ entails a main issue: how to identify the optimal number k of centroids (see 2004).

Spectral clustering

Spectral clustering 2007 algorithms use the spectrum of the similarity matrix to identify relevant clusters (the generic element of a similarity matrix measures the similarity between the corresponding data). These methods allow to perform dimensionality reduction, so that clustering can be enforced along fewer dimensions. Similarity matrices can be generated in different ways –e.g., ϵ-neighborhood graph, k-nearest neighbor graphs and fully connected graph. The main tools for spectral clustering are graph Laplacian matrices. In particular, in this work we used the unnormalized graph Laplacian matrix defined as:

$$L = D - W \tag{7}$$

where D is the degree matrix (i.e., a diagonal matrix with the degrees $d_1, \ldots, d_n$ on the diagonal) and W is the adjacency (or similarity) matrix of the similarity graph. The following algorithm has been used to perform unnormalized spectral clustering:

1. Generate the fully connected similarity graph and let W be its adjacency matrix;
2. Compute the unnormalized Laplacian L;
3. Compute the first k eigenvectors $u_1, \ldots, u_k$ of L;
4. Let $U \in \Re^k$ be the matrix containing the eigenvectors $u_1, \ldots, u_k$ as columns;

5. For $i = 1, \ldots, n$, let $y_i \in \Re^k$ be the vector corresponding to the i-th row of U;
6. Cluster the points $(y_i)_{i=1,\ldots,n}$ in $\Re^k$ with the k-means algorithm into clusters $C_1, \ldots, C_k$.

Notably, also in this case the number k of cluster is required as input.

Comparative results

We run the k-Means algorithm (using the Euclidean metric) and the spectral clustering algorithm on the selected datasets –with the goal of getting new insights on the results of the partitioning procedure defined in *DAN*. Both algorithms used for comparison purposes have been run using the optimal values of k identified by means of multiresolution analysis. The comparison has been performed considering the distortion J computed for each solution. Figure 2 reports comparative results and clearly shows that, in around 72.2 percent of the cases, *DAN* achieves the best result. These results highlight the validity of the proposed framework, also considering that *DAN* computes partitions without any a priori knowledge about the datasets, as the optimal (or suboptimal) number of clusters is typically found by applying the previously described pattern. Although k–*Means* is faster than *DAN*, it is important to stress that its results, at each attempt, depend tightly on the initial position of the k centroids. Hence, in absence of a strategy for identifying the initial disposal of centroids, k–*Means* should be (and it is in fact) run several times –the solution with the smaller distortion being selected as optimal. The spectral clustering algorithm showed its effectiveness many times, although bad solutions have been computed with datasets 2 and 3 of *TS/2*, characterized by 20 and 30 clusters, respectively.

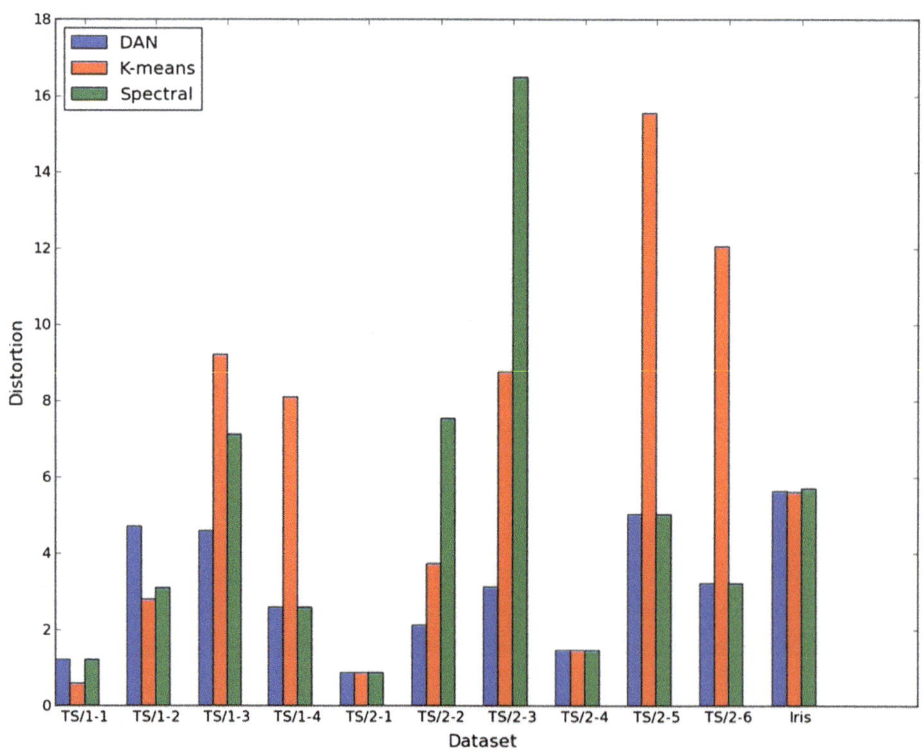

Figure 2 **Comparison, in terms of distortion, among solutions achieved by *DAN*, blue bars, *k*–*Means*, red bars and spectral clustering, green bars (the lesser the better).**

Conclusions

In this paper, a method for clustering multidimensional datasets has been described, able to find the most appropriate number of clusters also in absence of a priori knowledge. We have shown that community detection can be effectively used also for data clustering tasks, provided that datasets are viewed as complex networks. The proposed method, called *DAN*, makes use of transformations between metric spaces and enforces multiresolution analysis. A comparative assessment with other well-known clustering algorithms (i.e., $k-Means$ and spectral clustering) has also been performed, showing that *DAN* often computes better results.

As for future work, we are planning to test *DAN* with other relevant datasets, in a comparative setting. Furthermore, we are planning to study to which extent one can rely on the decision pattern described in the paper, assessing its statistical significance over a large number of datasets.

Endnote

[a]As pointed out by Arenas et al. 2008, it may not appropriate to speak of correct vs. incorrect solutions for multiresolution analysis. In a context of community detection we deem more appropriate to speak of optimal or suboptimal solutions (see also 2011 for more information on this issue).

Competing interests

The authors declare that they have no competing interests.

Authors' contributions

Both authors had the first insight about using complex networks to perform clustering tasks. They investigated the properties of the corresponding models and have devised suitable experiments to assess them. Marco A. Javarone has performed all experiments, using Python as programming language. Both authors read and approved the final manuscript.

Acknowledgements

Many thanks to Alessandro Chessa and to Vincenzo De Leo (both from Linkalab). The former for his wealth of ideas about complex networks and the latter for the support given to install and run their Complex Network Library.

References

Albert, R, Barabasi A: **Statistical Mechanics of Complex Networks.** *Rev Mod Phys* 2002, **74:**47–97.

Alsabti, K: **An efficient k-means clustering algorithm.** In *Proceedings of IPPS/SPDP Workshop on High Performance Data Mining*; 1998.

Arenas, A, Fernandez A, Gomez S: **Analysis of the structure of complex networks at different resolution levels.** *New Journal of Physics* 2008, **10**(5):053039.

Blondel, VD, Guillaume JL, Lambiotte R, Lefebvre E: **Fast unfolding of communities in large network.** *Journal of Statistical Mechanics: Theory and Experiment* 2008. **P10008**.

Eick, C, Zeidat N, Zhao Z: **Supervised Clustering – Algorithms and Benefits.** In *Proc. of ICTAI*; 2004.

Frank, A, Asuncion A: **UCI Machine Learning Repository.** 2010. [http://archive.ics.uci.edu/ml].

Gudkov, V, Montealegre V, Nussinov S, Nussinov Z: **Community detection in complex networks by dynamical simplex evolution.** *Phys Rev E* 2008, **78:**016113.

Guimer, R, Danon L, Diaz-Guilera A, Giralt F, Arenas A: **Self-similar community structure in a network of human interactions.** *Phys Rev E Stat Nonlin Soft Matter Phys* 2003, **68:**065103.

Jain, AK: **Data clustering: 50 years beyond K-means.** *Pattern Recognition Letters* 2010, **31**(8):651–666.

Li, Z, Hu Y, Xu B, Di Z, Fan Y: **Detecting the optimal number of communities in complex networks.** *Physica A: Statistical Mehcanics and Its Applications* 2011, **391:**1770–1776.

Mark, HH, Yu B: **Model Selection and the Principle of Minimum Description Length.** *Journal of the American Statistical Association* 1998, **96:**746–774.

Mucha, P, Richardson T, Macon K, Porter M, Onnela J: **Community Structure in Time-Dependent, Multiscale, and Multiplex Networks.** *Science* 2010, **328**(5980):876–878.

Newman, MEJ, Girvan M: **Finding and evaluating community structure in networks.** *Phys Rev* 2004, **69:**026113.

Newman, M: *Networks: An Introduction.* New York, NY, USA: Oxford University Press, Inc.; 2010.

Sporns, O, Chialvo DR, Kaiser M, Hilgetag C: **Organization, development and function of complex brain networks.** *Trend in Cognitive Sciences* 2004, **8**(9).

Heimo, T, Kaski K, Kumpula JM, Saramaki J: **Detecting modules in dense weighted networks with the Potts method.** *Journal of Statistical Mechanics: Theory and Experiment* 2008, **08:**08007.

Tibshirani, R, Walther G, Hastie T: **Estimating the number of clusters in a data set via the gap statistic.** *Journal of the Royal Statistical Society - Series B: Statistical Methodology* 2001, **63**(2):411–423.

Toivonen, R, Kivela M, Saramaki J, Viinikainen M, Vanhatalo M, Sams M: **Networks of Emotion Concepts.** *PLoS ONE* 2012, **7**(1):e28883.

von Luxburg, U: **A Tutorial on Spectral Clustering.** *Statistics and Computing* 2007, **17**(4):395–416.

9

Adaptive pedestrian behaviour for the preservation of group cohesion

Giuseppe Vizzari*, Lorenza Manenti and Luca Crociani

*Correspondence:
giuseppe.vizzari@disco.unimib.it
Complex Systems and Artificial Intelligence research center, Università degli Studi di Milano–Bicocca, Milano, Italy

Abstract

Purpose: A crowd of pedestrians is a complex system in which individuals exhibit conflicting behavioural mechanisms leading to self-organisation phenomena. Computer models for the simulation of crowds represent a consolidated type of application, employed on a day-to-day basis to support designers and decision makers. Most state of the art models, however, generally do not consider the explicit representation of pedestrians aggregations (groups) and their implications on the overall system dynamics. This work is aimed at discussing a research effort systematically exploring the potential implication of the presence of groups of pedestrians in different situations (e.g. changing density, spatial configurations of the environment).

Methods: The paper describes an agent-based model encompassing both traditional individual motivations (i.e. tendency to stay away from other pedestrians while moving towards the goal) and an adaptive mechanism representing the influence of group presence in the simulated population. The mechanism is designed to preserve the cohesion of specific types of groups (e.g. families and friends) even in high density and turbulent situations. The model is tested in simplified scenarios to evaluate the implications of modelling choices and the presence of groups.

Results: The model produces results in tune with available evidences from the literature, both from the perspective of pedestrian flows and space utilisation, in scenarios not comprising groups; when groups are present, the model is able to preserve their cohesion even in challenging situations (i.e. high density, presence of a counterflow), and it produces interesting results in high density situations that call for further observations and experiments to gather empirical data.

Conclusions: The introduced adaptive model for group cohesion is effective in qualitatively reproducing group related phenomena and it stimulates further research efforts aimed at gathering empirical evidences, on one hand, and modelling efforts aimed at reproducing additional related phenomena (e.g. leader-follower movement patterns).

Keywords: Agent-based modelling and simulation, Crowd simulation, Adaptive behaviours

Background

Crowds of pedestrians are generally recognised as a form of complex system (Batty 2001): even without making a serious attempt of providing a formal definition of the term crowd, and adopting a simplistic and common sense intuitive notion of "(too) many people in (too) little space" (Kruse 1986), the dynamics that can be identified in a crowded environment, in which several pedestrians move towards their own goals, are good indicators supporting this statement. Pedestrians share the same environment and they generally compete for the space resource; nonetheless, they can also exhibit collaborative patterns of interactions, for instance respecting (written or non-written) shared rules like giving way to passengers getting off a train before getting on board, or even respecting cultural dependant rules (e.g. gallantry). There are evidences of imitation among pedestrians whenever they need to cross a road (see, e.g., Helbing et al. 1997), but basic proxemic considerations indicate that normal behaviour includes a tendency to stay at a distance, to preserve a personal space (Hall 1966). Pedestrians continuously adapt their behaviour to the contextual conditions, considering the geometry of the environment but also social aspects. The overall crowd system, whose evolution depends on the individual decisions of pedestrians, shows several examples of self-organised behaviours, from the formation of lanes to oscillatory changes in the walking directions at narrow passages.

Computer models for the simulation of crowds are growingly investigated in the academic context and these efforts led to the implementation of commercial off-the-shelf simulators often adopted by firms and decision makers[a]. Models and simulators have proved their adequacy in supporting architectural designers and urban planners in their decisions by creating the possibility to envision the behaviour of crowds of pedestrians in specific actual environments and planned designs, to elaborate what-if scenarios and evaluate their decisions with reference to specific metrics and criteria.

Despite the substantial amount of results and efforts this area is still quite lively and we are far from a complete understanding of the complex phenomena related to crowds of pedestrians in the environment: one of the least studied and understood aspects of crowds of pedestrians is represented by the implications of the presence of groups (Challenger et al. 2009). Pedestrians, in fact, exhibit a substantially different behaviour in a given scenario if friends or family members, for instance, are present in the same environment at the same time: they will try to reach the desired destination but they will also try to preserve a limited distance from the other group members, also temporarily neglecting the tendency to move towards their goals, in an interesting form of *adaptive behaviour*. In most current models, instead, pedestrians simply interpret the presence of other individuals as a sort of *moving obstacle* or simply as openers of a potential route to follow (in case of imitative behaviours). The research that is summarised in this paper is aimed at systematically evaluate the impact of the presence of groups in a population of pedestrians, also considering relatively large groups potentially structured into smaller sub-groups, performing the most appropriate form of validation (either quantitative or qualitative) against real data, coming from new observations or already present in the literature. While, in fact, some of the implications of the presence of groups, for instance on the walking speed of its members, have already been analysed, at least in low density situations (Federici et al. 2012; Schultz et al. 2010; Willis et al. 2004), the overall impact on other observable metrics such as space utilisation is still not clear. As previously suggested, we will talk about *simple* groups when referring to small sets of pedestrians bound by a strong relationship (e.g. friends, family members): the agents representing members of this kind of group are characterised by an adaptive behavioural mechanism aimed at preserving its cohesion, even in situations of high local density and presence of obstacles or counter flows of other pedestrians. The model we will introduce also considers larger groups, sometimes of "artificial nature" (e.g. groups of tourists), that can in turn be made up of other smaller groups (either structured or simple) or just by individuals: agents representing members of this kind of group are characterised by a tendency to stay close to other members but this tendency is not as strong as for simple groups and it does not necessarily prevent the fragmentation of groups.

The paper breaks down as follows: the following section will briefly report the most relevant related works, while a description of the introduced adaptive model for pedestrian behaviour encompassing the effects of group presence will follow. Section "Simulation results" will present the scenarios in which the model has been applied and the achieved results. Conclusions and future developments will end the paper.

Related works

We will not provide here a comprehensive overview of the different approaches and models for the simulation of pedestrian and crowd dynamics: scientific interdisciplinary workshops and conferences are in fact specifically devoted to this topic (see, e.g., the proceedings of the first edition of the International Conference on Pedestrian and Evacuation Dynamics (Schreckenberg and Sharma 2001) and consider that this event has reached the sixth edition in 2012) and it would be impossible to summarise all the relevant works present in the literature. On the other hand, we propose a schema classifying the different current approaches based on the way pedestrians are represented and managed. From this perspective, pedestrian models can be roughly classified into three main categories that respectively consider pedestrians as *particles subject to forces*, particular *states of cells* in which the environment is subdivided in Cellular Automata (CA) approaches, or *autonomous agents* acting and interacting in an environment.

The most widely adopted particle based approach is represented by the *social force model* (Helbing and Molnár 1995), which implicitly employs fundamental proxemic (Hall 1966) concepts like the tendency of a pedestrian to stay away from other ones while moving towards his/her goal. Proxemics essentially represents a fundamental assumption of most modelling approaches, although very few authors actually mention this anthropological theory (notably CA based models (Ezaki et al. 2012; Was 2010) and an agent-based model (Manenti et al. 2010)). Recent works also extend the social force model for specific applications to evacuation scenarios in panic situations (Shiwakoti et al. 2196).

CA based approaches are based on a discrete representation of the simulated environment. Each cell represents a portion of the space and it can be either vacant, occupied by an obstacle or a single pedestrian (although there are models relaxing this constraint which allow more than a single pedestrian to be situated in a given cell at the same time). The management of system evolution is also based on a discrete representation of time, discretised in equally sized intervals of time (steps). A uniform transition rule guides the evolution of the system; the rule determines the next state of each cell considering its current state and the states of nearby ones (where nearby depends on a specific notion of neighbourhood, that is a function mapping every cell to a set of "visible" nearby cells). CA approaches can be roughly classified in ad-hoc models, that represent an effective and efficient solution but just for specific situations (such as the case of bidirectional flows at intersections described in (Blue and Adler 1999), in which there are specific "lane direction" rules) and more general models, that can be adopted for representing any kind of environment and pedestrian movement tendency, whose main representative is the floor-field approach (Schadschneider et al. 2002), in which the cells are endowed with a discretised gradient guiding pedestrians towards potential destinations.

While particle and CA based approaches are mostly aimed at generating quantitative results about pedestrian and crowd movement, agent based models are sometimes aimed at the generation of effective visualisations of believable crowd dynamics, and therefore the above approaches do not necessarily share the same notion of realism and validation. Works like (Bandini et al. 2004; Henein and White 2005) essentially extend CA approaches, separating the pedestrians from the environment and granting them a behavioural specification that is generally more complex than what is generally represented in terms of a simple CA transition rule, but they essentially adopt similar methodologies. Other approaches like (Musse and Thalmann 2001; Shao and Terzopoulos 2007) are more aimed at generating visually effective and believable pedestrians and crowds in virtual worlds. Finally, works like (Paris and Donikian 2009), employ cognitive agent models for different goals, but they are not generally aimed at making predictions about pedestrian movement for sake of decision support.

A relatively small number of recent works represent a relevant effort towards the modelling of groups, respectively in particle-based (Moussaïd et al. 2010; Xu and Duh 2010) (extending the social force model), in CA-based (Sarmady et al. 2009) (with ad-hoc approaches) and in agent-based approaches (Manzoni et al. 2011; Qiu and Hu 2010; Rodrigues et al. 2010; Tsai et al. 2011) (introducing specific behavioural rules for managing group oriented behaviours). All the above mentioned approaches interpret the impact of groups by means of additional contributions to the overall pedestrian behaviour representing the tendency to stay close to other group members. However, the above approaches mostly deal with small groups in relatively low density conditions; those dealing with relatively large groups (tens of pedestrians) were not validated against real data. The last point is a crucial and critical element of this kind of research effort: computational models represent a way to formally and precisely define a computable form of theory of pedestrian and crowd dynamics. These theories must be validated employing field data, acquired by means of experiments and observations of the modelled phenomena, before the models can actually be used for sake of prediction. The scarcity of data specifically characterising the behaviour of pedestrians in the presence of groups hinders this validation activity: for the present work we adopted the choice of validating the model in absence of groups with available data from the literature and exploring the effect of the introduction of groups with mechanisms and parameters that satisfied a (qualitative) *face validation* (Klügl 2008) against available video footages of groups of pedestrians (Federici et al. 2012) and preliminary data from experiments in controlled situations (Vizzari et al. 2012).

Methods

This section formally introduces a model representing pedestrian behaviour in an environment, considering the impact of the presence of simple and structured groups in the simulated scenario. The model is characterised by a discrete representation of the environment and time evolution, essentially based on floor-field CA approaches. Nonetheless, the pedestrian behavioural specification is so articulated, encompassing even an adaptive mechanism for the preservation of group cohesion, to the point that the model is more properly classified as agent-based. The different elements of the model will now be introduced, starting from the adopted representation of the environment and simulation evolution strategy. Then groups, pedestrians and the mechanism for the evaluation of available actions will be described. Finally, the adaptive mechanism for the preservation of group cohesion will be introduced.

Representation of the environment

The physical environment is represented in terms of a discrete grid of square cells: $Env = c_0, c_1, c_2, c_3, \ldots$ where $\forall c_i : c_i \in Cell$. The size of every cell is $40cm \times 40cm$ according to standard measure used in the literature and derived from empirical observation and experimental procedure (Fruin 1992; Weidmann 1993). Every cell has a row and a column index, which indicates its position in the grid: $Row(c_i), Col(c_i) : Cell \rightarrow \mathbb{N}$. Consequently, a cell is also identified by its row and column on the grid, with the following notation: $Env_{j,k} = c : (c \in Env) \wedge (Row(c) = j) \wedge (Col(c) = k)$.

Every cell is linked to other cells, that are considered its neighbours according to the Moore neighbourhood, that is all the cells surrounding the cell being considered, even in diagonal directions: it is possible to express Moore neighbourhood as joining of orthogonal cells, also called Von Neumann neighbourhood, and diagonal cells.

$$\begin{array}{ll} N(Env_{j,k}) = Env_{j+1,k}, & S(Env_{j,k}) = Env_{j-1,k}, \\ E(Env_{j,k}) = Env_{j,k+1}, & W(Env_{j,k}) = Env_{j,k-1}, \\ NE(Env_{j,k}) = Env_{j+1,k+1}, & SE(Env_{j,k}) = Env_{j-1,k+1}, \\ NW(Env_{j,k}) = Env_{j+1,k-1}, & SW(Env_{j,k}) = Env_{j-1,k-1} \end{array}$$

$$\begin{array}{l} VonNeumanNeighbours(c) = \{N(c), S(c), E(c), W(c)\} \\ DiagonalNeighbours(c) = \{NE(c), SE(c), NW(c), SW(c)\} \end{array}$$

$$neighbours(c) = VonNeumanNeighbours(c) \cup DiagonalNeighbours(c)$$

Every cell in the environment can be in three possible states: *free, occupied by an obstacle*, or *occupied by a pedestrian*. In the third case the cell contains also a reference to the specific pedestrian occupying it: $State(c) = s : s \in FREE, OBSTACLE, PEDESTRIAN_i$. In addition to the potential presence of physical objects (pedestrians and obstacles) each cell is also linked to additional structures that contain information useful to support pedestrian movement.

Definition of spatial markers

Space can be annotated at design-time with different markers, a set of cells that play particular roles in the simulation. Three kinds of marker are defined in the model:

- *start areas*, places (sets of cells) were pedestrians are generated: they contain information for pedestrian generation both related to the type of pedestrians and to the frequency of generation. In particular, a start area can generate different kinds of pedestrians according to two approaches: (i) *frequency-based generation*, in which pedestrians are generated during all the simulation according to a frequency distribution; (ii) *en-bloc generation*, in which a set of pedestrians is generated at once in the start area when the simulation starts;
- *destination areas*, final places where pedestrians want to go;
- *obstacles*, non-walkable cells defining obstacles and non-accessible areas.

Space annotation allows the definition of virtual grids on the environment, as containers of information for agents.

Definition of floor fields

Adopting the approach of the *floor field* model (Nishinari et al. 2004), the environment of the basic model is composed also of a set of superimposed virtual grids, structurally identical to the environment grid, that contains different floor fields that influence pedestrian behaviour.

The goal of these grids is to support long range interactions by representing the state of the environment (namely, the presence of pedestrians and their capability to be perceived from nearby cells) in terms of field modifications. In this way, a local perception for pedestrians actually simply consists in gathering the necessary information in the relevant cells of the floor field grids. In other works, the concept of perception and its practical implementation is more sophysticated, starting from the definition of the *field of view* mechanism in humans: in (Paris and Donikian 2009) pedestrian's perception is investigated from a cognitive point of view, while a more physical approach is adopted in (Shao and Terzopoulos 2007). Nonetheless, in CA based approaches such a precise perception model is rarely employed and still the achieved results are often extremely interesting, therefore we decided to employ a simple perception model and evaluate its adequacy.

Some of the floor fields are *static* (created at the beginning and not changing during the simulation) or *dynamic* (changing during the simulation). Three floor fields are considered in the model:

- the *path field* assigned to each destination area, that indicates for every cell the distance from the destination, acting thus as a potential field that drives pedestrians towards it (static floor field);
- the *obstacles field*, that indicates for every cell the distance from an obstacle or a wall (static floor field);
- the *density field* that indicates for each cell the pedestrian density in the surroundings at the current time-step (dynamic floor field).

All these fields can be seen as grids identical to the environment grid: a function that extracts the values of the fields for the given cell is defined as follows:

$$Val(f, c) : Field \times Cell \rightarrow \mathbb{R}$$

The following notation will be used to indicate these grids, assuming that pedestrians know only the path field associated to their own destination:

$$\begin{aligned} PathF_{j,k} &= Val(PathF, c) : ((c \in Env) \ \wedge (Env_{j,k} = c)) \\ ObsF_{j,k} &= Val(ObsF, c) : ((c \in Env) \ \wedge (Env_{j,k} = c)) \\ DensF_{j,k} &= Val(DensF, c) : ((c \in Env) \ \wedge (Env_{j,k} = c)) \end{aligned}$$

The definition of every type of floor field is now illustrated.

Path field

Each destination is associated to a *path field* indicating the shortest path between each cell in the environment and their destination. These floor fields act as a potential, driving pedestrian towards the destination (one floor field exists for every destination): starting from every destination defined in the scenario, the information are spread into the environment according to a particular method. In (Kretz et al. 2010) authors analysed different methods for the calculation of the distance potential field: starting from the considerations in this work, we decided to apply the *chessboard* metric to manage the Moore neighbourhood using the $\sqrt{2}$ variation over corners.

For every path field and for every cell, the value of the distance is calculated with this metric and it is associated to every cell: value increases if the distance increases.

Figure 1 proposes a graphical representation of the floor fields used in the model in a simple corridor scenario: in Figure 1a and 1b the path fields associated to destination areas (in blue) are proposed: darker tonalities indicate the approaching to the destination.

Obstacle field

This floor field contains all the information related to the position of obstacles in the scenario: just one grid exists for all the non-walkable areas in the environment. Chessboard metric with $\sqrt{2}$ variation over corners is used also to produce the spreading of the information in the obstacle field: a particular radius r_{ob} is considered as the border of propagation of information about the presence of the obstacle.

The algorithm works as follows: after the initialisation of all the cell with a null value, the chessboard metric with $\sqrt{2}$ variation for all the cells that lie under the radius r_{ob} from the obstacle is calculated; r_{ob} represents therefore the maximum distance for which an obstacle generates a repulsive effect. A non-null value is then associated to these cells (i.e. higher values represent cells nearest to the obstacles) according to the following functions:

$$Val(ObsF, c) = max\{0, r_{ob} - dist(c_{obs}, c)\}$$

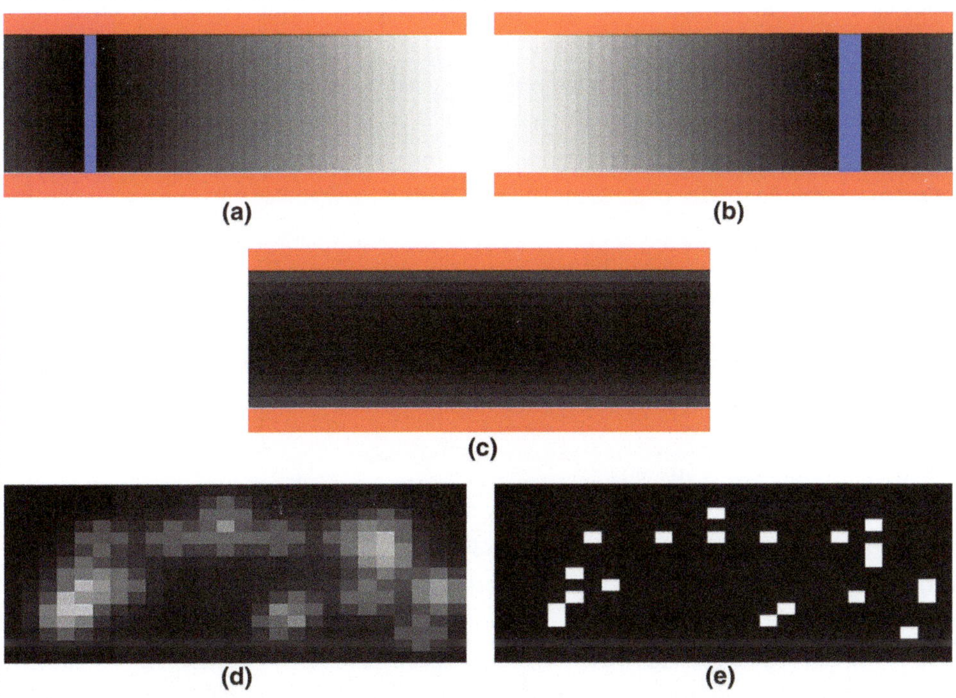

Figure 1 Graphical representation of different floor fields in a simple corridor scenario: Figures 1a and 1b represent two path fields associated to different destination areas. Figure 1**c** represents the obstacle fields provided by walls. Figure 1**d** and 1**e** show the density field with the related pedestrian position at a precise time step.

Figure 1c represents obstacle field in which lightest values indicate the approaching to the obstacles (in red).

Density field

The grid associated to this field contains information necessary to the management of interaction between pedestrians and to the calculation of statistics related to the local densities in the environment.

In general, the concept of *density* is usually related to the number of persons in a fixed portion of space: despite that, density can be measured in different ways and the concept of mean density in a particular cell of the scenario has to be analysed more in detail. In the pedestrian dynamics literature, the concept of *cumulative mean density* (CMD) indicates the density experienced by pedestrians in a cell: the concept was first introduced in (Still 2000), where the author defined that CMD is measured only when a pedestrian passes over there, by counting the number of people in the surroundings of the pedestrian (given a distance range). At the end of simulation, average on all the measurements is computed for each of the cells of the space. It is a local and pedestrian-based concept of density, that gives information about how pedestrian experiences the nearby space, ignoring the time periods in which a portion of space has been empty.

In (Castle et al. 2011), CMD evaluation in LEGION[b] and STEPS[c] software is analysed and compared: in LEGION, the surrounding of a pedestrian is identified into a radius $r = 1.5$ m with an area[d] of $Area = r^2 * \pi = 7.07m^2$. STEPS adopts a similar approach calculating the density in a discrete way using the cells that fall within $r = 1.25$ m of each persons: the area considered is $Area = r^2 = 6.25m^2$.

In our model, the management of density field is a bit more complicated with respect to previous cases. It is modified according to the following mechanism: when a pedestrian p moves in a cell c, the density field is modified in the grid adding 1 to the cell in which he/she moves, and subtracting 1 from the cell he/she just left. The modification is applied also to neighbour cells in a range given by a radius $r = 2$ m (equal to five cells from c, considering the proposed scale of discretisation), but the value added/subtracted decreases with the inverse of the square of the distance between the cell and p:

$$v = \frac{1}{d^2}$$

Figures 1d and 1e represent the density field (in which lightest values that indicate highest values of the density) and an instant position of agents, respectively.

Simulation time and update strategy

Simulation time is modelled in a discrete way by dividing time into steps of equal duration: we assume that a pedestrian moves exactly 1 cell per time step. The average velocity of a pedestrian, which can be estimated through observations or experiments (Fruin 1992) in about 1.2 ms^{-1}, will thus determine the duration of the each time step: considering that the size of the cell is $40cm \times 40cm$, the duration of each time step is thus $0.33s$.

Note that in this way, the maximum of velocity allowed in this model is $1.2ms^{-1}$: different works (Kirchner et al. 2004; Weng et al. 2006) investigated how variations in the pedestrian velocity ($1.0ms^{-1}$ and $1.5ms^{-1}$) can be modelled with CA approach and how these choices influence simulation results. It must be emphasised that, however, this parameter is not strictly embedded in the model and it could be changed, essentially modifying only the analysis and results interpretation phases.

When running a CA-based pedestrian model, three update strategies are possible (Klüpfel):

- *parallel* update, in which cells are updated all together;
- *sequential* update, in which cells are updated one after the other, always in the same order;
- *shuffled sequential update*, in which cells are updated one after the other, but with a different order every time.

The second and third update strategies lead to the definition of asynchronous CA models (see (Bandini et al. 2012) for a more thorough discussion on types of a-synchronicity in CA models).

In crowd simulation CA models, parallel update is generally preferred (Schadschneider et al. 2009), even if this strategy can lead to conflicts that must be solved. Some works (Kirchner et al. 2003) claim even that simulations are more realistic if the conflicts that arise are not solved, but to prevent the movement of all pedestrians involved in a conflict with a certain probability.

In their pioneering work, (Gipps and Marksjö 1985) used a sequential update, despite that (Blue et al. 1999) point out that "*with sequential updates the order of each move becomes unrealistically important, since as each entity moves, the next entity re-positions in relation to the previous entity. Thus, the first entity would act the position of all entities over the whole lattice*".

Nonetheless, we chose to investigate the effects of allowing the possibility of this form of micro coordinated movements and therefore we adopted a shuffled sequential update scheme for the activation of agent behaviours according to the fact that one of the elements involved in the prediction of movement is the previous position of the pedestrian in the environment and that conflicts may be represented by proxemic separation, rather than space exclusion.

Please note that the structure of the model and the defined mechanisms remain valid in case of a parallel update scheme: the model would only need the definition of a mechanism and strategy to manage conflicts to support this schema. Of course, this different choice would have an impact on the simulation results and thus on the calibration phases and a comparison among these different approaches can be pointed out.

Groups

As suggested in the introduction, we focus on two types of group: simple and structured. Simple or informal groups are generally made up of friends or family members and they are characterised by a high cohesion level, moving all together towards the same goal due to shared goals and to a continuous mechanism of adaptation of the chosen paths to try to preserve the possibility of performing non-verbal communication (Costa 2010). Structured groups, instead, are more complex entities, usually larger than simple groups (more than 4 individuals) and they can be considered as being composed of sub-groups that can be, in turn, either simple or structured. Structured groups are often artificially defined with the goal of organising and managing the movement (or some kind of other operation) of a set of pedestrians.

Groups can be formally described as:

$$Group_j = \langle Id, [\,Group_1, \ldots, Group_m], [\,Ped_1, \ldots, Ped_n] \rangle$$

Structured groups include at least one subgroup, while simple groups only comprise individual pedestrians. We will refer to the group an agent *a directly* belongs to as G_a, that is also the smallest group he belongs to; the *largest* group an agent *a* belongs to will instead be referred to as $\bar{G_a}$. It must be noted that $\bar{G_a} = G_a$ only when the agent *a* is member of a simple group that is not included in any structured group.

Pedestrians

In this model, a pedestrian is defined as an utility-based agent with state. Functions are defined for utility calculation and action choice, and rules are defined for state-change. Pedestrians are characterised as:

$$Pedestrian : \langle Id, GroupId, State, Actions, Destination \rangle$$

where:

1. $Id \in \mathbb{N}$ is the agent identification number;
2. $GroupId \in \mathbb{N}$ is the identification number of the group to which the pedestrian belong to; for pedestrians that are not member of any group this value is null.
3. *State* that represents the state of the agent related to its position in the space and to its attitude with respect to the simulated scenario. It is defined as:

   $$State : \langle Position, PrevDirection \rangle$$

   where *Position* indicates the current cell in which the agent is located, and *PrevDirection* is the direction followed in the last movement;
4. *Actions* is the set of possible actions that the agent can perform. Possible actions are movements in one of the eight neighbor cells (indicated as cardinal points), plus the action of remaining in the same cell (indicated by an '*X*'):

   $$Actions = \{N, S, W, E, NE, SE, NW, SW, X\}$$

   Admissible actions $AdmAct_p$ ($\subseteq Actions$) are all the actions that move the pedestrian p from cell c in cells that are free at the moment it is updated:

   $$AdmAct_p = \{a : a \in Actions \wedge State(a(c)) = FREE\}$$

   The effect of each action is to move the pedestrian p in the direction indicated. This means that when an action a is chosen (for example, *N*), the new cell is calculated as follows:

   $$newCell = a(oldCell)$$

   When the movement is completed, the cell is marked as occupied, and the old cell is marked as free:

   $$State(newCell) = PEDESTRIAN_p$$
   $$State(oldCell) = FREE$$

   The last effect of an action is to update the density field by reducing density in the surroundings of *oldCell* and increasing it in the surroundings of *newCell*;

5. *Destination* is the goal of the agent in terms of destination area. This term identifies the current destination of the pedestrian: in particular, every destination overlaps with a set of cells that are defined as destination areas by means of the appropriate spatial marker. *Destination* is used to identify which path field is relevant for the agent:

$$currentPathField = PathField(Destination)$$

where *PathField* is the precise path field associated to *Destination* and *currentPathField* is the path field relevant for the agent.

All these elements take part in the mechanism that manages the movement of pedestrians: as previously introduced, they are essentially utility-based agents. To every movement in the cell neighbourhood a value of utility is associated, according to a set of factors that concur in the overall dynamics.

Mechanism of action evaluation

In Algorithm 1 the agent life-cycle during all the simulation time is proposed: every time step, every pedestrian perceives the values of path field, obstacle field and density field for all the cells that are in its neighbourhood. On the basis of these values and according to different factors, the agent evaluates the different cells around him, associating an utility value to every cell and selects the action for moving into a specific cell.

Algorithm 1 Agent life-cycle

```
for all timestep ∈ SimulationTime do
    for all p ∈ Pedestrian do
        Utility[ ]
        for all c ∈ neighbours(Position) do
            pf ← Val(PathF, c)
            of ← Val(ObsF, c)
            df ← Val(DensF, c)
            Utility[c] ← Evaluation(pf, of, df)
        end for
        a = Choice(Utility[ ])
        Move(a)
    end for
end for
```

As previously suggested, the action selection strategy starts gathering the value of floor fields in cells included in the neighbourhood of agent's current position. The obtained values will be used in the evaluation of the movement towards the related cell.

After acquiring the perceived information from the environment, the agent elaborates a desirability value for each of the admissible actions (movements), according to several factors and more precisely:

- the desire to move towards a *goal*, a destination in the environment;
- the tendency to stay at a distance from the *obstacles* (e.g. walls, columns), that are perceived as repulsive;

- the desire to stay at a distance from other individuals, especially those that are not members of the same simple group, an effect of proxemic *separation*;
- a *direction inertia* factor, increasing the desirability of performing straight forms of movement;
- the penalisation of those movements that cause an *overlapping* event, the temporary sharing of the same cell by two distinct pedestrians;
- two contributions related to the tendency to preserve *group cohesion*, respectively devoted to simple and structured groups.

All the above contributions, which will be more thoroughly described in the following part of this section, are considered by the overall utility function $U_a(c)$ of a destination cell c which corresponds to an action/direction for agent a, that takes the form of a weighted sum of components associated to the above factors:

$$U_a(c) = \frac{\kappa_g G(c) + \kappa_{ob} Ob(c) + \kappa_s S(c) + \kappa_d D(c) + \kappa_{ov} Ov(c) + \kappa_c C_a(c) + \kappa_i I_a(c)}{d}$$

where d is the distance of the new cell from the current position, that is 1 for cells in the Von Neumann neighbourhood (vertically and horizontally neighbour cells) and $\sqrt{2}$ for diagonal cells: the factor is introduced to penalise the diagonal movements. Note that $\kappa_g, \kappa_{ob}, \kappa_s, \kappa_d, \kappa_{ov}, \kappa_c, \kappa_i \in [0, 100]$: the use of these parameters, in addition to allowing the calibration and the fine tuning of the model, also supports the possibility of describing and managing different types of pedestrian, or even different states of the same pedestrian in different moments of a single simulated scenario.

Given the list of possible actions and associated utilities, an action is chosen with a probability proportional to its utility. In particular, the probability for an agent a of choosing an action associated to the movement towards a cell c is given by the exponential of the utility, normalised on all the possible actions the pedestrian can take in the current turn:

$$p_a(c) = N \cdot e^{U_a(c)}$$

where N is the normalisation factor and c is the currently considered destination cell.

Every element that contributes to the utility calculation will now be formally described.

Goal attraction

Agents are driven towards their goal using information derived from the related path field, calculating the distance between their current cell and the destination area. The function that manages the goal attraction evaluates the reduction of distance moving from cell *Position* in cell c where $x = Row(Position)$, $y = Col(Position)$, $i = Row(c)$ and $j = Col(c)$:

$$G(c) = \frac{PathF_{x,y} - PathF_{i,j}}{\sqrt{2}}$$

where $\forall c \in Cells, G(c) \in [-1, 1]$.

Obstacle repulsion

The interaction between agents and obstacles and non-accessible areas in the environment has negative impact towards the movement, because of the tendency of pedestrians to manage the available space without walking too close to obstacles and walls: for instance, considering the scenario of a corridor, pedestrians tend to stay in the centre instead of close to the walls. Information for the location and influence of obstacle are associated to the obstacle field introduced in Section "Obstacle field". Considering a cell c

with $i = Row(c)$ and $j = Col(c)$,

$$Ob(c) = -\frac{ObsF_{i,j}}{r_{ob}}$$

where $\forall c \in Cells, Ob(c) \in [-1, 0]$ and r_{ob} is the maximum distance for which an obstacle generates a repulsive effect.

Proxemic separation

All the interactions among pedestrians are subjected to the proxemic separation relationship, stating that all the persons tend to maintain a certain distance with respect to others and according to the local density calculated, stored and maintained into the grid of the density field. According to Proxemic theory (Hall 1966) the public distance among pedestrians is between 3.0 m and 6.0 m. Considering a cell c with $c = Env_{i,j}$:

$$S(c) = -\frac{DensF_{i,j}}{MaxDensity}$$

where $DensF_{i,j}$ is the value of the density field for the cell c and $MaxDensity$ is the maximum global density value that the density field can assume, according to the radius r used in its definition. Output of S function are in the $[-1, 0]$ range. In our case, the value of $MaxDensity$ that can be reached with a discretisation of $40\ cm \times 40\ cm$ is equal to $6.25\ m^{-2}$ with $r = 1$.

Direction inertia

This factor represents the fact that pedestrians tend to maintain the direction during movement towards a destination. Unexpected changes of direction are usually avoided by pedestrians: for this reason, a value is added in the case that cell c is located in the same direction with respect to the previous movement (except the case in which agent remained in the same cell, i.e. $PrevDirection = X$):

$$D(c) = \begin{cases} 1\ if\ PrevDirection = D(c) \\ 0\ otherwise \end{cases}$$

where $D(c)$ is the direction in which the agent must move to reach cell c from the current position.

Overlapping extension

CA-based models are characterised by a limit to the maximum manageable density that also limits their possibility to effectively represent highly crowded situations. A relaxation to the non-interpenetration principle, allowing the overlapping of two pedestrians in a single cell, was investigated as a means to overcome this limit in a controlled and systematic way (Klüpfel). We adopted this approach, allowing pedestrians to transiently overlap with a small probability: at each time step (a maximum of) two pedestrians are allowed to stay on each cell.

$$State(c) = s : s \in \{FREE, OBSTACLE, ONE\_PED_i, TWO\_PEDS_{i,j}\}$$

State $TWO\_PEDS_{i,j}$ indicates that cell c is occupied at the same time by agent i and agent j. The set of admissible actions $AdmAct$ is modified allowing that also cells already occupied by one other pedestrian are admissible cells, for $j \in Pedestrians$:

$$AdmAct_p = \{a : a \in Act \wedge (State(a(c)) = FREE \vee State(a(c)) = ONE\_PED_j)\}$$

With this extension, we want to model the fact that in some situations, especially in high densities, pedestrians rotate their body to pass in tight spaces. So, densities higher than the limit of 6.25 m^{-2} are allowed, because of the maximum possible density is 12.5 m^{-2}, thought those parameters must be finely calibrated to prevent unreasonable (and not justified by empirical evidences) crowding conditions. Overlapping also influences the calculation of utility function $U_a(c)$, assigning a penalty if the overlapping occurs:

$$Ov(c) = \begin{cases} -1 & \textit{if } State(c) = ONE\_PED \\ 0 & \textit{otherwise} \end{cases}$$

Because of overlapping event can happen just in particular situation of densities, a trade-off function on the basis of the density value in the scenario is defined, managing the calibration of overlapping event k_{ov} according to contextual factors (essentially the local density):

$$Balance_{ov}(c) = \begin{cases} k_{ov} & \textit{if } DensF_{i,j} \geq \delta_{high} \\ k_{ov} + \delta_{high} - DensF_{i,j} & \textit{if } \delta_{low} \leq DensF_{i,j} < \delta_{high} \\ 0 & \textit{otherwise} \end{cases}$$

where $i = Row(c)$, $j = Col(c)$, $DensF_{i,j}$ is the value of density field in the cell c, δ_{high} and δ_{low} are the two density thresholds that regulate the activation of overlapping. Please note that a zero value for the κ_{ov} parameter does not mean that the overlapping comes without costs, on the contrary it means that the overlapping is *not* allowed. Therefore, the overall *Balance* function gradually makes the overlapping phenomenon more likely with the growth of the local density.

Cohesion for simple groups

A positive contribution to the evaluation of the utility of a given cell can be assigned whenever the movement towards that point of the environment is able to reduce the perceived distance from other members of a simple group. The overall reduction of distance is essentially an aggregation of the reduction of distance from all other members of the group; more formally the value of the C_a function for a given cell c, an agent a member of group G_a is defined as follows:

$$C_a(c) = \left[\left(\eta \cdot \sum_{a_i \in G_a} (DistFunction_{a,a_i}(c)) \right) \cdot 2 \right] - 1$$

where η is a normalisation factor that, along with numerical values, allows to translate the cohesion value into the range $[-1, 1]$, and *DistFunction* is a function that represents the gain of agent a with respect to agent a_i belonging to the same group G_a, moving into cell c. In the case of the evaluation of group cohesion, the perception of agents is expanded: every agent is able to perceive the members of the same group considering a distance parametric value g_d. *DistFunction* is so defined as:

$$DistFunction_{a,a_i}(c) = \frac{distance(Position(a), Position(a_i)) - distance(c, Position(a_i))}{size(G_a) - 1}$$

representing the gain that agent a obtains moving in a particular cell c with respect to agent a_i.

Inter group cohesion

The role of this component of the utility associated to a movement is to increase the desirability of choices that reduce the distance from the members of the structured group the agent belongs to (if any). The form of the function is therefore relatively similar to the one associated to the cohesion of simple groups, with a significant difference: the larger a group is the more difficult is to perceive its "centre" and also its direction. Moreover, ties between members of a large, possibly artificial group, are plausibly less influential than those binding members of a simple group. Therefore we decided to reduce the overall effect of cohesion for very large groups. We also support the definition of a hierarchical structure of groups and we exploit this structure when computing the value for this contribution to the overall utility of a given movement.

Considering an agent a, let us remind that by $\bar{G}_a$ we denote the largest group agent a belongs to; it could be a highly structured group including a subgroup, simple or structured, to which a directly belongs to, or it could even be equal to G_a, should a be member of a simple group not belonging to any structured group. Considering this notation, an $I_a(c)$ function for a given cell c, is defined as follows:

$$I_a(c) = \left[\left(\eta \cdot \sum_{a_i \in \bar{G}_a} (DistFunctionI\bar{G}_{a,a_i}(c)) \right) \cdot 2 \right] - 1$$

where function $DistFunctionI\bar{G}_{a,a_i}$ works on the tree-structure of the macro-group, identifying the proximity of two sub-groups G_a and G_{a_i} (i.e. the groups agents a and a_i directly belong to) in the tree of the group structure by means of the detection of the nearest common root of the two groups in $\bar{G}$.

More formally:

$$DistFunctionI\bar{G}_{a,a_i}(c) = \frac{1}{distance(c, Position(a_i))} \cdot \frac{1}{(Size(mcg(G_g, G_{a_i})) - 1)}$$

where mcg is the smallest sub-group of $\bar{G}_a$ including both G_a and G_{a_i}.

Adaptation mechanism for group cohesion preservation

While the above elements are sufficient to generate a simple pedestrian model that considers the presence of groups, even structured ones, the introduced mechanisms are not sufficient to preserve group cohesion, as discussed in a previous work adopting a very similar approach (Bandini et al. 2011). This is mainly due to the fact that in certain situations pedestrians adapt their behaviour in a more significant way than what is supported by simple and relatively small modifications of the perceived utility of a certain movement. In certain situations pedestrians perform an adaptation that appears in a much more decisive way a *decision*: they can suddenly seem to temporarily loose interest in what was previously considered a destination to reach and they instead focus on moving closer to (or at least do not move farther from) members of their group, generally whenever they feel that the distance from them has become excessive. In the following, we will discuss a metric of group dispersion that we adopted to quantify this perceived distance and then we will show how it can be used to adapt the weights of the different components of the movement utility computation to preserve group cohesion.

Group dispersion metrics

Intuitively, the dispersion of a group can be seen as the degree of spatial distribution of their members. In the area of pedestrian modelling and simulation, the estimation of different metrics for group dispersion has been discussed in (Bandini et al. 2011) in which different approaches are compared to evaluate the dispersion of groups through their movement in the environment. In particular, two different approaches are compared here: (i) dispersion as occupied area and (ii) dispersion as distance from the centroid of the group. This topic was also considered in the context of computer vision algorithms (Schultz et al. 2012), in which however essentially only *line abreast* patterns were analysed. Therefore we will focus on the former approach.

Formally, the above introduced formulas of group dispersion for each approach are defined as follows:

$$Disp(Group) = \frac{Area(Group)}{Size(Group)} \quad \text{(Area method)}$$

$$Disp(Group) = \frac{\sum_{i=1}^{Size(Group)} distance(centroid, a_i)}{Size(Group)} \quad \text{(Centroid method)}$$

with *Area*(*Group*) as the area occupied by the group, *Size*(*Group*) as the number of its members, *centroid* as its centroid. Results underline that the second approach suffers the effect of particular configurations in which the value of cohesion appears as low while a face validation of the situation indicates a good group cohesion. These wrong evaluations are detected in particular in medium and high-density situations in which groups tend to stretch themselves to walk through bottlenecks or narrow walkable areas. The centroid method identifies groups as highly disperse under these conditions, because some pedestrians can be far from the centre of the group.

Differently, the first metric, that can appear as more simple, defines the dispersion of the group as the portion of space occupied by the group with respect to the size of the group. Figure 2 illustrates how this metric works: the first step works on all the vertices (i.e. the members of the group, see Figure 2a), building a convex polygon with the minimum number of edges that contain all the vertices. The second step works on this output, calculating the area of the convex polygon (see Figure 2b). The dispersion value is calculated as the relationship between the polygon area and the size of the group.

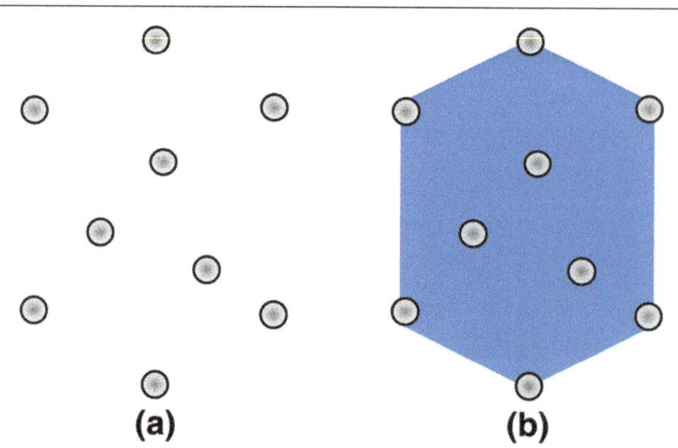

Figure 2 Graphical representation of a group composed of nine members (a) and the area of the convex polygon that contains all the group members (b).

Trade-off analysis

A trade-off between the goal attraction and the intra-inter group cohesion is necessary to preserve group cohesion: in the situation in which the spatial dispersion value is low, the cohesion tendency has to be less influential than the goal attraction on the overall pedestrian behaviour. On the contrary, if the level of dispersion of group is high, the cohesion component is more important than the goal attraction. An adaptation of the two related parameters in the utility computation is necessary, by means of a *Balance*(k) function that can be used to formalise these requirements:

$$Balance(k) = \begin{cases} \frac{1}{3} \cdot k + (\frac{2}{3} \cdot k \cdot DispBalance) & \text{if } k = k_c \\ \frac{1}{3} \cdot k + (\frac{2}{3} \cdot k \cdot (1 - DispBalance)) & \text{if } k = k_g \vee k = k_i \\ k & \text{otherwise} \end{cases}$$

where k_i, k_g and k_c are the weighted parameters $U_a(c)$ and

$$DispBalance = tanh\left(\frac{Disp(Group)}{\delta}\right)$$

is another function that works on the value of group dispersion as the relationship between the area and the size of the group, applying on it the hyperbolic tangent (assuring values in the $[0, 1)$ interval for the considered range of values for group dispersion). The value of δ is a constant that essentially represents a threshold above which the adaptation mechanism starts to become more influential; after a face validation phase, we set this value to 2.5, allowing the output of *DispBalance* function in the range $[0, 1]$ according to all elements in $U_a(c)$. The hyperbolic tangent approaches value 1 when $Disp(Group) \geq 6$ (values ≥ 6 indicate a high level of dispersion for small-medium size groups (1-4 members)).

A graphical representation of the trade-off mechanism is shown in Figure 3: red and green boxes represent the progress of parameter k_c and parameter k_g (k_i is treated analogously), respectively. Note that the increasing of the dispersion value produces an increment of k_c value and a reduction of k_g parameter.

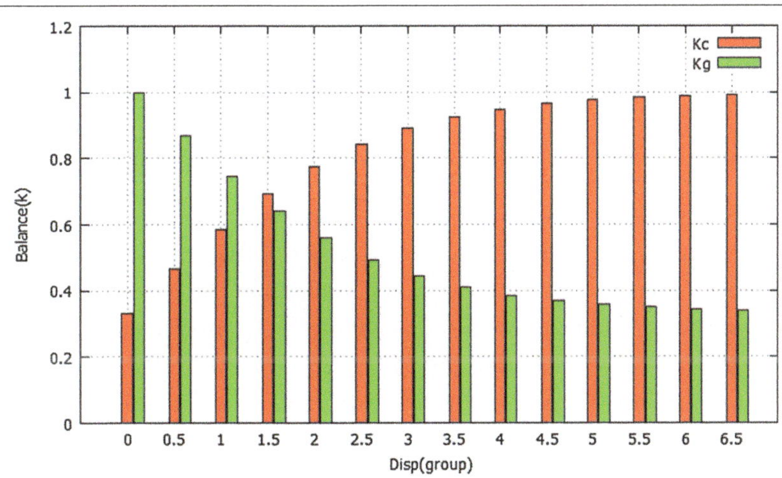

Figure 3 Graphical representation of *Balance(k)*, for $k = 1$ and $\delta = 2.5$.

Furthermore, according to (Xu and Duh 2010), the value of separation among group members has to be modified, on the basis of the assumption that pedestrians within a group allow to stay more close each other with respect to strangers: more in detail, the value of separation in a group is equal to the half among strangers.

The S function must therefore be substituted by a S_a function, considering these social aspects. In particular, this function is defined as follows:

$$S_a(c) = -\frac{DensF_{i,j} - SepGroup_a(c)}{MaxDensity}$$

where $SepGroup_a(c)$ provides the value to discount to the separation repulsion on the basis of the group to which the agent a belongs to (the case in which agent does not belong to a group is also expressed):

$$SepGroup_a(c) = \begin{cases} \sum\limits_{a_i \in G-\{a\}} \frac{1}{distance(a_i, c)^2} \cdot 0.5 & if\ a \in G \\ 0 & otherwise \end{cases}$$

It must be emphasised the fact that this adaptive balancing mechanism and the current values for its parameters were heuristically established and they actually require a validation (and plausibly a subsequent calibration) by comparing results achieved with this configuration and relevant empirical data about group dispersion gathered from actual observations and experiments in controlled situations.

Results and discussion

This section describes the results of a simulation campaign carried out to evaluate the performance of the above described model that had mainly two goals: (i) *validate* the model, in situations for which the adaptation mechanism was not activated (i.e. no simple groups were present in the simulated population), (ii) *evaluate* the effects of the introduction of simple groups, performing a qualitative face validation of the introduced adaptation mechanism considering available video footages of the behaviour of groups in real and experimental situations.

The chosen situations are relatively simple ones and they were chosen due to the availability of relevant and significant data from the literature. In particular, the first one is a linear scenario, a *corridor* in which we test the capability of the model to correctly reproduce a situation in which two groups of pedestrians enter from one of the ends and move towards the other. This situation is characterised by a counterflow causing local situations of high density and conflicts on the shared space. We essentially evaluate and validate this scenario by means of a fundamental diagram (Schadschneider et al. 2009): shows how the average velocity of pedestrians in a section (e.g. one of the ends of a corridor) varies according to the density of the observed environment. Since the flow of pedestrians is directly proportional to their velocity, this diagram is sometimes presented in an equivalent form that shows the variation of flow according to the density. In general, we expect to have a decrease in the velocity when density grows; the flow, instead, initially grows, since it is also directly proportional to the density, until a certain threshold value is reached (also called *critical density*), then it decreases. Despite being of great relevance, different experiments gathered different values of empirical data. Consensus on the shape of the function is wide, but the range of the possible values has even significant differences from different versions, as you can see in Figure 4 showing a set of different diagrams, both lines related to design manuals (namely Predtechenskii and Milinskiĭ 1978; Weidmann 1993) and also data points related to experimental measurements (respectively carried out in the context of the Hajj in Saudi Arabia by (Helbing et al. 2007) and in the city of Osaka, in Japan, by (Mori and Tsukaguchi 1987)).

The second scenario in which the model has been tested is instead associated to a situation in which pedestrians have to perform a turn and two different flows have to merge: geometrically, it is represented by a T-Junction also depicted in Figure 5, in which two branches of a corridor meet and form a unique stream. Also in this kind of situation it is not hard to reach high local densities, especially where the incoming flows meet to turn and merge in the outgoing corridor. The main goal of the analysis of the model behaviour in this scenario was the evaluation of its capability to generate patterns of spatial utilisation (i.e. aggregated density maps resulting from the single individual trajectories adopted by the pedestrians) that are in good agreement with those resulting from the actual observations available in the literature (Zhang et al. 2011).

We investigated both scenarios with a significant number of simulations, varying the level of density by adjusting the number of pedestrians present in the environment, so as to analyse different crowding situations. For every scenario, in terms of environmental configuration and level of density, a minimum of 3 and a maximum of 8 simulations were executed, according to the variability of the achieved results (more simulations were run when the variability was high, generally around levels of density close to the critical thresholds). Every simulation considered at least 1800 simulation turns, corresponding to 10 minutes of simulated time. The rationale was to observe a good number of complete paths of pedestrians throughout the environment, that was configured to resemble

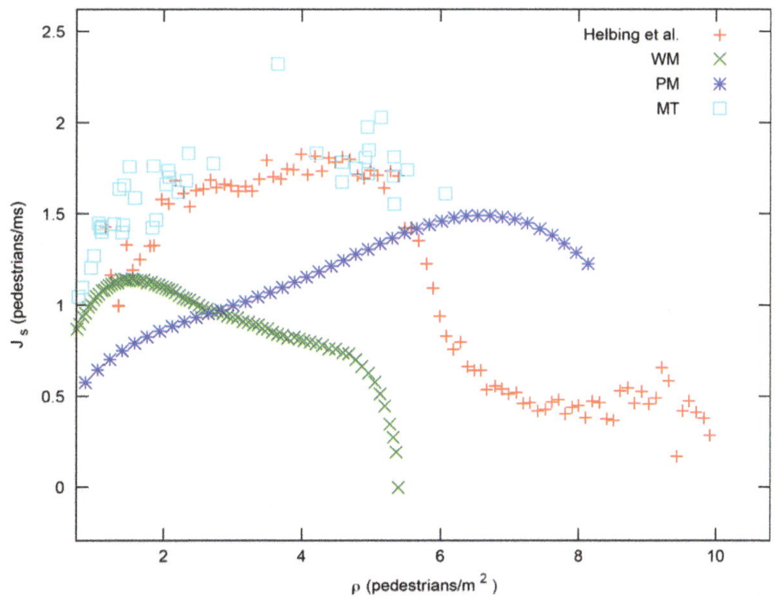

Figure 4 Empirical fundamental diagrams for pedestrian movement in planar facilities as reported by design manuals ((Weidmann 1993) – WM – and - (Predtechenskii and Milinskiĭ 1978) – PM) and also data points related to experimental measurements (respectively carried out in the context of the Hajj in Saudi Arabia by (Helbing et al. 2007) – Helbing – and in the city of Osaka, in Japan, by (Mori and Tsukaguchi 1987) – MT).

a torus (e.g. pedestrians completing a movement through it re-entered the scenario from their starting point), therefore simulations of situations characterised by a higher density were also set to last longer.

As suggested above, for both the scenarios, we adopted two different experiment settings: in the first one, the individual pedestrians belonging to a given flow (i.e. all the pedestrians entering the corridor or the T-junction from one end) are represented as members of a large structured group, but no simple groups are present. This first part of the experimentation was also necessary to perform a proper calibration of the model, for the parameters not involved in simple group modelling. In the second experimental setting, we included a variable number of simple groups (based on the total number of pedestrians in the environment and according to available data on the frequency of groups of different size in a crowd (Federici et al. 2012; Moussaïd et al. 2010))[e] first of all to calibrate and qualitatively validate the adequacy of the adaptation mechanism and then to explore its implications on the overall crowd dynamics.

The model has been implemented as an improvement of MAKKSim (Bonomi et al. 2011), an already existing simulation platform consisting of an extension to the Blender 3D modelling and rendering system. MAKKSim extends Blender by means of a set of Python scripts realising the model and simulation engine, in addition to support tools like algorithms for the semi–automatic generation of the fields starting from a CAD file and some specific spatial annotations defined through Blender's user interface. A complete description of this system and its performance is out of the scope of this paper and it is object of current and future works: we just conclude this introduction to the achieved results by clarifying that the performance of the MAKKSim simulation engine is strongly bound to the number of simulated pedestrians. Situations characterised by a low level of density required less than 5 minutes to be completed, whereas high density situations required more than two hours.

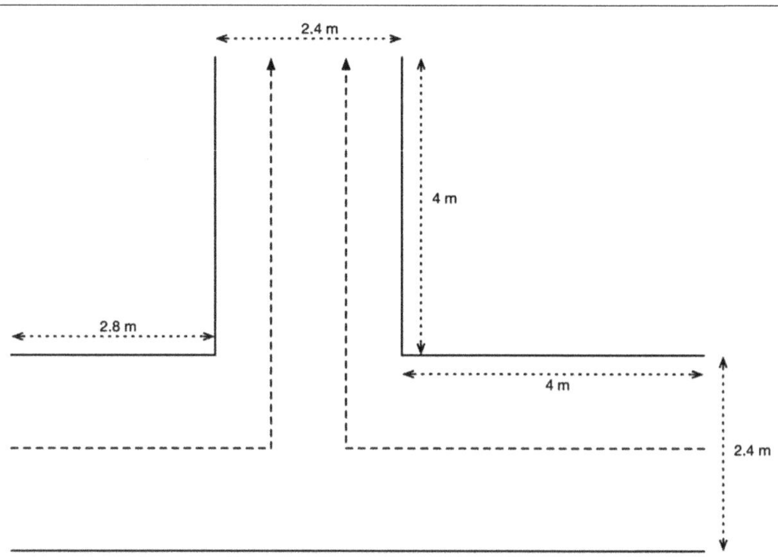

Figure 5 T-junction scenario.

Linear scenarios

This scenario proposes the analysis of the model performance when a bidirectional flow of pedestrians crossing a corridor is simulated. More in detail, the analysis of the flow is performed on three different linear scenarios with a variation of size in terms of width and height. The configurations taken into account are related to a linear corridor with size 2.4 m × 20 m (A), 3.6 m × 13.3 m (B) and 4.8 m × 10 m (C). Note that the variation in terms of width and height were applied according to the choice of maintaining the total area equal to 48 m^2 in every scenario. In particular, between the first and the second configuration the values of width and height were respectively multiplied and divided by 1.5, while between the first and the third configuration the values where respectively multiplied and divided by 2.

Figures 6, 7 and 8 show three fundamental diagrams, one for each environmental configuration, in which blue and red points respectively represent pedestrian flow without and with groups. Each point is associated with a simulation run of at least 10 minutes; in each run a pedestrian performs the crossing of the corridor a number of times, but when the density grows, in order to preserve the number of crossings, we need to increase the simulated time. Therefore, runs performed with higher density conditions had to be longer than those representing low density situations. Finally, we performed more runs in the density situations close to the critical threshold, since in those situations the results were characterised by the highest level of variability.

All the diagrams properly represent changes in the flow with respect to the level of density. The range for the critical value of density belongs to the interval [1.8 − 2.3] m^{-2} in the situation without group, in tune with experimental results and empirical data from the literature.

About the impact of groups, a variation of flow in case of groups with respect to the case without group has to be analysed. Considering charts in Figures 6, 7 and 8, it is possible to note that the level of critical density reached by the flow without groups is higher with respect to the flow with groups: in the latter, the value of critical density belongs to the

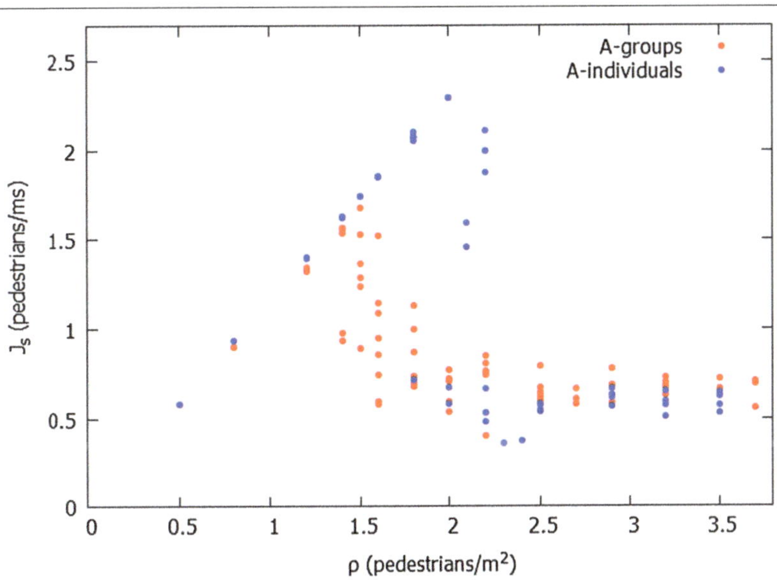

Figure 6 Fundamental diagrams with and without groups in corridor A.

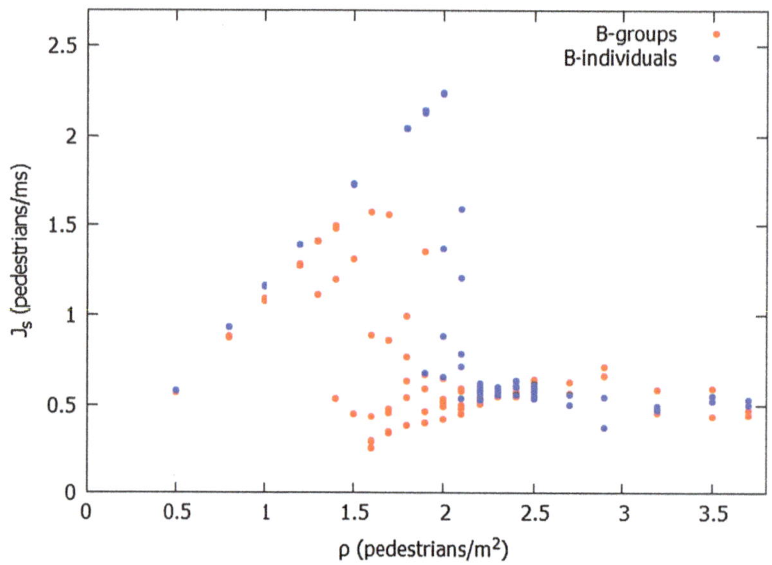

Figure 7 Fundamental diagrams with and without groups in corridor B.

interval [1.5 – 1.8] m^{-2}. This means that the flow without groups increases until values in the interval [2.0 – 2.2] ms^{-1} while the flow with groups until value in [1.5 – 2.0] ms^{-1}.

Considering these analyses, the presence of groups can be interpreted as a *negative factor* on the flow dynamics. This trend is maintained for a level of density $< 2.5m^{-2}$.

Differently, for higher densities (from $2.5m^{-2}$ to $4.0m^{-2}$), the presence of groups has a little impact on pedestrian flow: in Figures 6 and 7 it is possible to note that higher level of flow are assigned to situations with groups with respect to situations without groups. The behaviour in terms of speed and the detection of lanes explains the variation in the fundamental diagrams on the overall density interval. Also preliminary analyses on experimental data discussed in (Manenti et al. 2011) support these results, that might

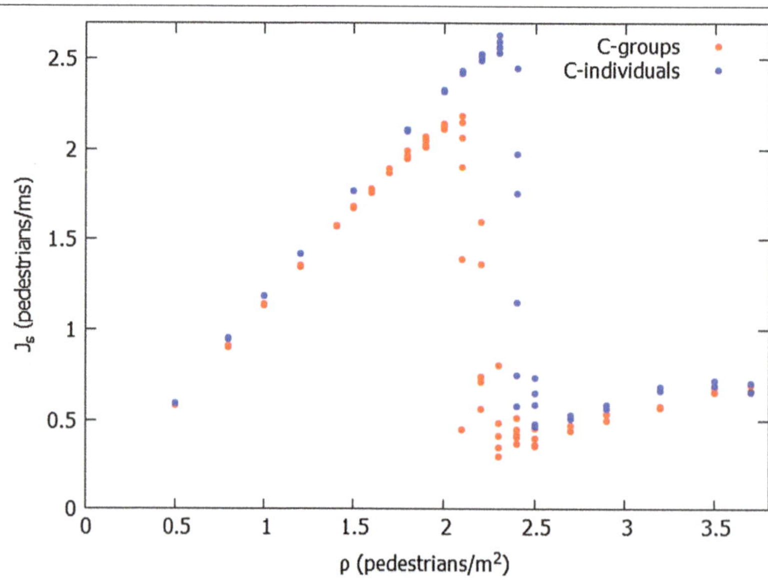

Figure 8 Fundamental diagrams with and without groups in corridor C.

seem counterintuitive, showing that the presence of groups and, in particular, of couples, can positively influence the pedestrian flow.

The impact of groups can be analysed more in detail considering the singular influence of every type of group (according to its size) to the pedestrian flow with the goal to understand if a relationship exists between the size of the group and its (negative) contribution to the overall pedestrian dynamics.

To achieve this goal, data related to the different types of simulated groups were aggregated, and a comparison among the related fundamental diagrams was performed. As summary, Figure 9 represents on the same chart all group contributions: unlike the previous diagrams, the depicted points (and the connecting line) represent the average flow achieved for that kind of group in the total simulated time.

The choice of evaluating the influence of groups in different linear scenarios was also inspired by (Zhang et al. 2011), in which a comparison in terms of pedestrians flow from experimental data among three corridors of width 1.8 m, 2.4 m and 3.0 m is presented. In this case, authors show that, in conformance with (Hankin and Wright 1958), above a certain minimum of about 1.22 m, the maximum flow is directly proportional to the width of the corridors.

We have chosen to expand the width of the corridors considering the comparison among sizes of 2.4 m, 3.6 m and 4.8 m. Figure 10 represents the three different fundamental diagrams in case of groups: these results are in tune with the above mentioned theory about the dependency between maximum reachable flow and the width of a corridor. In particular, in the case of corridor C, relevant variations can be detected around the critical density value. After that value, the influence of width on the pedestrian flow seems to be not relevant, probably due to the activation of overlapping extension to deal with high density situation (Section "Overlapping extension") and to the phenomenon of lane formation.

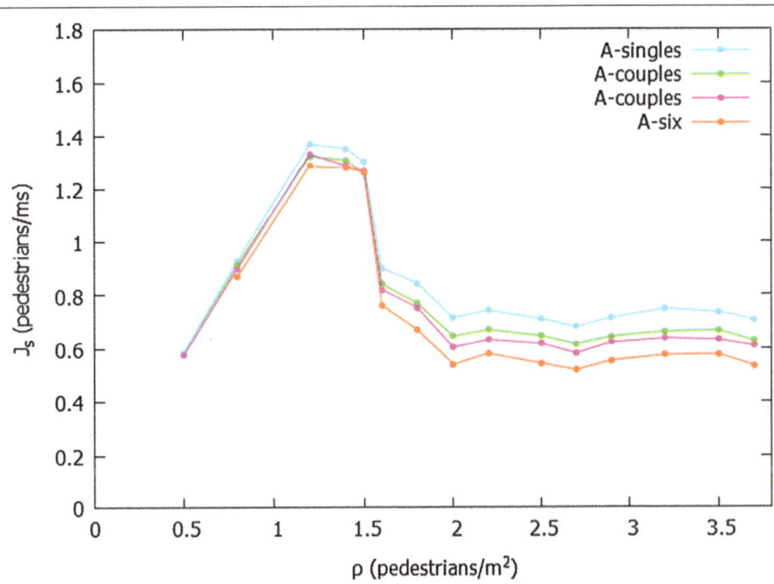

Figure 9 Comparison among fundamental diagrams on aggregate data with respect to groups of different size in corridor A.

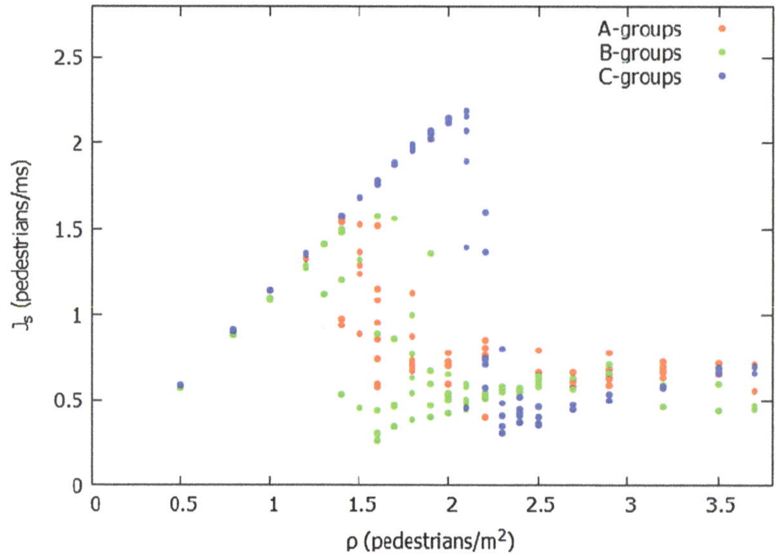

Figure 10 Comparison among corridor widths: fundamental diagrams in case of groups in corridors A, B and C.

Both in the linear and corner scenarios an analysis on the effectiveness of the mechanism to manage group cohesion is required: as previously introduced in Section "Trade-off analysis" a balance mechanism regulating the tendency to reach the goal and the tendency to stay cohesive with group members works on a metrics by evaluating group dispersion as ratio between the area occupied by the group and its size.

Finally, we analysed the relationship between the increase of the density and the level of group dispersion. The adopted indicator for the dispersion of groups is the one introduced in Section "Group dispersion metrics", that is, the area actually occupied by the group, computed as shown in Section "Group dispersion metrics". We found out that the density does not significantly influence the group cohesion: dispersion in couples is always in the interval $[0.2 - 0.4]\, m^2$, dispersion in triples in the interval $[0.9 - 1.3]\, m^2$ and dispersion in groups of six in the interval $[5.0 - 5.6]\, m^2$. This means that the mechanism to manage group cohesion works well, decreasing the tendency to reach the goal to preserve group cohesion. Figure 11 shows the comparison between dispersion values in groups of six in the three different linear scenarios. While in corridor B and C the dispersion of this kind of group is comparable, in corridor A the width definitely impacts on the variability on the dispersion of groups that belong to the interval $[5.1 - 7.1]\, m^2$. It must be noted that, in order to effectively validate the group cohesion mechanism we would need empirical data on the level of dispersion of groups in this kind of scenario and in different situations.

T-junction scenario

The central result of this scenario is not represented by fundamental diagram data, but rather by an indication on how the pedestrians employed the available space throughout the simulation. In particular, we adopted a metric is called *cumulative mean density* (CMD) (Castle et al. 2011), a measure associated to a given position of the environment indicating the average density perceived by pedestrians that passed through that point. It is quite straightforward to compute this value in a discrete approach like the one

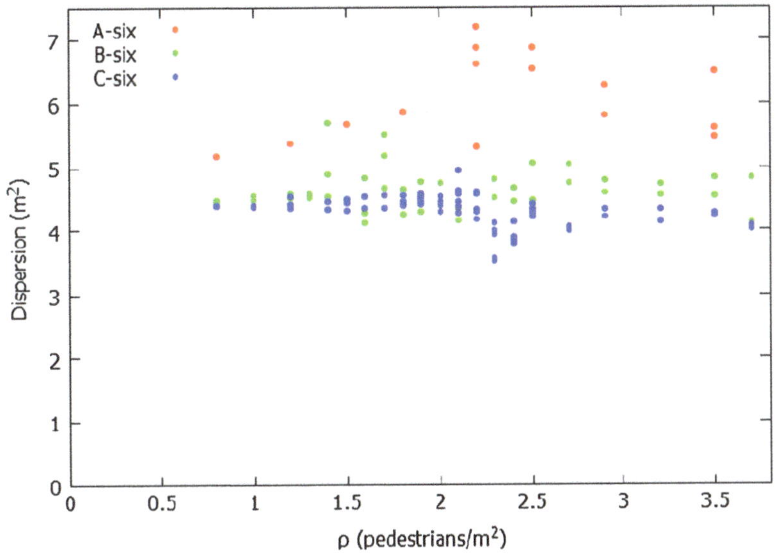

Figure 11 Comparison among the level of dispersions for group of six members in corridor A, B and C.

described in this work. As suggested above, we wanted to evaluate the capability of the model to reproduce patterns of spatial utilisation that are in good agreement with those resulting from the actual observations available in the literature (Zhang et al. 2011). In this work an analysis on spatial dynamics of the motion of pedestrians in an experimental observation carried out on a T-junction scenario is performed by means of topographical information for density profile according to CMD values. The density in the T-junction is not homogeneous and a higher density region appears near the junction. The lowest density region is located at a small triangle area, where the left and right branches begin to merge. The density in the branches (near to starting areas) are not uniform and are higher over the inner side. From this point of view, pedestrians prefer to move along the shorter and smoother path.

Our model is able to reproduce the phenomenon: a comparison between the two different density profiles from one run of a simulation with an overall density equal to $2.5m^{-2}$ is shown in Figure 12 (a). It can be noted that our model represents well the use of the space by pedestrians and their trajectories. The maximum CMD reaches is the same in the two density profile, and equal to $4.5m^{-2}$.

In Figure 12 (b), instead, we propose the results of the simulations carried out in the same scenario but introducing simple groups, in the same proportion of the previous

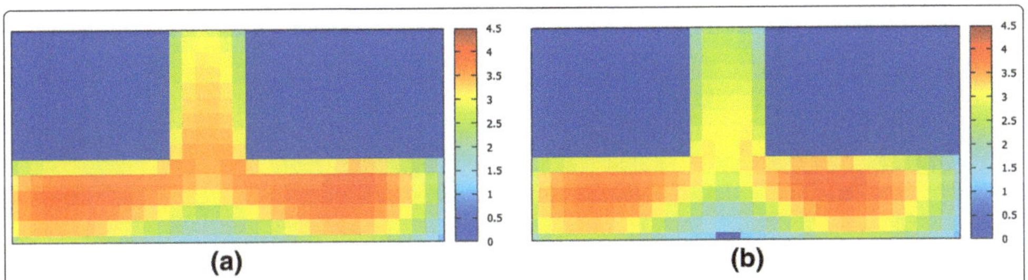

Figure 12 Density profile for simulations in the T-junction scenario without simple groups*(a)* and in presence of simple groups *(b)*.

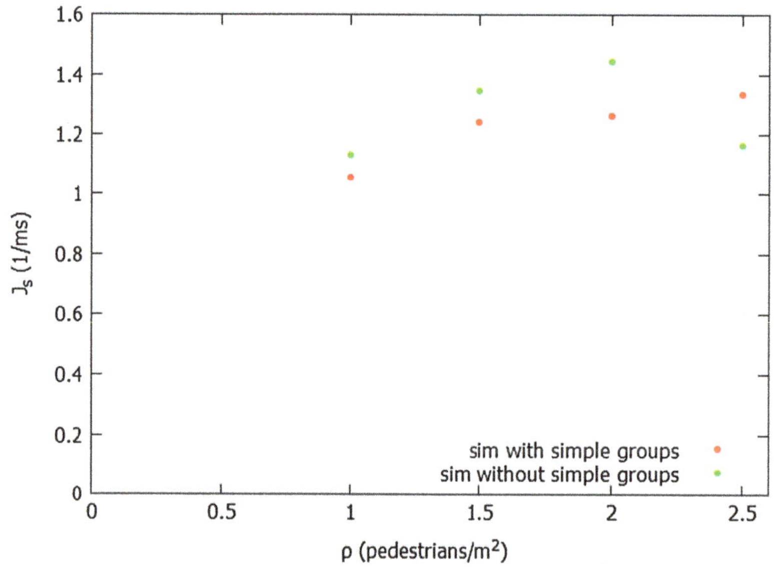

Figure 13 Comparison among the fundamental diagrams of experiments not including simple groups and including a set of simple groups of variable size.

scenario: while qualitatively the diagram looks similar to the previous one, the highest level of density is actually sensibly lower and in general the CMD is lower in most parts of the environment. This result is interesting because the lower level of CMD suggests that in turning and merging situations characterised by an overall high level of density the impact of simple groups could be significant and beneficial for the overall flow. We further analysed this phenomenon by performing a set of simulations varying the density of pedestrians in the environment and the resulting fundamental diagram is shown in Figure 13: in low to moderate densities the presence of simple groups slightly reduces the

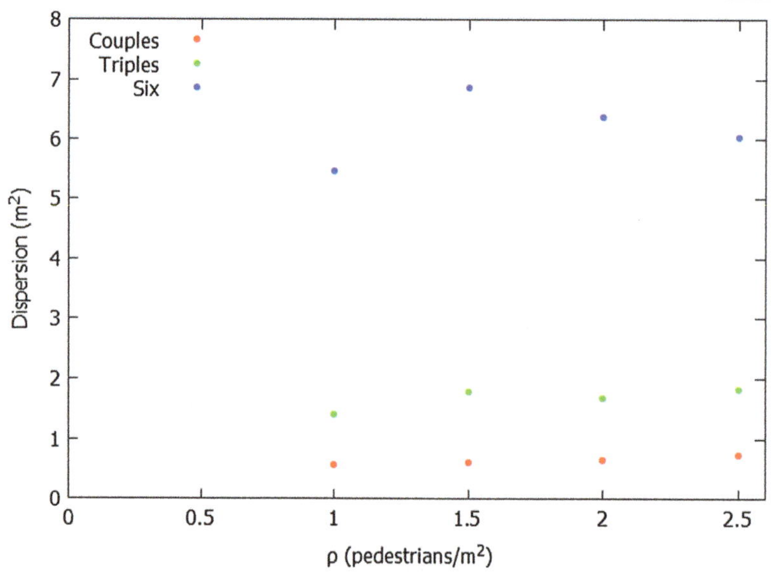

Figure 14 Comparison among the level of dispersions for groups of different size in the T-junction scenario in different density conditions.

overall flow of pedestrians, but in high density conditions the presence of simple groups actually makes the flow of pedestrian smoother than in the situation only including large structured groups. However, once again, it must be stressed that, in order to effectively validate the group cohesion mechanism, we would need empirical data also on the spatial patterns of movement in this kind of scenario and in different situations.

Finally, in Figure 14 we also report a graph showing the variations in the level of group dispersion in different situations characterised by a different level of density. As for the linear scenario, the level of dispersion does not seem to present a straightforward relationship with the level of density: it seems to increase at density levels close to the critical density, but then it seems to decrease to physiological levels (considering the size of the group).

Conclusions

The paper presented a model for the adaptive behaviour of pedestrians that are members of simple groups, that is, groups of friends and/or relatives. These pedestrians need to adapt their behaviour, namely reducing their goal orientation in order to preserve the cohesion of the group. The paper has formally introduced the model and discussed results that were compared to the available empirical data and relevant literature in absence of effects from the newly introduced mechanism and it has explored the effects of its introduction, after a face validation of its adequacy. The simulation results suggest that the effect of the presence of groups in the simulated population is not so simple to evaluate, since in the analysed situations the pedestrian flow is reduced due to the presence of groups only in low to moderate levels of density, while in high density situations the presence of groups might be even beneficial, smoothening the overall pedestrian flow.

Future works are aimed, on one hand, at achieving empirical data supporting an empirical validation and subsequent calibration of the group cohesion mechanism and then at further extending the model, for instance for representing the movement of elderly and special groups (like elderly and accompanying person), but also leader/follower schemes.

Endnotes

[a]see http://www.evacmod.net/?q=node/5 for a significant although not necessarily comprehensive list of simulation platforms. The list includes over 60 models, commercial and academic, general purpose or specifically targeted on certain situations and scenarios, maintained or discontinued.

[b]LEGION is the most famous and used commercial software for pedestrian dynamics simulation, see http://www.legion.com/legion-software.

[c]STEPS is an agent-based micro-simulation tool developed by Mott MacDonald for the simulation of pedestrian movement under both normal and emergency conditions, see http://www.steps.mottmac.com/.

[d]Note that LEGION works on continuous representation of environment, so the area is calculated as the circular area around the pedestrian.

[e]Groups of size 2 include about 28% of the total number of pedestrians, groups of size 3 about 24% and groups of size 6 about 12%.

Competing interests

The authors declare that they have no competing interests.

Authors' contributions

The work presented in this paper was carried out in collaboration between all authors. GV defined the research theme, suggested the basic approach to be adopted and extended in order to define the new model for pedestrian and crowd simulation in presence of simple and structured groups; he also supervised the project. GV, LM and LC designed the model. LC extended the MAKKSim simulation system to implement the defined model. GV, LM and LC designed the experiments, carried out the simulation campaign, analysed the data, interpreted the results and wrote the paper. All authors have contributed to, read and approved the final manuscript.

Acknowledgements

The authors would like express their sincere appreciation to Stefania Bandini, Katsuhiro Nishinari, Andreas Schadschneider and Armin Seyfried for many fruitful discussions and valuable hints. The authors would also like to thank the reviewers for for their careful reading of the paper and their helpful comments.

References

Bandini, S, Bonomi A, Vizzari G: **An analysis of different types and effects of asynchronicity in cellular automata update schemes.** *Nat Comput* 2012, **11**(2):277–287.

Bandini, S, Manzoni S, Vizzari G: **Situated cellular agents: a model to simulate crowding dynamics.** *IEICE Trans Inf Syst Spec Issues Cell Automata* 2004, **E87-D**(3):669–676.

Bandini, S, Rubagotti F, Vizzari G, Shimura K: **An agent model of pedestrian and group dynamics: experiments on group cohesion.** In *AI*IA Volume 6934 of* Lecture Notes in Computer Science. Edited by Pirrone, R, Sorbello F: Springer; 2011:104–116.

Batty, M: **Agent based pedestrian modeling (editorial).** *Environ Plann B: Plann Des* 2001, **28:**321–326.

Blue, VJ, Adler JL: **Cellular automata microsimulation of bi-directional pedestrian flows.** *Transportation Res Rec* 1999, **1678:**135–141.

Bonomi, A, Manenti L, Manzoni S, Vizzari G: **MAKKSim: Dealing with pedestrian groups in MAS-based crowd simulation.** In *WOA, Volume 741 of* CEUR Workshop Proceedings. Edited by Fortino, G, Garro A, Palopoli L, Russo W, Spezzano G; 2011:166–170. CEUR-WS.org.

Castle, C, Waterson N, Pellissier E, Le Bail S: **A comparison of grid-based and continuous space pedestrian modelling software: analysis of two UK train stations.** In *Pedestrian and Evacuation Dynamics*. Edited by Peacock, RD, Kuligowski ED, Averill JD. US: Springer; 2011:433–444.

Challenger, R, Clegg CW, Robinson MA: *Understanding Crowd Behaviours: Supporting Evidence*. University of Leeds: Tech. rep; 2009.

Costa, M: **Interpersonal distances in group walking.** *J Nonverbal Behav* 2010, **34:**15–26. [http://dx.doi.org/10.1007/s10919-009-0077-y]

Ezaki, T, Yanagisawa D, Ohtsuka K, Nishinari K: **Simulation of space acquisition process of pedestrians using proxemic floor field model.** *Physica A* 2012, **391**(1–2):291–299. [http://www.sciencedirect.com/science/article/pii/S0378437111006054]

Federici, ML, Gorrini A, Manenti L, Vizzari G: **Data collection for modeling and simulation: case study at the university of Milan-Bicocca.** In *ACRI Volume 7495 of* Lecture Notes in Computer Science. Edited by Sirakoulis, GC, Bandini S: Springer; 2012:699–708.

Fruin, J: **Designing for pedestrians.** In *Public Transportation United States*. Englewood Cliffs: Prentice Hal; 1992.

Gipps, P, Marksjö B: **A micro-simulation model for pedestrian flows.** *Math Comput Simul* 1985, **27**(2-3):95–105.

Hall, ET: *The Hidden Dimension*. New York: Anchor Books; 1966.

Hankin, B, Wright R: **Passenger flow in subways.** *J Oper Res* 1958, **9**(2):81–88.

Helbing, D, Johansson A, Al-Abideen HZ: **The dynamics of crowd disasters: an empirical study.** *Phys Rev E* 2007, **75:**046109.

Helbing, D, Molnár P: **Social force model for pedestrian dynamics.** *Phys Rev E* 1995, **51**(5):4282–4286.

Helbing, D, Schweitzer F, Keltsch J, Molnár P: **Active walker model for the formation of human and animal trail systems.** *Phys Rev E* 1997, **56**(3):2527–2539.

Henein, CM, White T: **Agent-based modelling of forces in crowds.** In *Multi-Agent and Multi-Agent-Based Simulation, Joint Workshop MABS 2004, New York, NY, USA, July 19, 2004, Revised Selected Papers, Volume 3415 of* Lecture Notes in Computer Science. Edited by Davidsson, P, Logan B, Takadama K: Springer–Verlag; 2005:173–184.

Kirchner, A, Klüpfel H, Nishinari K, Schadschneider A, Schreckenberg M: **Discretization effects and the influence of walking speed in cellular automata models for pedestrian dynamics.** *J Stat Mech: Theory Exp* 2004, **2004**(10):P10011.

Kirchner, A, Nishinari K, Schadschneider A: **Friction effects and clogging in a cellular automaton model for pedestrian dynamics.** *Phys Rev E* 2003, **67:**056122. [http://link.aps.org/doi/10.1103/PhysRevE.67.056122]

Klügl, F: **A validation methodology for agent-based simulations.** In *Symposium on Applied Computing*. Edited by Menezes, R, Viroli M: ACM Press; 2008:39–43.

Klüpfel, H: **A cellular automaton model for crowd movement and egress simulation.** *PhD thesis*, University Duisburg-Essen; 2003.

Kretz, T, Bönisch C, Vortisch P: **Comparison of various methods for the calculation of the distance potential field.** In *Pedestrian and evacuation dynamics 2008*. Edited by Rogsch, C, Klingsch W, Schadschneider A: Springer; 2010:335–346.

Kruse, L: *Changing Conceptions of Crowd Ming and Behaviour*. Berlin: Springer; 1986. chap. Conceptions of crowds and crowding.

Manzoni, S, Vizzari G, Ohtsuka K, Shimura K: **Towards an agent-based proxemic model for pedestrian and group dynamics: motivations and first experiments.** In *Proc. of 10th Int. Conf. on Autonomous Agents and Multiagent*

Systems – Innovative Applications Track (AAMAS 2011). Edited by Tumer, K, Yolum P, Sonenberg L, Stone P; 2011:1223–1224.
Manenti, L, Manzoni S, Vizzari G, Ohtsuka K, Shimura K: **Towards an agent-based proxemic model for pedestrian and group dynamic.** In *WOA , Volume 621 of CEUR Workshop Proceedings*. Edited by Omicini, A, Viroli M, CEUR-WS.org; 2010. [http://ceur-ws.org/Vol-621/paper02.pdf]
Manenti, L, Manzoni, S, Vizzari G, Ohtsuka K, Shimura K: **An agent-based proxemic model for pedestrian and group dynamics: motivations and first experiments.** In *MABS, Volume 7124 of Lecture Notes in Computer Science*. Edited by Villatoro, D, Sabater-Mir J, Sichman JS: Springer; 2011:74–89.
Mori, M, Tsukaguchi H: **A new method for evaluation of level of service in pedestrian facilities.** *Transportation Res Part A* 1987, **21**(3):223–234.
Musse, SR, Thalmann D: **Hierarchical model for real time simulation of virtual human crowds.** *IEEE Trans Vis Comput Graph* 2001, **7**(2):152–164.
Moussaïd, M, Perozo N, Garnier S, Helbing D, Theraulaz G: **The walking behaviour of pedestrian social groups and its impact on crowd dynamics.** *PLoS ONE* 2010, **5**(4):e10047. [http://dx.doi.org/10.1371%2Fjournal.pone.0010047]
Nishinari, K, Kirchner A, Namazi A, Schadschneider A: **Extended floor field CA model for evacuation dynamics.** *IEICE Trans Inf Syst* 2004, **87**(3):726–732.
Paris, S, Donikian S: **Activity-driven populace: a cognitive approach to crowd simulation.** *IEEE Comput Graph Appl* 2009, **29**(4):34–43.
Predtechenskii, V, Milinskiĭ: *Planning for Foot Traffic Flow in Buildings*. New York: Amerind Publishing; 1978.
Qiu, F, Hu X: **Modeling group structures in pedestrian crowd simulation.** *Simul Model Pract Theory* 2010, **18**(2):190–205.
Rodrigues, RA, de Lima Bicho A, Paravisi M, Jung CR, Magalhães LP, Musse SR: **An interactive model for steering behaviors of groups of characters.** *Appl Artif Intell* 2010, **24**(6):594–616.
Sarmady, S, Haron F, Talib AZH: **Modeling groups of pedestrians in least effort crowd movements using cellular automata.** In *Asia International Conference on Modelling and Simulation*. Edited by Al-Dabass, D, Triweko R, Susanto S, Abraham A: IEEE Computer Society; 2009:520–525.
Schadschneider, A, Kirchner A, Nishinari K: **CA approach to collective phenomena in pedestrian dynamics.** In *Cellular Automata, 5th International Conference on Cellular Automata for Research and Industry, ACRI 2002, Volume 2493 of* Lecture Notes in Computer Science. Edited by Bandini, S, Chopard B, Tomassini M: Springer; 2002:239–248.
Schadschneider, A, Klingsch W, Klüpfel H, Kretz T, Rogsch C, Seyfried A: **Evacuation dynamics: empirical results, modeling and applications.** In *Encyclopedia of Complexity and Systems Science*. Edited by Meyers, RA: Springer; 2009:3142–3176.
Schreckenberg, M, Sharma SD (Eds): *Pedestrian and Evacuation Dynamics*. Berlin Heidelberg: Springer; 2001.
Schultz, M, Rößger L, Fricke H, Schlag B: **Group dynamic behavior and psychometric profiles as substantial driver for pedestrian dynamics.** In *Pedestrian and Evacuation Conference (PED2012)*; 2012. [http://arxiv.org/abs/1210.5553]
Schultz, M, Schulz C, Fricke H: **Passenger dynamics at airport terminal environment.** In *Pedestrian and Evacuation Dynamics*. Edited by Klingsch, WWF, Rogsch C, Schadschneider A, Schreckenberg M. Berlin Heidelberg: Springer; 2010:381–396.
Shao, W, Terzopoulos D: **Autonomous pedestrians.** *Graphical Models* 2007, **69**(5-6):246–274.
Shiwakoti, N, Sarvi M, Rose G, Burd M: **Biologically inspired modeling approach for collective pedestrian dynamics under emergency conditions.** *Transportation Res Rec* 2196, **2010:**176–184.
Still, G: **Crowd dynamics.** 2000. *PhD Thesis* [http://www.gkstill.com/CV/PhD/]
Tsai, J, Fridman N, Bowring E, Brown M, Epstein S, Kaminka GA, Marsella S, Ogden A, Rika I, Sheel A, Taylor ME, Wang X, Zilka A, Tambe M: **ESCAPES - evacuation simulation with children, authorities, parents, emotions, and social comparison.** In *Proc. of 10th Int. Conf. on Autonomous Agents and Multiagent Systems – Innovative Applications Track (AAMAS 2011)*. Edited by Tumer, K, Yolum P, Sonenberg L, Stone P; 2011:457–464.
Vizzari, G, Manenti L, Ohtsuka K, Shimura K: **An agent-based approach to pedestrian and group dynamics: experimental and real world scenarios.** In *Proceedings of the 7th International Workshop on Agents in Traffic and Transportation*; 2012. [http://www.ia.urjc.es/att2012/papers/att2012_submission_1.pdf]
Was, J: **Crowd dynamics modeling in the light of proxemic theories.** In *ICAISC (2) Volume 6114 of Lecture Notes in Computer Science*. Edited by Rutkowski, L, Scherer R, Tadeusiewicz R, Zadeh LA, Zurada JM. Berlin Heidelberg: Springer; 2010:683–688.
Weidmann, U: **Transporttechnik der Fussgänger - Transporttechnische Eigenschaftendes Fussgängerverkehrs (Literaturstudie).** 1993. Literature Research 90, Institut füer Verkehrsplanung, Transporttechnik, Strassen- und Eisenbahnbau IVT an der ETH Zürich. [www.ped-net.org/uploads/media/weidmann-1993_01.pdf]
Weng, WG, Chen T, Yuan HY, Fan WC: **Cellular automaton simulation of pedestrian counter flow with different walk velocities.** *Phys Rev E* 2006, **74:**036102.
Willis, A, Gjersoe N, Havard C, Kerridge J, Kukla R: **Human movement behaviour in urban spaces: implications for the design and modelling of effective pedestrian environments.** *Environ Plan B* 2004, **31**(6):805–828.
Xu, S, Duh HL: **A simulation of bonding effects and their impacts on pedestrian dynamics.** *IEEE Trans Intell Transportation Syst* 2010, **11:**153–161.
Zhang, J, Klingsch W, Rupprecht T, Schadschneider A, Seyfried A: **Empirical study of turning and merging of pedestrians streams in T-junction.** *arXiv preprint arXiv:1112.5299* 2011. [http://arxiv.org/abs/1207.5688]
Zhang, J, Klingsch W, Schadschneider A, Seyfried A: **Transitions in pedestrian fundamental diagrams of straight corridors and T-junctions.** *J Stat Mecha: Theory Exp* 2011, **2011**(06):P06004.

Optimal surveillance network design: a value of information model

Matteo Convertino[1,2,3,4*], Yang Liu[1] and Haejin Hwang[1]

*Correspondence: matteoc@umn.edu
[1] School of Public Health, Environmental Health Sciences Division and Public Health Informatics Program, HumNat Lab, University of Minnesota, Minneapolis, MN, USA
[2] Institute on the Environment, University of Minnesota, Minneapolis, MN, USA
Full list of author information is available at the end of the article

Abstract

Purpose: Infectious diseases are the second leading cause of deaths worldwide, accounting for 15 million deaths – that is more than 25% of all deaths – each year. Food plays a crucial role, contributing to 1.5 million deaths, most of which are children, through foodborne diarrheal disease alone. Thus, the ability to timely detect outbreak pathways via high-efficiency surveillance system is essential to the physical and social well being of populations. For this purpose, a traceability model inspired by wavepattern recognition models to detect "zero-patient" areas based on outbreak spread is proposed.

Methods: Model effectiveness is assessed for data from the 2010 Cholera epidemic in Cameroon, the 2012 foodborne Salmonella epidemic in USA, and the 2004-2007 H5N1 avian influenza pandemic. Previous models are complemented by the introduction of an optimal selection algorithm of surveillance networks based on the Value of Information (VoI) of reporting nodes that are subnetworks of mobility networks in which people, food, and species move. The surveillance network is considered the response variable to be determined in maximizing the accuracy of outbreak source detections while minimizing detection error. Surveillance network topologies are selected by considering their integrated network resilience expressing the rewiring probability that is related to the ability to report outbreak information even in case of network destruction or missing information.

Results: Independently of the outbreak epidemiology, the maximization of the VoI leads to a minimum increase in accuracy of 40% compared to the random surveillance model. Such accuracy is accompanied by an average reduction of 25% in required surveillance nodes with respect to random surveillance. Accuracy in systems diagnosis increases when system syndromic signs are the most informative in a way they reveal linkages between outbreak patterns and network transmission processes.

Conclusions: The model developed is extremely useful for the optimization of surveillance networks to drastically reduce the burden of food-borne and other infectious diseases. The model can be the framework of a cyber-technology that governments and industries can utilize in a real-time manner to avoid catastrophic and costly health and economic outcomes. Further applications are envisioned for chronic diseases, socially communicable diseases, biodefense and other detection related problems at different scales.

Keywords: Value of information; Surveillance network; Transmission network; Network resilience; Topology; Uncertainty; Biosafety; Biodefense; Outbreak sources

Background

Surveillance and uncertainty

Increased movements of people, and expansion of international trade in food and other commodities, enhanced by social and environmental changes linked to urbanization, are all manifestations of the rapidly-changing nature of the world we live in. Such changes contribute to the rapid adaptation and movement of microorganisms, which has facilitated the return of old communicable diseases, the emergence of new ones, and the evolution of antimicrobial resistance. Food, is in many aspect the "connectome" of most issues related to infectious diseases and for this motivation experts are now talking about "emerging foodborne infectious diseases" whose complex dynamics embraces a larger spectrum of than the one traditionally considered for infectious diseases. The majority of the pathogens causing the significant foodborne disease burden - estimated as -1.9 million people annually at the global level - are now considered to be zoonotic (Schlundt et al. 2004). For this motivation a one-health approach is now embraced (NAS 2012). To put it more simply, food is in many cases involved in many infectious disease propagations as the vehicle of pathogen spread in the transmission chain (e.g., environment-animal-food-humans) leading to infections (e.g. for cholera and tuberculosis) beyond the classical foodborne illnesses where there is no human-human transmission. As for foodborne diseases, in USA result in 37.2 million illnesses, 228,744 hospitalizations, and 2,612 deaths each year (Batz et al. 2011; Hanson et al. 2012; NAS 2012; Scallan et al. 2011a, 2011b). Foodborne illnesses can be caused by a variety of microbial pathogens, chemicals, and parasites that contaminate food at different points in the food production and preparation process. Although most of these diarrheal deaths occur in poor countries, foodborne diseases are neither limited to developing countries nor to children. In general, infectious diseases other than foodborne are a huge problem also for creating susceptibility to chronic diseases. Because of the necessity to decrease such burden, surveillance is the primary public health strategy to emphasize, but a set of questions arise spontaneously.

What is the optimal set of surveillance nodes that best inform about the spread of an epidemic accurately? Is it possible to gauge the shape of transmission networks based on epidemic outbreaks? The accuracy in such estimations is a function of the source of information reliability. This means that a node, or a set of nodes performing surveillance, can report or not report whether a case of an infectious disease case is observed. Such information tremendously affects the estimation of outbreak path and sources. Thus, the functioning of surveillance networks is crucial for early detection and rapid response of epidemics which is possibly the only way to respond to emerging/re-emerging infectious diseases for which we are unprepared (Stachenko 2008). The function of surveillance networks is the ability to observe epidemic information (outbreak occurrence) moving along the mobility network. Therefore, the design of surveillance networks via the maximization of reporting accuracy (i.e., the accuracy in surveillance function) is a key process for saving lives of human and animal populations (Helbing et al. 2014). Economical and environmental outcomes are also associated to such optimal design: for instance industries and government costs related to commodity recalls and health care, and footprints on the environment related to pathogen and/or chemical spills can be highly reduced. To face such interconnected problems systemic risk approaches are welcome in order to design globally resilient systems (Haldane and May 2011; Helbing 2013; Helbing et al. 2014; Park et al. 2012).

For surveillance design, the schematization of surveillance as a network is very convenient and realistic (Figure 1). Nodes are sites where information about cases is observed and reported and these nodes may share information in real-time. The network scheme is related to the simplification of the system into a physical structure (the network) that is much simpler and controllable than the whole "3D domain" that is unnecessary to simulate. The mathematical problem of network search is easier than the one in which nodes are considered as separate nodes. The realism of the network scheme is related to the fact that surveillance - that is a network function - is occurring on a subset of the transmission network. Yet, surveillance and transmission networks are multiplex networks (Estrada and Gómez-Gardeñes 2014; Newman 2003) in which information and pathogens are spreading, respectively. The backbone of such networks are mobility networks of people (e.g., subway, train, airline networks, etc), food (e.g., supply chains), wild animals (e.g., migration routes), and environmental vehicle networks (e.g., watercourses and urban drainage networks) that connect portions of ecosystems and countries worldwide by contributing to the spread of infectious diseases (den Broeck et al. 2011; Goncalves et al. 2013; Knobler et al. 2006; On Effectiveness of National Biosurveillance Systems: BioWatch C, the Public Health System NRC 2011; Pastore Y Piontti et al. 2014). Thus, it is important to implement optimal surveillance on mobility networks (that can be transmission networks when a pathogen spread) via personnel and/or sensors whose reported information can be integrated into a cyber-infrastructure that supports public health authorities and industries.

The consideration of deep uncertainty in reported outbreaks is a very rarely addressed topic in epidemiology. Recently large attention has been placed on detecting outbreak sources (Convertino and Hedberg 2014a; Convertino and Hedberg 2014b; Pinto et al. 2012). However, in these studies uncertainty was neglected in reported outbreaks. Convertino and Hedberg (Convertino M, Hedberg C: Multisite outbreak detection: how many foodborne outbreak are attributable to farmerõs markets?, submitted) considered

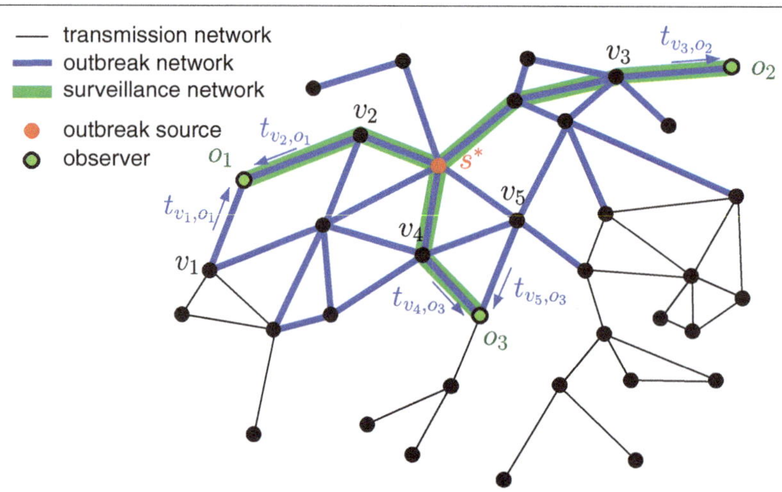

Figure 1 Schematic framework of the network based surveillance system. The reporting and non-reporting vertices are indicated with *o* and *v*, respectively. *s** is the outbreak source. Note that not all nodes in the surveillance network need to have reporting ability. *t* is the traveling time of the outbreak among nodes. The ideal situation is to have an anticipation of epidemic arrivals for rapid response to the epidemic rather than absent or late reporting of epidemic arrivals that translates into a certain, possibly high magnitude outbreaks. The underlying template network is taken from (Pinto et al. 2012).

large variability in food trade network factors to capture the deep uncertainty related to attribution of foodborne outbreaks to different local food establishments (versus food imported globally), with particular focus on farmers' markets. While this approach considered deep uncertainty, reported outbreaks were considered as "perfect information". Other studies, such as (Pinto et al. 2012) considered random and non-random placement of surveillance observers which partially captures reporting uncertainty; nonetheless these studies did not formulate any optimization algorithm maximizing detection accuracy even in presence of uncertainty.

Moreover, outbreak uncertainty is certainly biased, thus (Pinto et al. 2012)'s and others' approaches, based on regular schemes of surveillance underestimated such bias related to the anisotropic spreading of epidemics. In addition to such considerations, (Pinto et al. 2012)'s approach considered the problem of surveillance and outbreak source detection as coupled problems. In this study and in (Convertino and Hedberg 2014a) it is shown that such problem can be mathematically decoupled and the optimization of surveillance implies the optimal detection of outbreak sources but not vice versa. Pinto et al. (2012)'s model, analogous to the method used by telecommunication towers to pinpoint cell phone users, focused on arrival times of outbreaks - thought as information arrivals - that are correctly reproduced by the correct definition of outbreak sources. Here, we propose a simpler model, based on the original idea of (Brockmann and Helbing 2013) and subsequently developed by (Convertino and Hedberg 2014a; Convertino and Hedberg 2014b; (Convertino M, Hedberg C: Multisite outbreak detection: how many foodborne outbreak are attributable to farmerõs markets?, submitted)) and (Manitz et al. 2014), that considers the minimization of the error in the distance from real outbreak multi-sources with the most informative surveillance networks. All explored surveillance networks correspond to different reported outbreak patterns. Other approaches for designing surveillance have been developed in the past; for instance (Bajardi et al. 2012) developed a model for robust outbreak cluster detection based on network features. While such model has an optimization component, deep uncertainty in outbreak reporting (that is uncertainty related to report cases at a selected site and time) is not considered and the detection of clusters do not detect outbreak sources. The optimal surveillance network maximizes outbreak information, or equivalently minimizes the uncertainty related to the average distance between real and predicted outbreak sources.

As a side note we want to emphasize that outbreak sources are oftentimes confused with "onset areas" which are areas where first cases occur. Those areas are not necessarily where the first contamination of the environment and/or human infection occur considering environmental and social determinants that contribute to the spread of the pathogen in space. However, considering the typical short range of dispersal of most pathogens (Mundt et al. 2009), onset areas tend to coincide with outbreak sources. In some cases one of the onsets is the outbreak source and all other onsets are generated by epidemic propagation. For other outbreaks in which vehicle or vector (e.g, wild birds, food, and people via airplanes) of pathogens travel for long distances the proposed model works well because of the large resolution at which outbreaks are investigated and because of the uniqueness of the outbreak source. We argue that the algorithm of (Pinto et al. 2012) is more likely to fail in detecting real outbreak sources because it does not consider uncertainty in reported outbreaks, thus the implemented matching function of the observed delay in reported information detects onset areas. Certainly, the definition of outbreak sources is always dependent on the spatial and temporal scale of the system considered and care has to be placed when planning control strategies based on supposed outbreak sources.

Proposed approach

The model here proposed is explicitly based on a multiobjective optimization in which the value of information of reporting nodes is maximized. This maximization determines a minimization of the uncertainty related to the detection of the correct outbreak source. In order to test the model independently of the epidemiological dynamics determined by the interplay of socio-environmental and biological factors, we consider three well-known epidemics with different geographic range, pathogen features, and transmission routes. In this way we focus on macro-linkages between patterns and process on networks that allows us to maximize detection via an optimally design surveillance. Figure 2 shows the major steps of the proposed algorithm. Specifically, the novelties of the integrated model are the following.

- Global Sensitivity and Uncertainty Analyses (GSUA) (Convertino et al. 2014a; (Convertino M, S L, M A, Morris S: Importance, Interaction, and Scale-dependence of Cholera Outbreak Drivers: Metamodeling Predictions, submitted); Saltelli et al. 2008) as a method for evaluating the systemic uncertainty in reported outbreaks for the determination of outbreak sources, importance and interaction of transmission network variables. GSUA fully attributes uncertainty to any input factor of the model via probability distributions, and such uncertainty is propagated to study how it

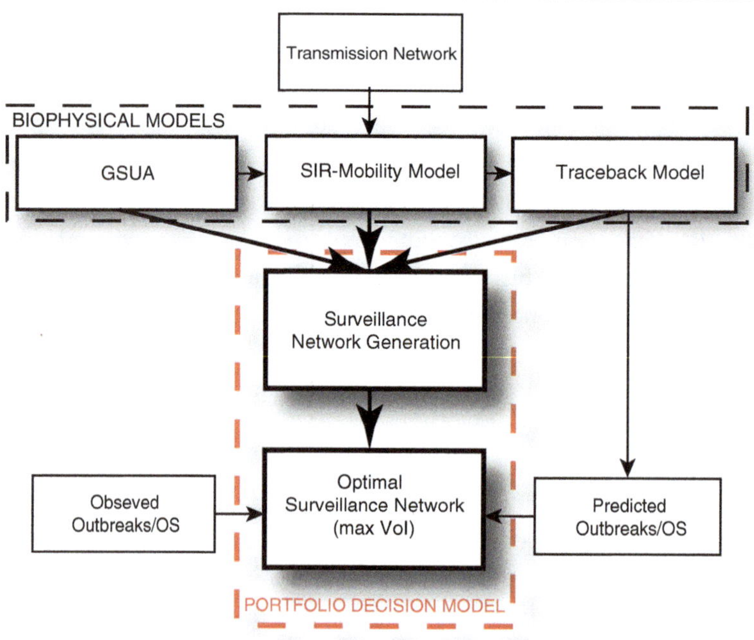

Figure 2 Integrated modeling framework. Transmission networks are identified and used in a coupled epidemiological-mobility model (where mobility can be of people, vehicles and vectors), the traceback model, and global sensitivity and uncertainty analyses model. These are models related to the physics of the problem. A portfolio decision model considers all potential surveillance network topologies and detects the one that maximize the VoI calculated based on the comparison of all predicted outbreaks and observed ones.

affects the uncertainty of the output that is the effective distance from observed outbreak sources (Brockmann and Helbing 2013).

- Value of Information (VoI) portfolio model (Convertino and Valverde 2013; Convertino et al. 2014c, 2014d; Trainor-Guitton et al. 2012) with Pareto optimization for the design of optimal surveillance networks by selecting observers in network topologies with the highest information for detecting outbreak sources.
- Network based (Brockmann and Helbing 2013; Convertino and Hedberg 2014a; Newman 2003) versus random node surveillance design, and effective distance based prediction of outbreak spread with source detection. The effective distance implementation (Brockmann and Helbing 2013) does require only the knowledge of outbreak occurrence versus the number of cases.

Materials

Case-study outbreaks

The choice about outbreaks considered in this study is related to the objective of capturing the salient links between epidemic patterns and processes regardless of disease etiology that is a result of peculiar epidemiological dynamics in terms of socio-environmental drivers, velocity, and geographical coverage (Convertino et al. 2009). This is with the aim of a general theory and cyber-infrastructure development for automated surveillance (Convertino and Hedberg 2014a). The goal is to show how observed outbreak patterns can be analyzed with the same physical-based models to detect outbreak sources and design surveillance networks considering the underlying transmission networks; thus, linking structure and functions of networks (Newman 2003) related to "invisible" outbreak dynamics over space and time. We chose among the fastest and high incidence rate epidemics: the 2004-2007 H5N1 pandemic (Kilpatrick et al. 2006), the 2010 cholera epidemic in Cameroon ((Convertino M, S L, M A, Morris S: Importance, Interaction, and Scale-dependence of Cholera Outbreak Drivers: Metamodeling Predictions, submitted); Guevart et al. 2006; Njoh 2010; Tatah et al. 2012), and the 2012 Salmonella epidemic in USA due to contaminated tuna (GATS 2014a). 34%, 27%, and 12% is the incidence rate for H5N1, cholera, and Salmonella, respectively. The estimated traveling velocity of these epidemics is 11, 22, and 122 km/month, respectively. The velocity is estimated using epidemic arrival times and distance covered over time. Additional file 1 report more information about the outbreak considered, and Figure 3 shows the observed outbreak patterns and transmission networks.

Mobility and surveillance networks

Mobility networks are the backbone of both transmission and surveillance networks that are both subnetworks of the mobility network. Mobility networks can be related to the mobility of vehicles, vectors, or pathogens themselves. For salmonella in tuna and H5N1 in humans we consider the food and human mobility network that are the primary networks responsible for the transmission of the pathogen. For cholera, previous studies such as (Pinto et al. 2012), have considered only the river network; however, considering the higher importance of human mobility in the transmission of cholera infections here we take into account human mobility and water-flow networks, simultaneously. However, only one network is enough for detecting outbreak sources. Surveillance networks

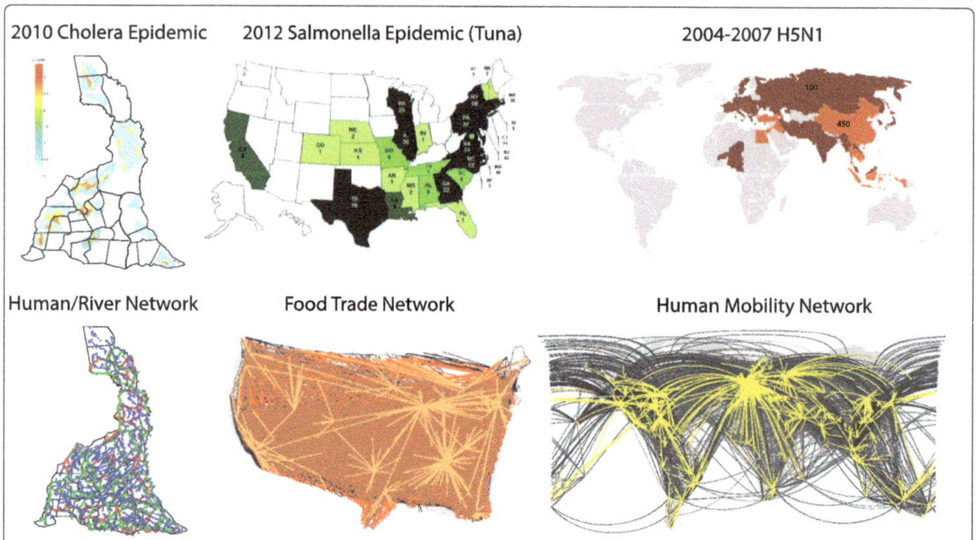

Figure 3 Epidemic considered for the validation of the surveillance system design algorithm. The top plots show the emerged predicted outbreak patterns determined by epidemic invasion fronts driven by food and human mobility in which food and people acted as a vehicles and vectors of transmission respectively. As for cholera (data from (Convertino M, S L, M A, Morris S: Importance, Interaction, and Scale-dependence of Cholera Outbreak Drivers: Metamodeling Predictions, submitted)) and H5N1 (WHO data), water and birds are also vehicle and vector, respectively, contributing to the spreading of the pathogen. Salmonella epidemic data are from CDC. Here, both water networks and bird mobility are neglected since the search is focused on human-manageable network on which perform surveillance. The bottom plots are the transmission networks of the epidemics considered. For cholera, transmission networks are based on (Convertino M, S L, M A, Morris S: Importance, Interaction, and Scale-dependence of Cholera Outbreak Drivers: Metamodeling Predictions, submitted). The food network is built using the radiation model (Convertino and Hedberg 2014a; Simini et al. 2012) and the human mobility network is based on airline data (www.oag.com).

can be though to coincide with human mobility networks rather than with environmental networks unless sensors are placed along those.

Human and food mobility network fluxes are estimated using the radiation model of (Simini et al. 2012) (Additional file 1) that is based on the population distribution. The population dataset is from the Web sites of the Gridded Population of the World and the Global Urban-Rural Mapping projects (CIESIN 2014; GRUMP 2014), which are run by the Socioeconomic Data and Application Center (SEDAC) of Columbia University. According to this dataset, the surface of the world is divided into a grid of cells that can have different resolution levels. Each of these cells has been assigned an estimated population value. Out of the possible resolutions, we have opted for cells of 15×15 minutes of arc to constitute the basis of our model. This corresponds to an area of each cell approximately equivalent to an area of 25×25 km^2 along the Equator. The dataset comprises 823,680 cells, of which 250,206 are populated. Since the coordinates of each cell and those of the airports in the World Airport Network are known, the distance between the cells and the airports can be calculated.

For the H5N1 we consider only the estimated mobility for the World Airport Network (WAN) (Colizza et al. 2006; den Broeck et al. 2011). WAN is composed of 3362 commercial airports indexed by the International Air Transport Association (IATA) that are located in 220 different countries. The database contains the number of available seats per year for each direct connection between two of these airports. The coverage of the dataset is estimated to be 99% of the global commercial traffic. The WAN can be seen as a weighted graph comprising 16,846 edges whose weight, represents the passenger flow between airports. The network shows a high degree of heterogeneity both in the number of destinations per airport and in the number of passengers per connection.

The International Agro-Food Trade Network (IFTN) (Convertino and Liang 2014; Ercsey-Ravasz et al. 2012) for the USA is built using data of (ComTrade 2014) for the worldwide linkages among countries, and data from the United States Department of Agriculture, Foreign Agricultural Service's Global Agricultural Trade System (GATS) (FAS 2014; GATS 2014b) for the trade network of imported food commodities in USA. Additional data is used from the Agricultural Marketing Service (AMS 2014). The use of both datasets for food allows one to cross check erroneous data and to complement missing data missing. Data of imported and produced food that is locally consumed is obtained from FAOSTAT food balance sheets (FAOSTAT 2014). Such information is useful for calibrating the model factor that determines how many individuals consume the contaminated food.

Methods

The integrated modeling used in this paper consists in the following main sequential steps.

1. Generation of potential surveillance networks within network topologies;
2. Model-based detection of onset/outbreak sources with uncertainty assigned to the reported outbreaks for all feasible surveillance networks from (1); the determination of outbreak sources is performed by minimizing the uncertainty in the effective distance estimations; and,
3. Determination of the most informative surveillance network by maximizing the VoI considering all combinations of feasible surveillance networks corresponding to potential reporting patterns.

Effective distance and traceback model

The model is based on the optimal inference of epidemiological factors that better explain the spatio-temporal occurrence of foodborne outbreaks for any potential outbreak location to which different food trade paths correspond. In this study, differently from (Convertino and Hedberg 2014a), Manitz et al. (2014), and (Brockmann and Helbing 2013), deep uncertainty in reported outbreaks is introduced. The possible outbreak sources are determined by perturbing the reported outbreaks considering different surveillance networks. All feasible outbreak sources are tested considering their likelihood to satisfy the relationship between outbreak arrival times and velocity at any time step of the epidemic and for all infected communities simultaneously.

We define the effective distance (Brockmann and Helbing 2013) of directly connected nodes as a function of the most probable food trajectories defined by the radiation model (Section S2.1 in Additional file 1) that can be derived from the connectivity matrix P that specifies which nodes are connected; specifically $d_{mn} = 1 - \log P_{mn}$ where for any couple $m - n$ there are multiple distances (Brockmann and Helbing 2013). For any candidate outbreak source we use the definition of "shortest path" effective distance D_{mn} (Brockmann and Helbing 2013) from an arbitrary reference node n to another node m in the network (not necessarily directed connected) as the length of the shortest path from n to m,

$D_{mn} = \min_\Gamma \lambda(\Gamma)$, where $\lambda(\Gamma)$ is the directed length of an ordered path $\Gamma = n_1, \ldots, n_L$ as the sum of effective lengths along the path Brockmann and Helbing (2013). The average shortest path tree is calculated among all effective distances that are associated to potential candidate outbreak sources. Thus, both average effective distance (d_{mn}) and average shortest path tree (that is the shortest path among many effective distances) responsible of outbreak spreading (D_{mn}) depend only on the static mobility matrix weighted by the food fluxes. There is a family of effective distances and there one of shortest path distances between community m and n. Theoretically, there can be more than one shortest path distance but the dimension of the ensemble of shortest path distances is always smaller (or equal) than the ensemble of effective distances.

The above topological considerations allow us to say that, on a spatial scale described by the metacommunity model, the complexity of spatio-temporal outbreak patterns is largely determined by the structure of the mobility network (in this case of food and people movement) and not by the nonlinearities of the pathogen/-food/-people couples and epidemiological factors. According to the second Newtons law the arrival time of outbreaks is defined as

$$T_a = D_e(P)/v_e(\alpha, R_0, \gamma, \epsilon)\ . \tag{1}$$

This equation states that effective outbreak distances D_e can be computed with high fidelity based on outbreak arrival times on and effective spreading speed v_e, and that each factor depends on different factors of the dynamical system considered (Brockmann and Helbing 2013). Note that here we indicate D_e as D_{mn} previously defined. The epidemiological factors associated with the classical SIR model determine the effective speed, whereas effective distance depends only on the topological features of the static underlying network, i.e., the matrix P. Because, T_a and v_e are know and estimated from data (that can be generated in real-time), respectively, it is easy to estimate D_e of an outbreak considering also the uncertainty in epidemiological dynamics and food trade network. Thus, Eq. 1 can be used as a test of hypotheses where hypotheses are about all candidate outbreak sources responsible for the observed outbreak patterns. The most likely geographical distance can be assessed after determining the effective distance and considering all potential food trade paths. The SIR dynamics (Section "Epidemiological dynamics and information preading: predictive metacommunity model") is calibrated on the range of the outbreak velocity and arrival times, and is simulated by maximizing the prediction accuracy of the observed outbreak patterns in terms of cumulative outbreaks. Such accuracy is maximized when the uncertainty about the effective distance from the real outbreak source is minimized, that determines the maximization of the concentricity of outbreak spreading waves (Section "Outbreak detection: linking patterns and processes"). The concentricity is maximized only when the accuracy in the reported outbreaks is maximized that is equivalent to the maximum of the value of information (Section "Value of information and Pareto optimization") considering any uncertainty source in outbreak patterns, supply chain and disease dynamics (Section "Global sensitivity and uncertainty analyses").

Epidemiological dynamics and information spreading: predictive metacommunity model

The metacommunity model is inspired by the metacommunity concept in (Convertino 2011; Convertino et al. 2009) and (Convertino M, S L, M A, Morris S: Importance, Interaction, and Scale-dependence of Cholera Outbreak Drivers: Metamodeling Predictions, submitted) and the mathematics in (Brockmann and Helbing 2013) and more recently in (Convertino and Hedberg 2014a). The metacommunity is an ensemble of metapopulations (not only individuals) that can be divided into different categories. The model consists in the coupling of a classical susceptible-infected-recovered (SIR) dynamic model with a radiation model (Simini et al. 2012) of food/human mobility that determines the interlinked variations of food/human mobility and population states for all communities of the system considered simultaneously. The infection of individuals occurs only via food-human and/or human-human interactions that define the infectious contact matrix. In general, the model is a reaction-diffusion transport model of information spreading with embedded bias transport in relation of the radiation model of mobility. The information is about the occurrence of a case related to the epidemic considered; yet, information can be a binary variable in which “1” stands for the occurrence of an case in a location within the system. Uncertainty in reporting means to have some areas that report the outbreak or not. Thus, such uncertainty affects the ability to reproduce correctly the observed outbreak pattern and to detect the correct outbreak source. The metacommunity model, whose mathematics is written as in (Brockmann and Helbing 2013), is

$$\begin{aligned} \delta_t j_n &= \alpha s_n j_n \sigma(j_n/\epsilon) - \beta j_n + \gamma \textstyle\sum_{m \neq n} P_{mn}(j_m - j_n) \\ \delta_t s_n &= -\alpha s_n j_n \sigma(j_n/\epsilon) + \gamma \textstyle\sum_{m \neq n} P_{mn}(s_m - s_n) \,, \end{aligned} \tag{2}$$

where $s_n = S_n/N_n$, $j_n = I_n/N_n$, and $r_n = 1 s_n j_n$. $R_0 = \alpha/\beta$ that is the ratio of mean infection and recovery rate. A detailed derivation of such equations is provided in (Brockmann and Helbing 2013). The mobility parameter γ is the average food/human mobility rate estimated in the data (Section “Mobility and surveillance networks”), i.e. $\gamma = \Phi/\Omega$, where $\Omega = \sum_n N_n$ is the total population in the system and $\Phi = \sum_{n,m} F_{nm}$ is the total food/human flux. This yields numerical values in the range $\gamma = 0.01150.0379$ day^{-1}. The matrix $\mathbf{P}$ with $0 \leq P_{mn} \leq 1$ quantifies the fraction of the food/human flux with destination m exported from node n, i.e., $P_{mn} = F_{mn}/F_n$, where $F_n = \sum_m F_{mn}$ (Brockmann and Helbing 2013). The sigmoid function $\sigma(x) = x^n/(1 + x^n)$ with gain parameter $\eta \gg 0$ accounts for the local invasion threshold ϵ and fluctuation effects for $j_n < \epsilon$ (Brockmann and Helbing 2013). Typical ϵ and η average values are $\eta = 4, 8, 10, 12, \ldots, \infty$ and -$\log_{10}\epsilon = 4, 6, 8, \ldots$. For human-driven contagions ϵ is related to the number of infected people that travel within the system. For food ϵ depends on how many individuals eat the contaminated food. Here to avoid the use of two factors, one for local consumption and another for consumption of contaminated food, we use ϵ as the only factor determining spread of contamination. This means that we just consider the portion of consumed food that is contaminated. The fraction of produced and imported food that is locally consumed is assessed from (FAOSTAT 2014). In general, it is possible to assume that consumption is proportional to the size of the population of each community. This is one of the motivations for which the radiation model works well in reproducing food trade patterns (Section “Mobility and surveillance networks”) (Brockmann and Helbing 2013). Considering any effective path as a linear system - discretized at regular intervals l along the transmission network - the set of Equations 2 can be rewritten as

$$\begin{aligned} \delta_t j_n &= \alpha s_n j_n \sigma(j_n/\epsilon) - \beta j_n + D\delta_\Gamma^2 j_n \\ \delta_t s_n &= -\alpha s_n j_n \sigma(j_n/\epsilon) + D\delta_\Gamma^2 s_n \,, \end{aligned} \tag{3}$$

where $D = l^2\gamma$ is the diffusion coefficient that is related to a classic form of a Fisher equation which has been deployed in mathematical epidemiology for describing wave spreading in reaction-diffusion system such as for epidemics (Belik et al. 2011; Campos and Mendez 2005). Note that $s_n = s_n(\hat{s}(O), t)$ and $j_n = j_n(\hat{s}(O), t)$ where $(\hat{s}(O), t)$ is the surveillance network in Equation 4.

For sufficiently localized initial conditions this systems exhibits traveling waves with speed $c = 2\sqrt{\alpha D(1-\beta/\alpha)} \sim \sqrt{\gamma}$ (Belik et al. 2011). Such velocity can consider bias transport as in (Bertuzzo et al. 2007) in case the bias is known and elevated in the transport process. The telegraph model introduced by (Holmes et al. 1993) and simulated by (Bertuzzo et al. 2007) for cholera possesses both diffusion and wave motions. The Fisher equation of diffusion works well when using the effective distance, while the telegraph model equation works well when using the geographical distance of the transmission network. GSUA is used here to consider uncertainty in the traceback model and to assess the relative importance and interactions of factors - epidemiological and mobility network factors (Section "Epidemiological dynamics and information spreading: predictive metacommunity model" and Additional file 1: Section S2.1, respectively) - leading to outbreaks. Interactions of causal factors are typically neglected as well as the determination of the causality of such factors. The traceback model is run multiple times according to the Sobol scheme (Additional file 1: Section S1.2) (Figure 2) sampling all model factors along their probability distributions (Additional file 1: Figure S4) estimated from data or assumed as uniform according to a maximum entropy principle (Convertino et al. 2014a).

Outbreak detection: linking patterns and processes

The simplest method to determine which food and which outbreak source are the most likely to determine the observed outbreak patterns is to test which feasible candidate source has the lowest variability in terms of mean and variance assessed from Eq. 1. This is essentially a portfolio problem (Convertino and Valverde 2013) in which a large set of outbreak source alternatives exist and the whole set is evaluated considering average risk and variance, where in this case the "risk" is related to the inability to predict the correct outbreak patterns in space and time. Thus, for each potential candidate outbreak location the model computes the effective distance to the subset of nodes in which an outbreak is observed. This is done for any time step of the epidemic. On the basis of this set of effective distances (denoted by different outbreak sources), we compute the mean $\mu(D_e)$ and standard deviation $\sigma(D_e)$ of the effective distance. As shown by (Brockmann and Helbing 2013), concentricity of epidemic waves increases with a combined minimization of mean and standard deviation of the estimated effective distance. In other words, such approach minimizes the deviation from the expected relationship arrival times-distance, or velocity-distance, equivalently. Here we perform a minimization of the sum $\sqrt{\mu(D_e)^2 + \sigma(D_e)^2}$ as in a portfolio model where the objective function is the Euclidian distance built with the variables to minimize (Convertino and Valverde 2013). The correct outbreak source is the one that satisfies the minimum of the aforementioned sum for the whole epidemic duration. It is possible to visualize the dynamics of epidemic spread and concentricity of outbreak waves by plotting outbreak sources at the center of a network where all other nodes are placed around the central node at a radial distance equal to the effective distance. Correct outbreak sources determine cohesively evolving outbreak waves from the center to the nodes at the boundary from the beginning to the end of the epidemic.

Value of information and Pareto optimization

The Value of Information (VoI) is classically defined as the amount a decision maker would be willing to pay for information prior to making a decision (Convertino et al. 2014c; Keisler 2004). In our case we consider the set of observers $\hat{s}(O)$ as the economically and epidemiologically valuable features on which stakeholders may take a decision. In the real world such decision may be related to early control strategies of the epidemic that have economic and health repercussions. In order to calculate the VoI, or the "value of spatial information" more precisely because $\hat{s}(O)$ is a spatially explicit variable (Trainor-Guitton et al. 2012), we assess uncertainty reduction in the detection of outbreak sources when extra knowledge about outbreaks is available. The coverage of the surveillance network determines where we have information about cases. Thus, that is a macroscale uncertainty whose magnitude can be high or low, and there is an optimal coverage determined by the optimal surveillance network the proposed model aims to determine. The uncertainty in case reporting is assessed as distance $d(\hat{s}(O))$ from the outbreak source (Eq. 1 and Section "Outbreak detection: linking patterns and processes"), that is related to the variability in the reported cases expressed by the function of cumulated cases $C(t)$ over the epidemic. Note that in general, d can be both effective and geographical distance. The evaluation of the VoI is done by comparing the uncertainty before and after "pinching" the input, i.e., replacing it with another input (of different value) and a lower degree of uncertainty, or with no uncertainty (Oakley 2009). The former scenario is when a set of monitored communities is considered for any network topology (random, high degree, and maxVoI), and the latter is when all the monitoring communities that are reporting cases are considered. In this case study we consider such scenarios with complete presence (as reported) or absence of reported cases in USA for all states involved. The VoI in the latter scenario is commonly known as "value of perfect information" that is assumed in this study equal to the information gathered from all data. By considering the above reasoning and $\hat{s}(O)$ with its probability structure in space and time, the VoI is defined as:

$$VoI(t)_{i,n} = \left[d(\hat{s}(O),t)_{i,n=1} - d(\hat{s}(O),t)_{i,n=M}\right] \propto \left[C_{i,n=1}(t) - C_{i,n=M}(t)\right] , \qquad (4)$$

where $C(t)$ and $d(\hat{s}(O),t)$ is evaluated for all possible networks n (among all topologies (Additional file 1: Figure S1)), communities i, and time steps t; M is the total number of considered networks. $\propto$ stands for "proportional to". Hence, the VoI is the difference of $d(\hat{s}(O))$ estimated for different networks and all communities of the system. The VoI (Eq. 4) can be theoretically defined by the difference between two estimates of $d(\hat{s}(O))$ for different surveillance networks (Eq. 4) (i.e. a different set of observer nodes). VoI is always taken equal or greater than zero and a rational decision maker always chooses more valuable information. Note that here we do not consider any cost function in the VoI because no information is available; however, that information is easily included as function detracted to the predictive benefits in Eq. 4. In presence of a physical-based model reproducing outbreaks in space and time the VoI can also be measured as accuracy in predicting prevalence patterns. In fact, errors in detecting outbreak sources translate into errors in initial conditions of the model that reduce model predictive accuracy. Such predictions generated by models that take in input surveillance data are used for public health decision making and system design.

Global sensitivity and uncertainty analyses

GSUA involves five steps: (1) the probability distribution functions (pdfs) for each input factor are selected (epidemiological, mobility, and network topology factors); (2) sample points are generated on the input factor distributions using the Sobol method; (3) multiple model execution using each of the sample points and a set of outputs is generated; (4) global sensitivity analysis is performed (i.e., assessment of factor interaction and importance); and, (5) the important input factors are selected. Probability distributions of mobility network factors (link length and clustering coefficient) are derived from data as in (Convertino M, S L, M A, Morris S: Importance, Interaction, and Scale-dependence of Cholera Outbreak Drivers: Metamodeling Predictions, submitted). Epidemiological factors (α, and β) and contamination factors (σ an ϵ) are assumed to be normally distributed within a normalized range [0,1]. γ is related to the range of variability of the total food and human mobility flux. As for human mobility, more than 60 million people travel billions of miles on more than 2 million international flights each week. GSUA is extremely important in complex systems that are dominated by factor interactions rather than an additive contribution of each single factor. In fact, classical sensitivity analysis fails to account for factor interactions and does not offer a full probabilistic investigation of system states; yet, classical one factor sensitivity analysis provides misleading conclusions. GSUA is rarely performed and classical one-factor at a time sensitivity approaches are used to evaluated uncertainty. GSUA first assign uncertainty and that drives sensitivity analysis via variability of model output that accounts for non-linearity among model input factors and those and model output. The Sobol method (Sobol 1993, 2001) is a variance-based method that performs a quantitative analysis of model sensitivity based on the principle of variance decomposition. According to this principle, the full variance of the model output is given by the sum of the variances of all input factors (Saltelli et al. 2008). The Sobol method has the capacity to quantify the influence of the full range of variation of each input factor as well as the interaction effects among the input factors on the model output (Saltelli et al. 2008). The Sobol method estimates sensitivity measures which summarize the models behavior. The most common measure of sensitivity is the first-order sensitivity index, S_i, that represents the main effect (direct contribution) of each input factor to the variance of the output. It is expressed as $S_i = V_i/V$ where V_i is the part of the variance due to the input factor X_i, and V is the total variance of the model output. Another measure of sensitivity is the total factor sensitivity that includes the interactions. The total effect index, S_{T_i}, that is the result of the variance decomposition, accounts for the total contribution to the output variation due to factor X_i, i.e., its first-order effect plus all the higher order effects due to interactions (Saltelli et al. 2008). Thus, the total effect index of factor X_i can be expressed as $S_{T_i} = S_i + S_{i,i+1} + S_{i,i+1,i+2} + S_{i,i+1,i+2,i+3} + S_{i,i+1,i+2,i+3+...+N_{IF}} = S_i + S_{I_i}$, where N_{IF} is the number of input factors and S_{I_i} is the sum of all the interaction indexes. The Sobol pairwise interaction between factors can be calculated and we identify this with the notation $S_{I_{ij}}$ where i and j are the factors that are considered. As an example, in the case of three input factors S_{T_1} is the total sensitivity index of X_1, S_1 is the main effect of X_1, $S_{I_{12}}$ is the interaction effect between X_1 and X_2, and $S_{I_{123}}$ is the interaction effect between X_1, X_2, and X_3. Considering the previous expression, $S_{T_1} - S_1$ provides a measure of how much X_1 is involved in interactions with all other input factors (Saltelli et al. 2008). The sum of all S_i is equal to one for additive models and less than one for non-additive models. The difference $1 - S_i$ can be used as an indicator of the presence of interactions in the model and we indicate that as S_I. The number of simulations required for the Sobol method for a two-index analysis (i.e., for first order and total indices) is given as $N = M(2k + 2)$ where M is the sample size of each index (typically taken between 500-1000) and k is the number of uncertain input factors after the Morris screening. In this study because $k = 7$, and we take $M = 600$ we have 9,600 Sobol simulations. Since the Sobol method uses a pseudo-randomized multivariate sampling procedure, it can be used as a basis for a global uncertainty evaluation by constructing the pdfs and cumulative distribution functions for each of the selected outputs.

Results and discussion

The results of the study hold when surveillance works for any network topology. Here surveillance is seen as a function of the network to contrast the spreading of infectious diseases. The random surveillance design model, i.e. placing reporting nodes of outbreaks randomly along the transmission network produces errors in the distance from the real outbreak sources as shown Figure 4 for the 2010 cholera outbreak in Cameroon. Table 1 shows the surveillance node coverage necessary for achieving an outbreak source detection higher than 90% also for other surveillance network models. The higher the percentage the higher is mismatch between the surveillance network model and the epidemic transmission process. For instance, for cholera 20% of random node coverage is sufficient for detecting precisely the outbreak source - possibly because of the extended human mobility and outbreaks - but for other highly heterogeneous epidemics such as H5N1 the percentage increase to almost 50%. For cholera, with a reporting node coverage of 20% the error is 2 hops (i.e., about 4.4. km) (Figure 4). With higher percentage of node coverage, from 40 to 80 %, the error is more or less invariant to 2 km that is a very small error considering the extension of the Far North Region of Cameroon ($\sim$ 241 km long and 145 km wide). For $N_0/N = 100\%$ the error goes to zero and there is almost perfect detection of the three main outbreak sources.

By selecting as surveillance nodes only the nodes with the highest degree, where the node degree is defined as the the number of connections, the coverage required for achieving the same detection accuracy (90%) is lower than for the random coverage. 5, 13 and 38% are the percentages required for cholera, Salmonella, and H5N1 (Table 1). The results are quite invariant considering different thresholds - from 5 to 10 - for establishing "high degree" nodes. We believe that the high degree surveillance scheme captures

Table 1 Surveillance node coverage for achieving an outbreak source detection higher than 90%

	Observers		
Outbreak (Network)	**Random**	**High degree**	**maxVol**
Cholera Cameroon 2010 (Tree)	20%	5%	3%
Salmonella USA 2012 (Tree)	37%	13%	5%
H5N1 2007 (Graph)	46%	38%	20%

Number are the percentage (N_o/N x 100) of observers necessary to achieve a probability of localization of outbreak sources equal to 90%. Such percentage is for the random, high degree, and maxVol surveillance network topology irrespectively of the outbreak type. The maxVol algorithm is the one used for all potential surveillance network topologies.

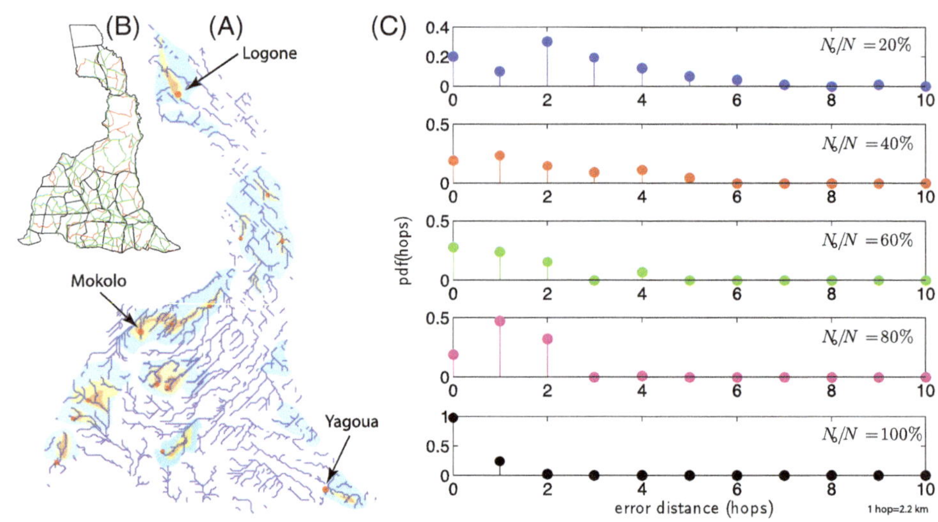

Figure 4 Cholera outbreak sources and error in detecting outbreak sources for random reporting node placement. (A) River networks and cumulated outbreak pattern of cholera in 2010 in the Cameroon Far North region. **(B)** Road connectivity influencing people mobility; in this case we use the radiation model for estimating mobility fluxes for consistency with the other epidemic considered. **(C)** $N_O/N \times 100$ is the percentage of randomly placed nodes for the cholera epidemic. The frequency distributions are about the "hop" that is the error distance from the detected to the real outbreak sources. The map is showing the detected sources (red dots) for $N_O/N \times 100 = 20\%$. For the cholera case 1 hop = 2.2 km.

better outbreak evolutions than other schemes because epidemics are very likely to pass through the most connected nodes. However, selecting only these nodes is not enough since many infections pass thought other nodes. The maxVoI model (Eq. 4) determines a drastic reduction in the number of surveillance nodes required to detect outbreak sources with an accuracy higher than 90%. 3, 5 and 20% are the percentages of nodes required for cholera, Salmonella, and H5N1 (Table 1). Such percentages correspond to errors

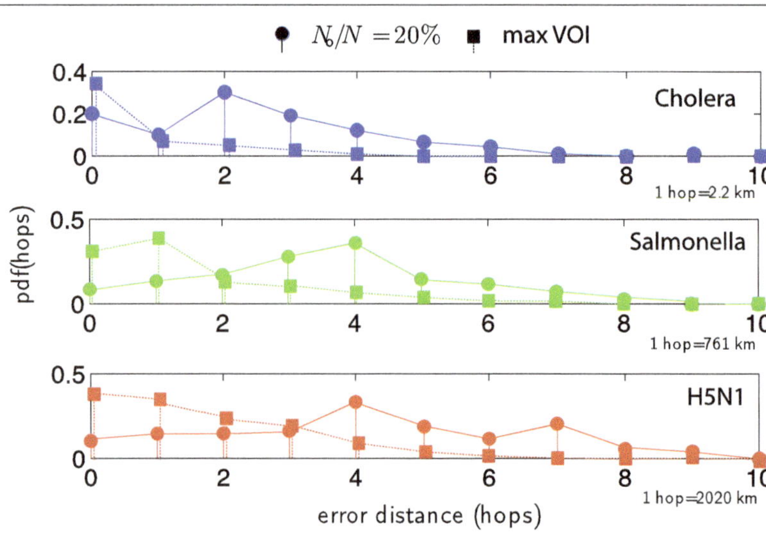

Figure 5 Error in detecting outbreak sources for value of information based and random surveillance design schemes. The frequency distributions are about the "hop" that is the error distance from the detected to the real outbreak sources. Results for random and maxVoI models are shown. The maxVoI is achieved for small-world surveillance networks.

of 0.5×2.2 km, 1.0×761 km, and 1.5×2020 km as distances from the correct outbreak sources (Figure 5). In case of random placement the error in km units is 2.5 ×2.2 km, 3.8×761 km, and 4.6×2020 km. The reduction in surveillance nodes determined by the maxVoI model is 17%, 32%, and 26% with an average 25% reduction. These differences creates differences in predictions of outbreak patterns larger than 40% in terms of cumulative cases. The magnitude of reduction is significative of the both interlinked epidemic dynamics and transmission network; more space filling epidemics such as cholera show a smaller advantage of the maxVoI model than more long-range and less space filling epidemics such as Salmonella and H5N1 for which the food/human mobility network does not covert the whole system.

The ability of the maxVoI model to detect outbreak sources with high accuracy and low surveillance nodes is related to determination of the most meaningful nodes (in terms of known information for describing outbreak patterns) via a portfolio model with Pareto optimization that explores all potential combinations of observers and corresponding outbreak information. The VoI is maximized, thus the difference between two estimated distances from the real outbreak sources is maximized. The set of observer nodes for which the VoI is the maximum allows to predict the real outbreak source and the observed outbreak pattern.

For the outbreak considered the error in the estimation of the distance, and thus in the outbreak velocity, determined by the uncertainty of reporting outbreak arrivals is minimized by the maximization of the VoI. Such optimization results maximized in terms of VoI for small-world networks that likely capture both local and large scale outbreak spreading regardless of epidemiological details. Figure 6 shows that there is clearly an association between network topology, maximum VoI and error distance regardless of the epidemic considered. Additional file 1: Figure S2 shows these findings in an alternative way by plotting the normalized probability distribution of estimated distances from outbreak sources and the Integrated Network Resilience (INR) (Halpern et al. 2012; Pandit and Crittenden 2012) (see Additional file 1, Section S2.3, and Additional file 1: Figure S1) of all surveillance network topologies. Random and regular networks with medium and high INR, seem limited in maximizing the VoI; yet, with such network it is hard to predict accurately outbreak sources, and the evolution of outbreaks over space and time. INR, that is manifesting the structural resilience of a network in terms of reconfiguration ability (see Additional file 1) determines also the ability to inform about outbreak evolution. Additional file 1: Figure S3 shows the error in epidemic spreading velocity as a function of the VoI for all surveillance network topologies. The variability of the velocity is higher than the distance, nonetheless even the VoI-velocity error pattern shows a neat dependency of the epidemic dynamics on the network topology (Additional file 1: Figure S3). Small world networks have the highest VoI, thus the ability to detect outbreak sources and velocities carefully. Outbreak waves for different time steps of the epidemics considered are estimated with data related to the maxVoI surveillance model (bottom plot of Additional file 1: Figure S3). Such epidemic waves are dependent on the surveillance network and their estimation is crucial for detecting epidemic spreading and outbreak sources rapidly and carefully, respectively. Ideally, the wish is to have a surveillance system that detects and stops the epidemic spreading after the first cases that correspond to the first epidemic wave. The celerity of epidemic waves can be derived from Eq. 3 or more simply with Eq. 1 in the

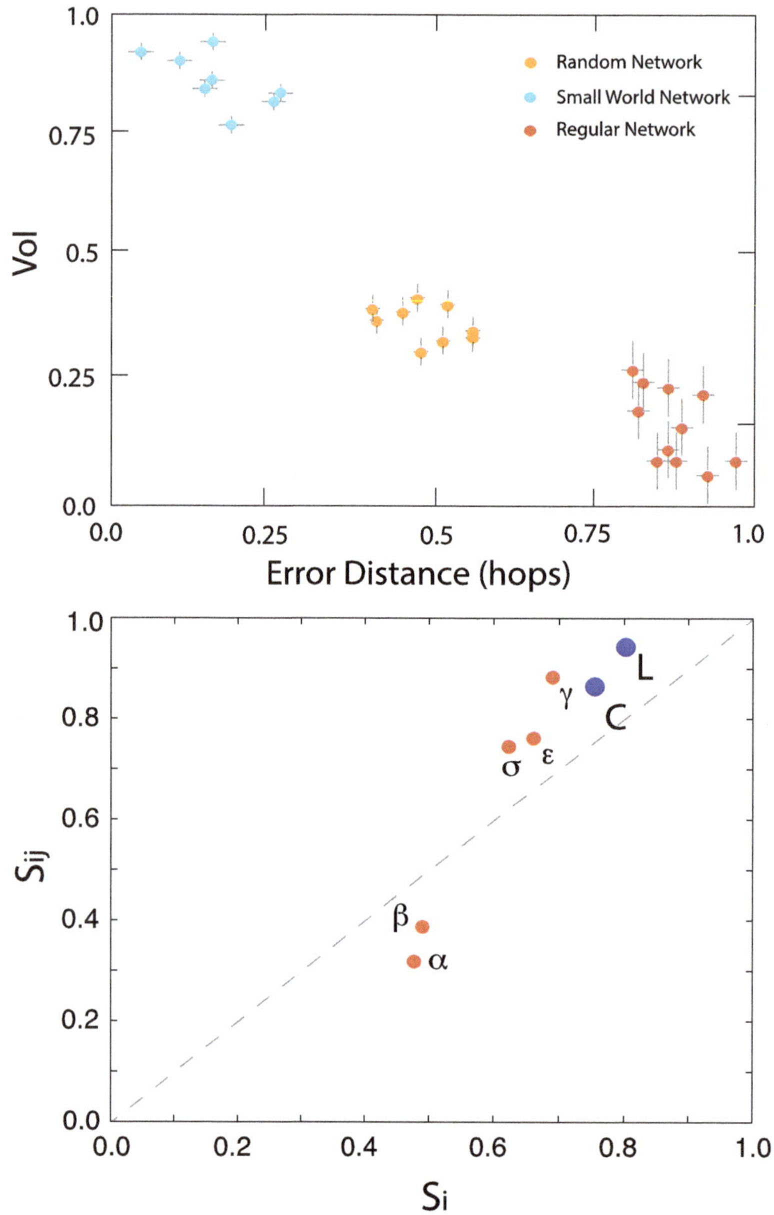

Figure 6 Value of information, prediction accuracy, and prediction determinants. The upper plot shows the VoI versus the error distance as a function of the network topology. The bottom plot is after Global Sensitivity and Uncertainty Analyses (GSUA).

effective distance domain. Wrong estimations of epidemic velocities are for inefficient outbreak reporting systems during the course of epidemics which makes very difficulty the detection of outbreak sources, attribution of outbreaks to food and/or infected individuals, and prediction of outbreak spread. The distance error from outbreak sources is anisotropic - in the sense that different errors are associated to different paths of the transmission network - because epidemic spreading is anisotropic in the geographical space. For this structured anisotropy related to the nonrandom and non regular structure of the mobility network, we believe that random and regular surveillance network are suboptimal surveillance networks. Vice versa, small-world surveillance networks such as power-law networks capture the small-world features of mobility networks (Bajardi et al. 2012), thus they are able to monitor outbreak evolution over space and time very accurately.

Figure 6 also shows the result of GSUA in which uncertainty is considered for SIR and mobility model input factors. Results show that topological factors of the mobility network, namely the average link length and the clustering coefficient L and C, are the most important factors in explaining outbreak patterns. The average mobility rate γ and the invasion threshold ϵ that defined the number of free-moving contaminated food and/or individuals are very important and interacting factors after topological factors. Epidemiological transition factors, α and β, are secondary importance and weakly interacting factors because they are strongly dependent on the aforementioned factors. These results underlines even more the importance of knowledge and controllability of the network on which epidemic may occur. Mobility networks define where outbreaks can occur but also where surveillance should be crucially designed. Because a fully extended surveillance network is likely unfeasible considering also the highly varying mobility network of food and people, an optimally designed surveillance system via maxVoI is ideal in terms of benefits and costs.

As for H5N1 the source is correctly identified in Vietnam (Hanoi as the outbreak source); subsequently the epidemic is predicted to move to Thailand, and within weeks to spread to ten countries and regions in Asia, including Indonesia, South Korea, Japan and China. As for the Salmonella outbreak in tuna, the outbreak source is detected in New York City if only the US food trade is considered, and in Kerala, India, if the whole worldwide food trade network is taken into account. For the cholera epidemic in Cameroon, the health districts of Logone (North), Mokolo (West), and Yagoua (South East) are detected as outbreak sources. Considering the epidemic starting some time before the peak of cases, the urban health district of Maroua is detected as outbreak source. This evidences that: (i) the cholera epidemic in Cameroon is very likely generated by infections in rural areas and urban cases are due to infected people mobility among cities where human-human transmission is higher (in this case Maroua that is the capital of the Far North Region); and (ii) the definition of outbreak sources depend on the time horizon considered. For instance, Maroua is a source only if previous cases are neglected. Equivalent considerations are related to the scale of analysis: outbreak sources are related to the scale of the system considered. For example, it is known that cholera is endemic in large Sub-Saharan countries and the migration of populations, in which there are also infected people, among such countries is very high. For Cameroon, a high migration is reported from Nigeria in the East and Chad in the North, thus real outbreak sources may be located outside Cameroon. These considerations highlight the need to be very careful when planning control strategies that may result very inefficient for decreasing the systemic risk of infection because of the wrong targeting of outbreak source areas.

We show that efficient surveillance in the human mobility network is sufficient to detect outbreak source rapidly; this is despite complications played by wild bird migratory pathways and infected domestic birds for H5N1, and water-driven pathogen mobility for cholera. The almost immediate detection of outbreak sources has been shown in previous papers (Brockmann and Helbing 2013; Convertino and Hedberg 2014a). Certainly, in general it is important to detect which transmission network causes the fastest spreading infections and we believe that in our cases, for both cholera an H5N1, the fastest infections are driven by human-human contacts related to the human mobility network as shown by epidemiological studies.

Conclusions

The design of optimal surveillance networks allows stakeholders to detect outbreak sources and paths correctly. Such detection is only possible if information about outbreak "fingerprints", i.e reported outbreaks, is accurate and timely. We show that independently of system scale and resolution, biological and socio-environmental dynamics the "unknowns" of epidemiological dynamics - such as epidemic spreading paths and sources - can be determined considering the same physics principles of reaction-diffusion processes and information theory. We believe that our results are robust independently of any other epidemic types, or it can provide the methodological framework to design epidemic-specific surveillance network. The following results are worth emphasizing.

- The structural resilience of the system is independent of epidemic dynamics and can be identified by the Integrated Network Resilience that combines two key network topological factors. Because of the spatio-temporal anisotropy of outbreaks the small-world network is the surveillance network with the highest value of information, that implies the most accurate outbreak source detection and outbreak pattern prediction. Structural resilience is just one component of the system to respond to outbreaks. A portion of system resilience has to be built considering training of personnel dedicated to surveillance.
- The knowledge of mobility networks and the value of information of surveillance networks as subsets of of the former is determinant for early detection and response of outbreaks. The use of effective distances allow to avoid the need to use topological features - necessary in the design process - that increase the computational complexity and uncertainty related to the estimation of the most likely transmission networks.
- Optimal dynamic design and real time surveillance are tasks that can be achieved using the proposed model. The design and outbreak source detection can be performed with a portfolio model with Pareto optimization that selects the most important observer networks by maximizing the VoI, and detects the most likely outbreak sources by minimizing the uncertainty in the effective distance from the real sources. The optimal design implies accuracy detection but not vice versa. Dynamic surveillance network in which surveillance change configurations, for instance taking advantage of smart sensors placed along the mobility network, are at the frontier of technological development and can implement the maxVoI model.

The model can work for other kinds of spreading processes (e.g, viruses in computer networks and biomarker activation in chronic diseases) in which information spreads on partially known networks, although networks can be back-inferred with a reverse application of the model where the question is the identification of the networks (among all candidate ones) that leas to the generation of observed epidemic patterns. For instance, an important application of the model is related to the determination of the most effective preparedness information, for example provided by the surveillance network or other information networks, with the aim to decrease the systemic risk of outbreaks in populations. Such information can be organized considering public media and/or online networks that inform people about performing tasks related to control outbreaks. Reversely, information networks can also be used as tools to more rapidly inform surveillance networks about the spread of outbreaks in populations.

Additional file

Additional file 1: Specifics of case-study outbreaks, supplementary methods (Radiation Model of Mobility Fluxes, Surveillance Network Topologies, and Integrated Network Resilience) and supplementary Figures.

Competing interests
The authors declare that they have no competing interests.

Authors' contributions
MC designed and performed the study. YL and HH helped in the data gathering and analysis. MC wrote the paper while YL and HH provided comments to the text. All authors read and approved the final manuscript.

Acknowledgments
M.C. acknowledges fundings form the Healthy Food Healthy Lives Institute (grant "Impact of Local and Organic Foods on Food Safety: Assessment of Supply Chain Controls and Value Outcomes for Quantitative-based Food Policies"), the Institute on the Environment "Discovery Grant" (grant "Food System Design for Resilient Population Health: The Minnesota Model"), and the Minnesota Discovery, Research and InnoVation Economy (MnDRIVE) Global Food Systems Faculty Scholar Start-Up Funding. Y.L. acknowledges the School of Public Health Dean Fellowship for the academic year 2014-2015.

Author details
[1]School of Public Health, Environmental Health Sciences Division and Public Health Informatics Program, HumNat Lab, University of Minnesota, Minneapolis, MN, USA. [2]Institute on the Environment, University of Minnesota, Minneapolis, MN, USA. [3]Institute for Engineering in Medicine, University of Minnesota, Minneapolis, MN, USA. [4]Academic Health Center, Mayo Memorial Building, Suite 1232 12th fl., MMC807, 420 Delaware St. SE, 55455 Minneapolis, MN, USA.

References
AMS: **Agricultural marketing service.** 2014. Tech. rep. [http://www.marketnews.usda.gov/portal/fv?&paf_gear_id=1200002&movNavClass=FVPHN&rowDisplayMax=25&repType=movementDaily&previousVal=&lastCommodity=&paf_dm=full&lastLocation=&volume=40000&locName=MEXICO&locAbr=MX&startIndex=51&dr=1; Accessed April 2014]

Bajardi P, Barrat A, Savini L, Colizza V: **Optimizing surveillance for livestock disease spreading through animal movements.** *J R Soc Interface* 2012.

Batz M, Hoffmann S, Morris JJ: **Ranking the risks, the 10 pathogen-food combinations with the greatest burden on public health.** 2011. Tech. rep., University of Florida, Emerging Pathogens Institute. [https://folio.iupui.edu/bitstream/handle/10244/1022/72267report.pdf; Date of access: 06/06/2013]

Belik V, Geisel T, Brockmann D: **Natural human mobility patterns and spatial spread of infectious diseases.** *Phys Rev X* 2011, **1:**011001.

Bertuzzo E, Maritan A, Gatto M, Rodriguez-Iturbe I, Rinaldo A: **River networks and ecological corridors: reactive transport on fractals, migration fronts, hydrochory.** *Water Resour Res* 2007, **43**(4).

Brockmann D, Helbing D: **The hidden geometry of complex, network-driven contagion phenomena.** *Science* 2013, **342**(6164):1337–1342.

Campos D, Mendez V: **Reaction-diffusion wave fronts on comblike structures.** *Phys Rev E* 2005, **71:**051104.

CIESIN: **The gridded population of the world version 3 (GPWv3) population grids.** 2014. Tech. rep. [http://sedac.ciesin.columbia.edu/gpw; Accessed April 2014]

ComTrade: **United Nations commodity trade statistics.** 2014. Tech. rep. [http://comtrade.un.org/db/; Accessed April 2014]

Colizza V, Barrat A, Barthlemy M, Vespignani A: **The role of the airline transportation network in the prediction and predictability of global epidemics.** *Proc Nat Acad Sci USA* 2006, **103**(7):2015–2020.

Convertino M: **Neutral metacommunity clustering and SAR: River basin vs. 2-D landscape biodiversity patterns.** *Ecol Model* 2011, **222**(11):1863–1879.

Convertino M, Valverde LJ Jr: **Portfolio decision analysis framework for value-focused ecosystem management.** *PLoS ONE* 2013, **8**(6):e65056.

Convertino M, Hedberg C: **Epidemic intelligence cyberinfrastructure: real-time outbreak source detection and prediction for rapid response.** *PLoS Curr Outbreaks* 2014a.

Convertino M, Hedberg C: **Intelli-food: cyberinfrastructure for real-time outbreak source detection and rapid response.** *Lecture Notes in Computer Science* 2014b, **8549:**181-196. [Smart Health book, Springer].

Convertino M, Liang S: **Probabilistic supply chain risk model for food safety.** *GRF Davos Planet@Risk* 2014. **2**. [Special Issue on One Health].

Convertino M, Muneepeerakul R, Azaele S, Bertuzzo E, Rinaldo A, Rodriguez-Iturbe I: **On neutral metacommunity patterns of river basins at different scales of aggregation.** *Water Resour Res* 2009, **45**(8).

Convertino M, Munoz-Carpena R, Chu-Agor M, Kiker G, Linkov I: **Untangling drivers of species distributions: Global sensitivity and uncertainty analyses of MaxEnt.** *Envir Model Softw* 2014a, **51:**296–309.

Convertino M, Foran CM, Keisler JM, Scarlett L, LoSchiavo A, Kiker GA, Linkov I: **Enhanced adaptive management: integrating decision analysis, scenario analysis and environmental modeling for the everglades.** *Sci Rep* 2014c, **2922:**.

Convertino M, Munoz-Carpena R, Kiker G, Perz S: **Design of monitoring networks by value of information optimization: experiment in the Amazon.** *Stochastic Environ Res Risk Assess* 2014d. In press.

den Broeck WV, Gioannini C, Goncalves B, Quaggiotto M, Colizza V, Vespignani A: **GLeaMviz computational tool, a publicly available software to explore realistic epidemic spreading scenarios at the global scale.** *BMC Infect Dis* 2011, **11:**.

Estrada E, Gómez-Gardeñes J: **Communicability reveals a transition to coordinated behavior in multiplex networks.** *Phys Rev E* 2014, **89**(4).

Ercsey-Ravasz M, Toroczkai Z, Lakner Z, Baranyi J: **Complexity of the international agro-food trade network and its impact on food safety.** *PLoS ONE* 2012, **7**(5):e37810. [http://dx.doi.org/10.1371%2Fjournal.pone.0037810]

FAOSTAT: **Food balance sheet website of the food and agriculture organization of the United Nations.** 2014. Tech. rep. [http://faostat.fao.org/site/368/default.aspx#ancor; Accessed April 2014]

FAS: **USDA foreign agricultural service.** 2014. Tech. rep. [http://www.fas.usda.gov/data; Accessed April 2014]

GATS: **Multistate outbreak of salmonella bareilly and salmonella nchanga infections associated with a raw scraped ground tuna product.** 2014a. Tech. rep. [http://www.cdc.gov/salmonella/bareilly-04-12/index.html?s_cid=cs_654; Accessed April 2014]

GATS: **USDA global agricultural trade system.** 2014b. Tech. rep. [http://www.fas.usda.gov/gats/default; Accessed April 2014]

Goncalves B, Balcan D, Vespignani A: **Human mobility and the worldwide impact of intentional localized highly pathogenic virus release.** *Scientific Reports* 2013, **3**(810).

GRUMP: **Global rural-urban mapping project alpha version population grids.** 2014. Tech. rep. [http://sedac.ciesin.columbia.edu/gpw; Accessed April 2014]

Guevart E, Noeske J, Solle J, Essomba J, Edjenguele M, Bita A, Mouangue A, Manga B: **Factors contributing to endemic cholera in Douala, Cameroon.** *Med Trop (Mars)* 2006, **66**(3).

Haldane AG, May RM: **Systemic risk in banking ecosystems.** *Nature* 2011, **469**(7330):351–355.

Halpern BS, Longo C, Hardy D, McLeod KL, Samhouri JF, Katona SK, Kleisner K, Lester SE, O/'Leary J, Ranelletti M, Rosenberg AA, Scarborough C, Selig ER, Best BD, Brumbaugh DR, Chapin FS, Crowder LB, Daly KL, Doney SC, Elfes C, Fogarty MJ, Gaines SD, Jacobsen KI, Karrer LB, Leslie HM, Neeley E, Pauly D, Polasky S, Ris B, St Martin K, et al: **An index to assess the health and benefits of the global ocean.** *Nature* 2012, **488:**615–620. [http://dx.doi.org/10.1038/nature11397]

Hanson L, Zahn E, Wild S, Dopfer D, Scott J, Stein C: **Estimating global mortality from potentially foodborne diseases: an analysis using vital registration data.** *Popul Health Metrics* 2012, **10**(5).

Helbing D: **Globally networked risks and how to respond.** *Nature* 2013, **497**(7447):51–59. [http://dx.doi.org/10.1038/nature12047]

Helbing D, Brockmann D, Chadefaux T, Donnay K, Blanke U, Woolley-Meza O, Moussaid M, Johansson A, Krause J, Schutte S, et al: **How to save human lives with complexity science.** *J Stat Phys* 2014. SSRN 2390049. DOI 10.1007/s10955-014-1024-9.

Holmes EE: **Are diffusion models too simple? A comparison with telegraph models of invasion.** *Am Nat* 1993, **142:**403–419.

Keisler J: **Value of information in portfolio decision analysis.** *Decis Anal* 2004, **1**(3):177–189. [http://da.journal.informs.org/content/1/3/177.abstract]

Kilpatrick AM, Chmura AA, Gibbons DW, Fleischer RC, Marra PP, Daszak P: **Predicting the global spread of H5N1 avian influenza.** *Proc Nat Acad Sci* 2006, **103**(51):19368–19373.

Knobler S, Mahmoud A, Lemon S, Pray L, on Microbial Threats EF: *The Impact of Globalization on Infectious Disease Emergence and Control: Exploring the Consequences and Opportunities, Workshop Summary - Forum on Microbial Threats*: The National Academies Press; 2006. [http://www.nap.edu/openbook.php?record_id=11588]

Manitz J, Kneib T, Schlather M, Helbing D, Brockmann D: **Origin detection during food-borne disease outbreaks a case study of the 2011 EHEC/HUS outbreak in Germany.** *PLOS Curr Outbreaks* 2014.

Mundt CC, Sackett KE, Wallace LD, Cowger C, Dudley JP: **Long?distance dispersal and accelerating waves of disease: empirical relationships.** *Am Nat* 2009, **173**(4):456–466. [http://www.jstor.org/stable/10.1086/597220]

NAS: *Improving food safety through a one health approach: workshop summary*. Washington (DC): National Academies Press (US); 2012. ISBN-13: 978-0-309-25933-0ISBN-10: 0-309-25933-9.

Newman M: **The structure and function of complex networks.** *SIAM Rev* 2003, **45**(2):167–256. [http://epubs.siam.org/doi/abs/10.1137/S003614450342480]

Njoh ME: **The cholera epidemic and barriers to healthy hygiene and sanitation in cameroon: a protocol study.** 2010. Tech. rep., Umea University. [http://bvs.per.paho.org/texcom/colera/MENjoh.pdf]

On Effectiveness of National Biosurveillance Systems: BioWatch C, the Public Health System NRC: *BioWatch and Public Health Surveillance: Evaluating Systems for the Early Detection of Biological Threats: Abbreviated Version*: The National Academies Press; 2011. [http://www.nap.edu/openbook.php?record_id=12688]

Oakley JE: **Decision-theoretic sensitivity analysis for complex computer models.** *Technometrics* 2009, **51**(2):121–129. [http://www.tandfonline.com/doi/abs/10.1198/TECH.2009.0014]

Pandit A, Crittenden J: **Index of Network Resilience (INR) for Urban Water Distribution Systems.** In *Nature*; 2012. [http://www.tisp.org/index.cfm?cdid=12519&pid=10261]

Park J, Seager TP, Rao PSC, Convertino M, Linkov I: **Integrating risk and resilience approaches to catastrophe management in engineering systems.** *Risk Anal* 2012, **33:**356–367. doi: 10.1111/j.1539-6924.2012.01885.x.

Pastore Y Piontti, A, Gomes Da Costa, M, Samay N, Perra N, Vespignani A: **The infection tree of global epidemics.** *Netw Sci* 2014, **2:**132–137. [http://journals.cambridge.org/article_S2050124214000058]

Pinto PC, Thiran P, Vetterli M: **Locating the source of diffusion in large-scale networks.** *Phys Rev Lett* 2012, **109:**068702. [http://link.aps.org/doi/10.1103/PhysRevLett.109.068702]

Saltelli A, Ratto TM, Andres, Campolongo F, Cariboni J, Gatelli D, Saisana M, Tarantola S: *Global Sensitivity Analysis - The Primer*. John Wiley and Sons; 2008.

Schlundt J, Toyofuku H, Jansen J, Herbst S: **Emerging food-borne zoonoses.** *Rev Sci Tech* 2004, **23**(2).

Scallan E, Griffin P, Angulo F, Tauxe R, Hoekstra R: **Foodborne illness acquired in the United states-Unspecified agents.** *Emerg Infect Dis* 2011a, **17:**.

Scallan E, Hoekstra R, Angulo F, Tauxe R, Widdowson S, Roy J, Jones P, Griffin P: **Foodborne illness acquired in the United StatesMajor Pathogens.** *Emerg Infect Dis* 2011b, **17**(7).

Simini F, González MC, Maritan A, Barabási AL: **A universal model for mobility and migration patterns.** *Nature* 2012, **484**(7392):96–100. [http://dx.doi.org/10.1038/nature10856]

Sobol I: **Sensitivity analysis for non-linear mathematical model.** *Math Comput Simul* 1993:407–414.

Sobol I: **Global sensitivity indices for nonlinear mathematical models and their Monte Carlo estimates.** *Math Comput Simul* 2001, **55:**271–280.

Stachenko S: **The role of surveillance and data use in the development of public health policies.** *Promot Educ* 2008, **15**(3):27–29. [http://dx.doi.org/10.1177/1025382308095654]

Trainor-Guitton WJ, Mukerji T, Knight R: **A methodology for quantifying the value of spatial information for dynamic Earth problems.** *Stochastic Environ Res Risk Assess* 2012:1–15. [http://dx.doi.org/10.1007/s00477-012-0619-4]

Tatah A, Pulcherie K, Mande N, Akum N: **Investigation of water sources as reservoirs of Vibrio cholerae in Bepanda, Douala and determination of physico-chemical factors maintaining its endemicity.** *Onderstepoort J Vet Res* 2012, **79**(2).

11

Complex adaptive systems modeling with Repast Simphony

Michael J North[1,2*], Nicholson T Collier[1], Jonathan Ozik[1,2], Eric R Tatara[1], Charles M Macal[1,2], Mark Bragen[1] and Pam Sydelko[1]

* Correspondence: north@anl.gov
[1]Decision and Information Sciences Division, Argonne National Laboratory, 9700 S. Cass Avenue, Argonne, IL 60439-4867, USA
[2]University of Chicago, Chicago, IL, USA

Abstract

Purpose: This paper is to describe development of the features and functions of Repast Simphony, the widely used, free, and open source agent-based modeling environment that builds on the Repast 3 library. Repast Simphony was designed from the ground up with a focus on well-factored abstractions. The resulting code has a modular architecture that allows individual components such as networks, logging, and time scheduling to be replaced as needed. The Repast family of agent-based modeling software has collectively been under continuous development for more than 10 years.

Method: Includes reviewing other free and open-source modeling libraries and environments as well as describing the architecture of Repast Simphony. The architectural description includes a discussion of the Simphony application framework, the core module, ReLogo, data collection, the geographical information system, visualization, freeze drying, and third party application integration.

Results: Include a review of several Repast Simphony applications and brief tutorial on how to use Repast Simphony to model a simple complex adaptive system.

Conclusions: We discuss opportunities for future work, including plans to provide support for increasingly large-scale modeling efforts.

Keywords: Complex adaptive systems modeling, Agent-based modeling, Java, Groovy

Background

Introduction

Complex adaptive systems (CASs) are composed of interacting, autonomous agents (Holland 2006). CAS agents have properties and behaviors. They interact with and influence each other, learn from their experiences, and adapt their behaviors so they are better suited to their environment(s). By modeling these agents individually, the full effects of the diversity that exists among agents with respect to their attributes and behaviors can be observed as they give rise to the dynamic behavior of the system as a whole.

Agent-based modeling provides a mechanism for modeling CASs (Bonabeau 2002; Macal and North 2010). Agent-based modeling has been used successfully to model complex adaptive systems in many disciplines, including archaeology, biology, ecology, supply chains, consumer market analysis, military planning, and economics (North and Macal 2007; North and Macal 2009). Some examples are reviewed in section three.

Repast Simphony is a widely used, free, and open source environment for agent-based modeling of CAS. The most recent version, 2.0, was released on March 5, 2012. Repast Simphony is a second-generation environment that builds upon the previous Repast 3 library described in North et al. (2006).

This paper describes Repast Simphony and shows how to apply it to modeling CASs. The next section discusses related work. Section three details the architecture of Repast Simphony. Section four uses a simple demonstration example to show how to model CAS with Repast Simphony. Conclusions are then presented.

Related work

This section reviews other free and open source modeling libraries and environments that are documented with published papers. The review includes NetLogo, StarLogo, Swarm, Repast 3, MASON, Ascape, and EcoLab.

StarLogo

StarLogo (StarLogo 2011) is a library and environment that uses a Java interpreter and interface. StarLogo is an educational system that leverages the Logo language (Feurzeig et al. 1970, Harvey 1997) to make it easier to learn the library and to use it to develop agent models. StarLogo is free and open source for non-commercial use. StarLogo substantially extended Logo by increasing the allowed number of agents (i.e., turtles) from tens to thousands by giving turtles much more interactive behaviors, and by transforming the turtle's environment from a rectangular array of pixels into a grid of first-class agents (i.e., patches). Resnick (1996) notes that StarLogo is designed around a single organizing principle. For StarLogo, that principle is modeling decentralized complex systems. There is no requirement to hide the complexity of foundational features, such as parallelism, if they embody the organizing principle.

NetLogo

NetLogo (Wilensky 2012) is a free and open source agent-based simulation environment that uses a modified version of the Logo programming language. NetLogo is available under the General Public License (GPL) (NetLogo User Manual – FAQ (Frequently Asked Questions)). NetLogo's builds on the StarLogo effort (Tisue and Wilensky 2004). NetLogo was designed to provide a basic laboratory for teaching complexity concepts. It can also be used to develop more sophisticated applications. NetLogo provides a graphical environment to create programs that control graphic turtles that reside in a world of patches, which are monitored by an observer. Links are also available to connect turtles to form networks. Wilensky (2012) and Tisue and Wilensky (2004) offer several lessons from the development of NetLogo. Agent-based simulation environments and languages should be simple enough to allow beginners to start quickly. Agent modeling libraries should, in principle, place no limits on what experienced users can do. Agent modeling libraries should also include a large number of example simulations to help beginning and experienced users alike.

For Wilensky (2012) and Tisue and Wilensky (2004), programming is not something to be avoided. The challenge is not to develop a programming-free way to specify models. Most people can learn to program. The challenge is to simplify programming syntax and semantics enough to make programming accessible. Visual programming is not necessarily the easiest way to specify models. They state that working with a carefully designed text-based language can sometimes be easier than a visual one.

Swarm

Swarm is a free and open source toolkit with both Objective-C and Java bindings (Minar et al. 1996). Swarm also uses a GPL license. Swarm was originally developed as a software toolkit for the creation of simulation models in the field of Artificial Life (A-Life) (Macal 2009). Swarm provides modelers with a flexible, nested approach to modeling interactions. Swarm was highly influential in the development of libraries that would arrive on the scene later, such as Repast 3 (North et al. 2006) and Repast Simphony. While discussing their design, (Minar et al. 1996) states that agent-based modeling toolkits or libraries are important in that they save scientists from wasting time on repetitive and error-prone programming. Discrete time scheduling is preferred over floating point scheduling, because discrete calculations do not suffer from rounding errors. Object orientation is also useful because of its close association with agents.

Minar et al. (1996) suggests that agents should be organized into containers with schedules of activities. These collections are called "swarms." Swarms should be interoperable to allow mixing and nesting of related models.

Repast 3

Repast 3 is a family of three free and open source agent-based modeling libraries (North et al. 2006). The three libraries are Java-based Repast J, C#-based Repast .NET, and NQP (Not Quite Python)-based Repast Py. Repast 3 uses a "new BSD" (Berkeley Software Distribution)-style license and includes third-party libraries with compatible licenses. Several conclusions were reached during the implementation of the libraries. Object-oriented design is essential to allow both the library and user models to be flexible. The use of design patterns is helpful for improving implementation quality. Environments should allow seamless programming in different languages to help simultaneously lower barriers to entry for new users and increase flexibility for experts. Choosing environments that are cross-library makes systems usable for a wider audience compared to single library approaches. Modular construction is essential. Library developers should be keenly aware of differences in the fundamental details of the candidate implementation environments, such as differences in mathematics libraries. Library developers should also consider the availability and capability of the ecosystem surrounding an implementation environment when making their selections.

MASON

MASON is a free and open source, agent-based modeling library that provides core services that can be mixed easily with other libraries (Luke et al. 2005). Luke et al. (2005) recommends that libraries should be developed from first principles derived from specific domains. Another key attribute is that libraries should be minimal and mix well with third-party libraries. Only essential features should be provided to increase interoperability and simplify learning. There should be complete separation between models, user interfaces, and offline storage. A large number of agents (e.g., millions) should be supported. All model runs should be reproducible. Furthermore, checkpointing (i.e., saving and restoring model runs midstream) should be supported.

Ascape

Ascape (Parker 2001) is a free and open source, agent-based modeling library that represents models as a complex series of nested "scapes" populated by agents whose behaviors are implemented with abstracted rules. Parker (2001) states that the abstraction of complexity is a powerful way to focus design and development efforts. Selective abstraction should be used to provide structure for users and reduce the complexity of user code. The abstractions need to be very carefully considered to avoid limiting flexibility.

For Parker (2001), space should be represented with nested containers (i.e., scapes). Scapes dynamically represent space, identity type, and group membership. Scapes should be allowed to act as agents. This treatment allows complete nesting of models. Agents, however, should not have to explicitly address the nature of scapes, which greatly simplifies agent specifications and reduces the chances for errors. Agent behavior specification and scheduling should occur only at the scape-level, never at the individual level. This treatment allows behaviors to be dynamically changed and permits sophisticated, behavior-based designs of experiments.

EcoLab

EcoLab (Standish 2008) is a general purpose C++ and TCL/Tk agent-based modeling library. Users write models in C++ and then invoke the models using TCL. Tk can be used as needed to develop user interfaces. Standish (2008) suggests that it is helpful to have standardized models with which to compare agent modeling libraries. Standish states that reflection is essential for agent models. Reflection is the ability to examine the type information of objects dynamically, that is, while a program is running. Standish also states that dynamic pointers that automatically track objects such as agents are important. These pointers should automatically note the type of object being tracked and should indicate whether an object reference has been deallocated from memory.

Methods

Repast Simphony

The free and open source REcursive Porous Agent Simulation Toolkit (Repast) was originally developed by Sallach, Collier, and others (Collier et al. 2003) at the University of Chicago in 2000, and was subsequently expanded by Argonne National Laboratory as a reusable software infrastructure that could support "rapid social science discovery" based on extensive computational experimentation (Sallach and Macal 2001). Successive releases of Repast have extended the system to handle large-scale agent simulation application development. Repast is managed by the not-for-profit volunteer Repast Organization for Architecture and Design (ROAD). ROAD is led by a board of directors that includes members from a range of government, academic, and industrial organizations. Argonne National Laboratory is especially active in maintaining and developing Repast. Many other groups contribute to Repast's development. For example, the University of Michigan recently teamed with Argonne to add more than 20 demonstration models to Repast Simphony with support from the 2011 Google Summer of Code. The Repast system, including the source code, is available directly from the Web at http://repast.sourceforge.net/ (Repast – The Repast Suite). All versions of Repast use a "new BSD"-style license and include third-party libraries with compatible licenses.

Repast Simphony builds on the lessons of Repast 3 but uses a new code base. Repast Simphony was designed from the ground up with a strong focus on well-factored abstractions. The resulting code has a highly modular plug-in architecture that allows individual components such as networks, logging, and time scheduling to be replaced as needed. A plug-in is a software module that extends the features of a host program using a well-defined programming interface. The Repast Simphony code also has multiple layers within each plug-in module to allow specific implementations at lower levels to be replaced by new implementations when required. As discussed later, this layering allowed the authors to replace Repast Simphony's logging and visualization implementations while maintaining backward compatibility. Repast Simphony also includes many advanced features, such as watchers, that were not possible in Repast 3. These features are discussed later.

The Repast community is large and growing. These users have applied Repast to a wide variety of applications that range from social systems, to evolutionary systems, market modeling, and industrial analysis. A sampling of Repast applications will be presented next.

Example Repast Simphony applications

This section briefly discusses several example Repast Simphony applications that illustrate the versatility of the system. The examples include applications that study phenomena related to consumer products, systems engineering, possible future hydrogen infrastructures, social science, and ancient pedestrian traffic, to name a few. Additional examples can be found in North and Macal (2009).

The Virtual Market Learning Lab (North et al. 2010) is a large-scale, Repast Simphony model of consumer markets co-developed by Argonne National Laboratory and Procter & Gamble (P&G). It represents the shopping behavior of consumer households and the business behavior of retailers and manufacturers in a simulated national consumer market. Argonne and P&G successfully calibrated, verified, and validated the resulting agent-based model using several independent, real-world data sets for multiple consumer product categories with more than 60 comparison tests per data set. First Repast and then later Repast Simphony were used to implement the model. P&G has successfully applied the model to several challenging business problems where it has directly influenced managerial decision-making and has produced substantial cost savings.

The Monitoring, Analysis, and Diagnosis of Distributed Processes with Agent-Based Systems (MADCABS) model is a Repast Simphony simulation framework for investigating the operation and control of complex networks (Artel et al. 2011; Tatara et al. 2006). MADCABS combines agent models with traditional differential equations models. MADCABS has been used for a variety of research pursuits, including to study the effectiveness of perceptron-based learning for the robust operation of distributed chemical reactor networks.

The Hydrogen Economy Model (Mahalik et al. 2007) is a Repast Simphony representation of the Los Angeles, California, metropolitan area, with 5,000 square miles of detailed and geographic information system (GIS)-sourced interstate highways and omnipresent local roads. The model has driver and investor agents. Driver agents use their cars to move between their demographically assigned home neighborhoods and their jobs. Drivers have a variety of characteristics including income levels, environmental concerns, risk aversion, and car type preferences (e.g., wanting conventional fuel cars versus hydrogen cars). Investor agents build, own, and operate hydrogen fuel stations based on the investor's estimates of the potential for profits at each available geolocated site.

The Endogenous Emergence of Coordination (EndEC) model (Ozik and North 2010) is a Repast Simphony simulation of the endogenous emergence of coordination within a group of social agents. The model demonstrates the use of a dynamic language, specifically Groovy, to represent the socially coordinated evolution of behavior. The EndEC sample model is inspired in part by Holland's work on the emergence of language (Ke and Holland 2006; Tao et al. 2005). The model consists of a set of agents each with an individual list of movement capabilities and utterances. The agent's movement capabilities and utterances are drawn from separate supersets that are available to the population as a whole. The agents each maintain their own individual associations of movements with utterances. As the simulation proceeds, agents make utterances to announce their movements and, over time, adjust their associations to match the observed behavior of their neighbors. These simple behavior rules result in the dynamic emergence of complex coordination between groups of agents.

SHULGI is a Repast Simphony model of ancient pedestrian traffic in urban environments (Branting et al. 2007). The model represents a wide range of issues that affect pedestrian route choices, including source and destination locations; roadway availability and quality; and the metabolic energy costs of candidate routes. The factors can be flexibly weighted as needed to address different questions of interest. The model has been successfully validated by using archaeological data from the ancient Turkish city of Kerkenes Dag.

Repast Simphony design goals

The design goals of Repast Simphony are as follows (North et al. 2005):

- All of the core features and capabilities of the previous versions of Repast should be available. These features are detailed in North et al. (2006). The lessons learned from developing the original Repast series of libraries are summarized above in the related work section.
- There should be a strict separation between model specification, model execution, model visualization, and data storage.
- All user model components should be plain, unadorned Java objects that are accessible to and replaceable with external software (e.g., legacy models and enterprise information systems).
- Tasks that are commonly performed by model developers should be automated when possible.
- Imperative 'boilerplate' code should be eliminated or replaced with declarative runtime configuration settings when possible. Much like boilerplate text in standard English, boilerplate code is moderate- to large-sized pieces of code that are used again and again in model after model. An example in Repast 3.1 is code to setup a master time schedule or configure a network display.
- Idiomatic code expressions should be simple and direct. Idiomatic code expressions are short pieces of code that act as slang terms or shorthand. An example is code to update the weight on a network edge between agents.

These design goals have been met by the Repast Simphony architecture.

Eclipse

Eclipse (Eclipse Home Page) is a widely used, free, and open source development environment that is multilingual and integrated. Eclipse can be used to develop code in many languages, including Java and C++. Java support is particularly strong.

The Repast Simphony system uses Eclipse as its primary development environment. Repast Simphony leverages Eclipse's plug-in architecture to provide a set of development options. It is important to note that while Eclipse's plug-in architecture is conceptually similar to that of Repast, Eclipse uses a different implementation.

The Repast Simphony Eclipse plug-ins provide tools, views (i.e., windows), and perspectives (i.e., selected sets of windows) for creating a range of Repast-specific model components, including general Repast projects, ReLogo projects, flowcharts, and ReLogo agents. The plug-ins also allow Repast models to be executed, debugged, and packaged into self-contained installers for deployment. An example flowchart that can be used to encode agent attributes and behaviors is shown in Figure 1.

Repast Simphony architecture

The design goal of maintaining a strict separation between model specification, model execution, data storage, and visualization has led to creation of the multilingual Repast Simphony model development environment in Eclipse, a runtime engine for model execution, a runtime interface for model visualization, and freeze dryers for data storage. Freeze dryers are an automated way to send and retrieve the state of a simulation to and from secondary storage, such as an XML (extensible markup language) file. To implement this function, Repast Simphony uses a highly modular architecture. Each module is an independent 'plug-in' that can be connected or disconnected with a few

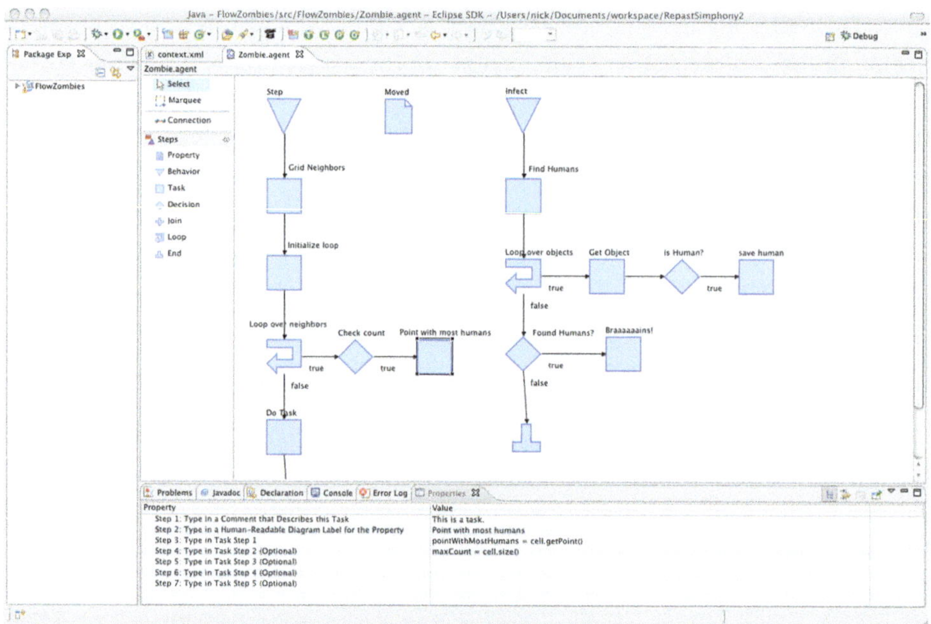

Figure 1 **An example Repast Simphony flowchart showing the behavior for a Zombie agent.**

lines of XML. Thanks to this architecture, Repast Simphony exists, in essence, as a user-selectable federation of plug-in components. Selected plug-ins and related sets of plug-ins are summarized in Table 1.

It should be noted that that Repast plug-ins use the Java Plug-in Framework (JPF) (Java Plug-in Framework (JPF) Project), which is separate from the Eclipse plug-in architecture. JPF is used at runtime by Repast models. Eclipse is used at development time by the Repast design tools.

The next section will discuss the plug-in system. The following sections will consider central plug-ins of particular interest.

Simphony application framework

Repast Simphony's modular architecture is built on top of the free and open source Simphony Application Framework (SAF). As an application framework, SAF is intended to simplify common application-building tasks, such as creating and configuring menus, toolbars, application window layout, and so on. SAF-based applications typically follow the model-view-controller design pattern with the addition of a class that mediates between various parts of the application. This mediator provides a link between the user interface and the application functionality triggered by the user. The Repast Simphony runtime interface implementation consists of three components: the user interface proper, the simulation engine infrastructure, and the application class that mediates the interaction between the two. The user interface is composed of other smaller model-view

Table 1 Selected repast simphony plug-ins and plug-in sets

Plugin or plugin sets	Function
Simphony Application Framework	Provides the basic runtime interface plug-in system and user interface tools
Eclipse Set	Provides model specification and programming tools
Core Set	Provides central simulation functions, such as scheduling
ReLogo Set	Provides a simple modeling language and structure
GIS	Provides GIS modeling and visualization
Freeze Dryer Set	Provides persistence in XML and text formats
Batch Run Set	Provides parameter sweeps and stochastic iteration
Deployment	Packages user models for self-contained release
Charts	Provides interactive model graphing in the runtime interface
Model Integration	Provides tools for embedding legacy models
2D Visualization Set	Provides tools for interactive two-dimensional (2D) model viewing
3D Visualization Set	Provides tools for interactive three-dimensional (3D) model viewing
Terracotta	Provides distributed model execution
System Dynamics	Provides tools for ordinary differential and difference equations
Third-Party Application Set	Supports a set of independent plug-ins for: • Geographic Resources Analysis Support System (GRASS), • Java Universal Network/Graph Framework (JUNG) network analysis, • *ORA network analysis, • Pajek network visualization, • R statistics, • iReport and Jasper Reports enterprise reporting, • Spreadsheets, • Structured Query Language (SQL) analysis within running simulations, • VisAD scientific visualization, and • Weka data mining.

-controller types of components that may or may not delegate some functionality to the application class.

SAF uses JPF to provide much of this functionality. JPF provides a runtime engine that dynamically discovers and loads plug-ins. A plug-in within the SAF context is a structured component that describes itself to JPF using a manifest, which is an XML document with a specific schema. This file contains information identifying the plug-in; what other plug-ins this plug-in depends on, if any; where its code is located; and where any third-party code that it depends on is located. JPF does not provide the plug-in content, but rather a plug-in discovery mechanism, a plug-in format, and an application programming interface (API) for working with such. SAF then uses these pieces to create an application framework.

This plug-in approach helps to make SAF applications modular. Specific bits of application functionality can be defined in separate plug-ins and loaded at runtime. Repast Simphony as a whole is simply a collection of plug-ins, allowing bits and pieces of Repast to be used as required. The Repast Simphony runtime interface also leverages this modularity in a variety of ways. For example, the integration with third-party applications, such as the R statistical package, that appear in the toolbar is accomplished by using the plug-in mechanism. Third-party application integration is discussed later.

In addition to promoting modularity, JPF also makes applications extensible by using plug-in extension points and extensions. An extension point functions as a generic interface that defines some specific functionality and how to invoke that functionality in an abstract way. An extension is an implementation of an extension point. SAF defines extension points appropriate for building applications, and the Repast Simphony runtime interface implements these as extensions. Both the Repast runtime interface and Repast Simphony define Repast-specific extension points and extension implementations. For example, the Repast Simphony runtime interface scenario tree that appears on the lower left side of Figure 2 contains various metadata (e.g., context loading, display definition, etc.) about the model and is itself a collection of extensions. In addition, the mechanisms by which these are manipulated (e.g., display wizards) are also extensions implementing extension points defined by the Repast Simphony runtime interface. In this way, Repast Simphony and the Repast runtime interface are modular and easily extensible. The value of Repast Simphony's modularity was demonstrated when the initial library used by SAF to provide docking windows needed to be replaced to meet increasing user requirements. The modular structure allowed the docking windows library to be swapped out while maintaining backward compatibility with existing user models.

Core

The Repast Simphony core plug-ins provide major simulation functions, including time scheduling, space management, behavior activation, and random number generation (North et al. 2007). Repast Simphony space management uses contexts and projections. Repast Simphony provides a discrete event time scheduler that uses double precision real numbers for event times. The scheduler also allows events with otherwise identical times to be differentiated with an optional priority ranking. The Repast scheduler offers imperative scheduling and declarative scheduling for both sequential and concurrent activities, as discussed later in this section.

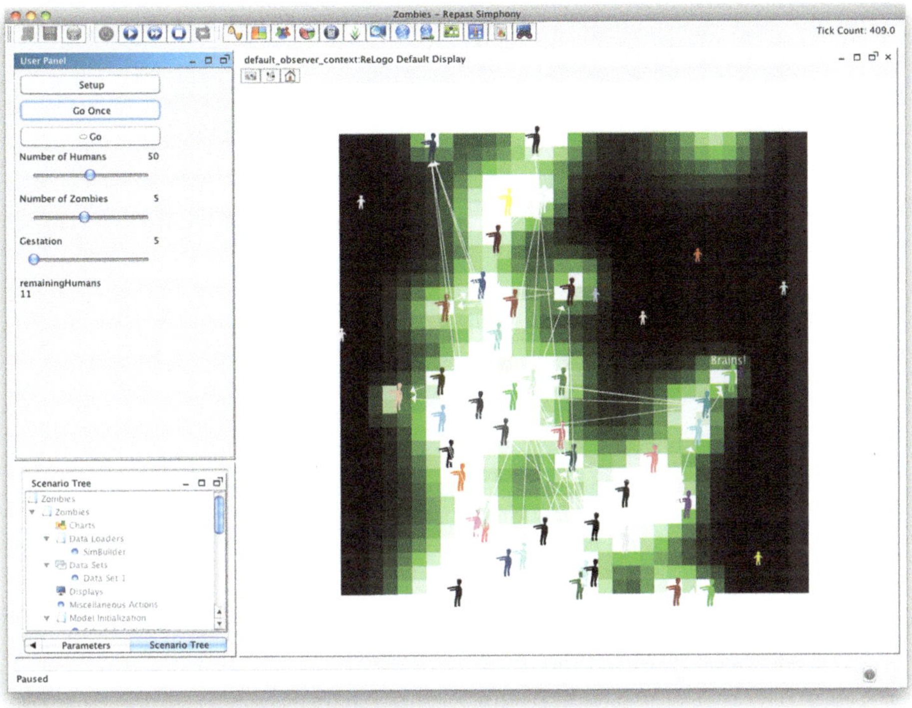

Figure 2 **The Repast Simphony zombie demonstration model.**

Repast Simphony contexts are hierarchically nested and named containers that hold model components. The model components can be any type of Java object, including other contexts, but are typically agent objects. It should be noted that contexts can be agents as well. Each model component can be present in as many contexts as the modeler desires. The hierarchical nesting means that a model component that is present in a context is also present in all of that context's parent contexts. Of course, the converse is not true in the general case. The hierarchical nesting structure can be declaratively or imperatively specified by the modeler. In addition, the contents of components within contexts (e.g., agent properties) can be declaratively logged at runtime.

In addition to supporting hierarchical nesting, contexts also support projections. Repast Simphony projections are named sets of relationships defined over the constituents of a context. For example, a Repast Simphony network projection stores a network or graph relationship between the members of its context. The members of this context can then ask to whom they are linked and who is linked to them. Similarly, the Repast Simphony grid projection stores a set of Cartesian coordinates for each member of the context. The members of this context can ask where they are located. Each context can support any mixture of projections. Also, projections can be declaratively visualized at runtime.

Contexts and projections can use watchers to activate behaviors. The following example shows an agent behavior that is activated at certain times when an appropriate social network neighbor changes its "contentment" attribute:

```
// This is an example watcher activation method.
@Watch(watcheeClassName =
```

```
"repast.user.models.ExampleSocialAgent",
watcheeFieldName = "contentment",
query = "linked_to 'neighbors' and not linked_from
  'work' and colocated",
triggerCondition = "$watchee.getContentment() !=
 $watcher.contentment",
whenToTrigger = WatcherTriggerSchedule.LATER,
scheduleTriggerDelta = 5,
scheduleTriggerPriority = 15)
public void neighborMoodUpdated(ExampleSocialAgent
  neighbor) {
  // Check the contentment level.
  if (neighbor.getContentment() > 0.5) {
   // Copy a neighbor.
   this.setContentment (neighbor.getContentment());
  } else {
   // Decide for ourselves.
   this.setContentment(RandomHelper.nextDouble());
  }
}
```

In this example, the "neighborMoodUpdated" method will be called after another ExampleSocialAgent's contentment changes if the current agent is linked to the watched agent in the neighbor network, is not linked from the watched agent in the work network, is colocated with the watched agent in a context, and has a contentment setting different from that of the watched agent's contentment. The "neighborMoodUpdated" method will be called five ticks after the change and the event will be given a scheduler priority of 15, which is substantially higher than the default value of zero. As shown in the example, contexts work directly with watchers by allowing watcher queries to use context names and properties. Similarly, projections work directly with watchers, allowing watcher queries to use projection names (e.g., "neighbors" and "work" in Figure 3), properties, and relationships (e.g., "linked_to," "linked_from," and "colocated" in Figure 3).

The watcher mechanism provides a listener-like design pattern for reactive agent behavior, which is computationally more efficient than polling a collection of agents at a fixed interval. Runtime agent class instrumentation is performed with the Javassist library (Chiba and Nishizawa 2003) when the agent classes are loaded by the Repast runtime, which occurs automatically for classes with @Watch annotations. Bozada et al. (2006) presented one of the first implementations the Repast watcher pattern in a large-scale, agent-based model of military deployment logistics. The execution speed of the logistics model implementation in Repast was more than an order of magnitude faster than the previous Java implementation with polling-based agent behaviors.

Annotations can also be used for tasks such as unconditional scheduling. In this usage, individual methods can be called at specified times without the need for explicit Java method calls to the Repast scheduling system. This approach has the potential to be both

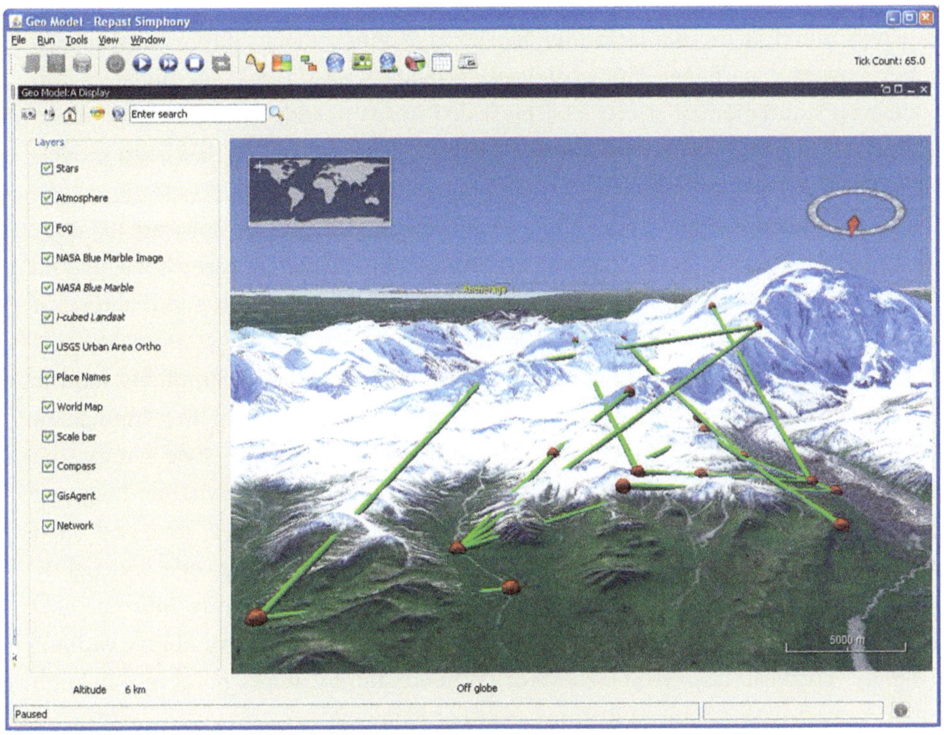

Figure 3 A Repast Simphony demonstration model that combines GIS with a network.

more convenient than the conventional Java method of call scheduling because it requires only the setting of a few annotation parameters rather than the making of a set of method calls. It is also better documented because it directly connects the scheduling information to the scheduled method itself rather than dividing this task among multiple locations in the model.

Both watch-based scheduling and unconditional scheduling allow concurrent or parallel code execution. This optional feature is activated by setting the duration of the activity being scheduled to a non-zero value. The duration is the length of time the activity is considered to take. Any activities that occur between the start of the activity with non-zero duration and the end of the duration are candidates to be concurrently executed with the activity having non-zero duration. The default duration is zero, indicating sequential execution. Whether the execution is simply concurrent or actually parallel is determined by the underlying hardware, the Java virtual machine in use, and the interrelated details of the activities being scheduled. This approach allows modelers to begin with simpler sequential execution and then scale up to concurrent or parallel execution as they continue to develop their models.

Many agent behaviors use random number generation for option selection. The Repast Simphony core module includes a full suite of random number generation tools. Repast uses the Colt library (Colt Home Page) to produce random numbers. Colt has been widely used for scientific computation and is well validated. Repast makes its Colt-based random number sources accessible through a "random helper" class with a supporting internal registry. The internal registry supports efficient reuse of random streams. It also allows all random streams to be simultaneously reset when a user changes the default random seed.

ReLogo

ReLogo wraps much of the Repast Simphony library's functionality into a semantically simple but powerful package. The goal is to provide a fast track so modelers can quickly develop simulations that can later be scaled up, as needed, using the features of the full Repast Simphony library. A tool for automatically converting NetLogo models into ReLogo models is also included.

ReLogo semantics are drawn from the venerable Logo lineage of software (Harvey 1997), informed especially by the previously discussed StarLogo (Resnick 1996) and NetLogo (Wilensky 2012) variants, while maintaining the object-oriented nature of Repast Simphony.

ReLogo provides a streamlined Eclipse perspective to create Repast Simphony models with turtles, links, patches, and observers. An example is shown in the Innovation model discussed later. ReLogo models are programmed using a Logo-style syntax (the ReLogo domain-specific language) based on the Groovy (Groovy: A dynamic language for the Java platform) dynamic language. Groovy itself is a widely used, free, and open source Java Virtual Machine language that both compiles to Java bytecode and tightly integrates with Java. Thanks to its Groovy foundations, ReLogo freely interoperates with Groovy and Java. ReLogo programmers can use any Java or Groovy library without special syntax. They can also write Groovy or Java code fluidly at any point in a ReLogo file to gain access to the advanced features of either language.

A ReLogo model is compiled by using the Groovy Eclipse joint compiler. The ReLogo code is first converted into its Abstract Syntax Tree (AST) representation. The AST is then repeatedly transformed by Repast Simphony to generate model-specific custom ReLogo code, as needed. The end result is Java byte code, which is run as a regular Repast Simphony model.

Data collection

The Repast Simphony data collection system is designed to gather and store information from a simulation while it is running. The system typically collects predefined sets of data from each agent at each time step: a particular piece of logged data represents the state of the simulation at a particular time. This logged data can contain aggregate and nonaggregate values. An aggregate data value represents the result of an aggregate operation (e.g., summation) performed over some collection of individual values. Typically, these individual values correspond to agent properties. A nonaggregate value is some individual value, again typically an agent property.

Data collection in Repast Simphony consists of three phases, namely initialization, recording, and termination. Initialization and termination occur once per run, whereas recording occurs at some regular interval (e.g., once every three ticks). Crucial to the proper functioning of data collection is specification of a set of notifications associated with these phases. For example, when writing a row of data in tabular delimited format, it is necessary to notify the writer that all of the data for that row has been collected and can now be written. Similarly, a histogram component requires notification that the record phase has ended and that the histogram should now be produced with the recorded data.

The first Repast Simphony data collection system was built upon Log4J (LOG4J – Logging Services). Although Log4J is a widely used for fast, minimal overhead logging, adapting it to the requirements of Repast Simphony's data collection functionality proved to be problematic. Log4J was difficult to adapt to the time-stamped and time-phased nature of Repast Simphony data collection. Log4J has no real user-facing mechanism for repeated initialization and termination from within a single application instance. Our incremental understanding of these requirements and adapting them to Log4J's system of logging levels and logging messages resulted in code that was inflexible and hard to manage. The isolation available in the multilayered Repast Simphony logging system made it easier for the Repast development team to transition from Log4J to custom code while largely maintaining backward compatibility.

Consequently, the data collection mechanism was replaced by a new one built on the following abstractions: data sources, data sinks, and data sets. Data sources are just that, sources of data. A nonaggregate data source receives a single object (e.g., an agent) and returns some data value, typically but not necessarily a property of that agent. An aggregate data source receives multiple objects, and performs some aggregate operation over all of the objects. It then returns the result of that operation. For example, it might calculate the mean value of some property over all of the objects. A data sink writes the output of these data sources, for example, to a file or a chart. A data set manages a collection of sources and sinks, providing the objects on which the data sources operate and directing the results to data sinks. Each data source can be thought of as a column in a tabular spreadsheet type of format, and each row is the result of recording data from each data source associated with a data set. Crucially, a data set also provides notifications to data sinks, notifying them that a new row of data is being started, that a row has ended, and that the entire record phase for the current time period has ended. It provides similar notifications to the data sources, for example, notifying an aggregate data source that the current row has started and thus any previous aggregate values should be reset. The new code embodying these abstractions is proving to be much more flexible than the previous Log4J code and has provided the basis for a more simple and more efficient user interface.

GIS

Geographic referencing is provided through geography projections that, like the other spatial projections in Repast Simphony, correlate the agents to positions in space. An agent's representation in a geography projection corresponds to a specific geographical feature, such as points, lines, and polygons. Repast Simphony uses GeoTools (GeoTools – The Open Source Java GIS Toolkit), an Open Geospatial Consortium (OGC) compliant library (OGC – Making location count), to provide support for the feature types described above, along with additional GIS data types and functions. Repast Simphony's GIS capabilities also include providing support for Environmental Systems Research Institute (ESRI) shapefiles (esri – Understanding our world) and a range of raster data files.

The geography projection is associated with a coordinate referencing system (CRS), which is based on the OGC standards and can be used to execute geographical queries on the topology of the agent features in the geography. Agents can query the geography to determine whether agent features overlap or are within a certain distance of, intersect with, or border other agent features in the geography. Features in the geography may have either static or dynamic positions. Typically, landmark features such as roads, buildings, and streams — represented by lines and polygons — are static. Mobile agent feature types are typically represented by point features, whose latitude/longitude coordinates can be updated dynamically during the course of a simulation.

2D and 3D visualization

The Repast Simphony 2D and 3D visualization modules provide tools for interactively viewing running models in 2D and 3D. A 2D ReLogo example is shown in Figure 2. A 3D GIS example is shown in Figure 3.

The initial 2D visualization code was based on the Piccolo2D graphical framework (Piccolo2D – A Structured 2D Graphics Framework). Piccolo2D provides a 2D scene graph that allows for excellent animation, layout, zooming, and user interaction. Unfortunately, it proved to be too slow. In particular, it did not provide adequate performance for turtle and patch visualization in our new ReLogo language. For efficient visualization, the Repast Simphony design calls for a two-phased approach where (1) all the visual representations of all of the agents are updated to reflect the current states of the agents, and then (2) the entire scene is rendered. The intention is that the first phase performs no graphical screen updates. It is only after this first phase that the actual display is updated to reflect the changes made in the first phase.

The new 2D visualization mechanism in Repast Simphony 2.0 uses OpenGL (OpenGL – The Industry's Foundation for High Performance Graphics) via the Java Binding for the OpenGL (JOGL) API library (jogl – Java Binding for the OpenGL API) and implements the two-phased approach described above. The abstraction provided by the multilayered Repast Simphony visualization system simplified transitioning from Piccolo2D to OpenGL while largely maintaining backward compatibility. As before, agent objects do not draw themselves, but rather are associated with particular styles and layouts. These styles and layouts are then applied to each agent to determine its display location and its visual representation. The actual 2D visualization is accomplished by using a lightweight 2D scene graph implemented on top of OpenGL. The scene graph consists of a hierarchically arranged tree of nodes where operations on parent nodes apply to the children of those nodes. This new library also allows the user to easily create a variety of geometric shapes, create custom shapes, import Scalable Vector Graphics-based geometry, import images, and style these in a variety of ways. It supports object selection, zooming, and panning via a camera implementation. This new implementation has proven to be many times faster than the original Piccolo-based approach when rendering simulations of similar visual complexity.

Both the original Piccolo-based 2D and the new OpenGL-based 2D, as well as the current 3D implementation, were guided by the need for the same agent to be displayed simultaneously in a variety of topologies (networks, grids, and so on) and a variety of styles. For example, a single agent could be appear in a 2D display as a red circle in a grid layout and simultaneously in a 3D display as a blue cube participating in a network with the corresponding links visualized. Previous experience with agent visualization has shown that this capability is best accomplished by separating the agent's implementation from the visual representation of that agent.

In Repast Simphony, style and layout are implemented as Java interfaces. Both the 2D and 3D style interfaces consist of methods that receive an agent as a parameter and return some value, typically based on some property of the agent. For example, there are methods that return the geometric representation (e.g., circle, cube, model-specific polygon) of the agent, the agent's color, rotation, scale, label, and so forth. An additional network style is used to style network edges. The layout interface works in a similar fashion. A location finding method takes an agent as a parameter and returns the location of that agent in either 2D or 3D space. Typical layouts compute the location for an individual agent relative to the topology in which it resides. The screen location of an agent in a network is determined by the incoming and outgoing links of that agent, whereas an agent in a grid topology is laid out according to its grid location. Consequently, the layout interface contains methods to associate it with the particular topology or topologies being displayed. Often, layouts need to be recalculated as agents are added and removed from the topology, and the layout interface reflects this need, as well. The implementation of style and layout as interfaces independent of agent implementation and the subsequent styling and layout of agents provide the required flexibility. Agents can be represented differently but simultaneously in 2D and 3D displays while participating in different types of topologies.

Both the 2D and 3D display implementations use a scene graph architecture. Individual agents are represented as geometric nodes in the scene. The appearance (color and so forth) of the object is determined by the style and applied to the object in some scene graph–specific way. For example, in a 3D display, the color of a node's surface appearance is determined by the color returned from the style. The node's ultimate screen location is the result of translating the node by the amount returned from the layout. Updates to the displays occur in two stages. Zooming, panning, and so forth are implemented as operations on the root node or camera of the scene. As mentioned above, display updates are a two-stage process. In the first, the scene graph is updated to reflect the latest styling and layout information, and then in the second, the scene graph is rendered.

GIS displays are available in both 2D and 3D implementations, both of which may use the same underlying geography projection. The 2D GIS display map is based on modified rendering utilities from GeoTools along with features from Piccolo2D. Agent features are visible in the 2D map according to the defined GIS style, which can be created and modified by the user at runtime via the built-in style editor, or imported from an OGC styled layer descriptor (SLD) file (OGC – Making Location Count: Standards). The 2D GIS display can be panned and zoomed in two dimensions.

An additional GIS display type is based on the NASA World Wind Java (WWJ) software development kit (SDK) (NASA World Wind) and shown in Figure 3. This WWJ display uses OpenGL to render an interactive globe on which geographical features are placed. The display behavior is similar to that of the 2D GIS display, such that features are represented by points, lines, and polygons and can be either static or dynamic. The default display contains several base raster layers for terrain and satellite data, which are fetched from public domain servers at runtime. Depending on the level of zoom magnification, finer and finer detail will be pulled automatically from the image servers. Agent feature layers are placed over the underlying raster image layers, and the extent of transparency, colors, and size of agent feature styles are fully customizable.

Freeze drying

Repast Simphony allows the user to serialize, or checkpoint, the state of the simulation for later retrieval and execution. Tabular and XML formats are supported. The tabular format writes the serialized objects to rows and columns, where each row represents an object and each column a property of that object. An additional column contains a unique identifier generated by Repast Simphony for each object. Objects are serialized by type, with a single table holding the data for all objects of that type. Additional tables record the state of the object tree. For example, a Map may contain agents, and those agents may themselves contain additional objects. These tree tables would then record the parent–child relationships between the map and the agents, and the agents and the objects they contain.

The motivation for the tabular format was two-fold. First, we wanted to be able to load data into a variety of standard databases easily, and a tabular format made that relatively straightforward. Second, we wanted the state of agents to be editable offline. When saved to a delimited file format, the state of the simulation and its agents can be loaded into common tools, such as Microsoft Excel, and easily edited. Moreover, additional objects (e.g., agents) can be created offline by adding rows to the relevant table.

This tabular serialization format has worked reasonably well, although it does present some maintenance difficulties. In particular, large changes to Repast Simphony classes, such as networks, grids, and so forth that are serialized often require changes to the serialization code. The implementation of tabular serialization also exposed some differences in Java reflection implementations across operating systems.

XML format serialization was introduced as a simpler and more extensible serialization solution for those applications that did not need the benefits of the tabular format or had issues with it. XML serialization uses the XStream (XStream Home Page) serialization library to serialize objects to and from XML. The default serialization provided by XStream can be extended and customized by using its converter mechanism. Repast Simphony takes advantage of this capability by implementing custom converters for its major components, providing more efficient serialization. It also allows users to provide their own custom converters for serializing their own models. This functionality allows the user to handle or work around any issues in serialization without requiring any time-consuming changes to Repast Simphony itself.

Third-party applications

The Repast Simphony runtime interface's plug-in architecture is open to allow almost any kind of software to be added to the system. We have used this feature to provide users with a default set of built-in tools for model results analysis. Some of these tools are directly embedded within the runtime interface, and others are externally activated.

The default-embedded tools are generally Java-based libraries with Repast Simphony–compatible licenses. Examples include the JUNG network analysis system (JUNG – Java Universal Network/Graph Framework) and the SQL for Java Objects (JoSQL) library (JoSQL – What is JoSQL?). Repast uses JUNG for network storage and statistical analysis and uses JoSQL to allow users to query paused simulations interactively using SQL.

The default external tools are either not Java based or use licenses that are free and open source but incompatible with the Repast Simphony license. These external programs are activated by wizards that go through the following sequence: (1) ask users which data sets should be selected for analysis, (2) export the data to a format that is readable by the external tool, (3) note the licensing terms of the external program, (4) start the external executable, (5) automatically load the exported data into the target application, and then (6) switch focus to the external program. The external tool loads into a separate memory space with no direct linkage to the initiating Repast model so there are no licensing conflicts. The experience is nonetheless seamless for the model user. Example default external tools include the R statistics system (http://www.r-project.org/), the GRASS GIS (GRASS GIS – The world's leading Free GIS software), the *ORA network analysis system (CASOS – *ORA), the Pajek network visualization system (Networks/Pajek: Program for Large Network Analysis), the iReport reporting tool (Jaspersoft Community – iReport Designer), the JasperReports reporting tools (Jaspersoft Community – JasperReports Library), common spreadsheets, the VisAD scientific visualization system (VisAD – McIDAS-V), and the Weka data mining system (WEKA – Machine Learning Group at the University of Waikato).

Results and discussion

CAS modeling example

This section uses a simple demonstration example to show how to develop models of complex adaptive systems with Repast Simphony. The approach presented here follows the methodology detailed in North and Macal (2007) and North and Macal (2009). The example scenario is intentionally simple so that it clearly shows the essential steps of the model design and development process. The scenario is an innovation diffusion environment in which a group of people randomly circulates. People occasionally communicate ideas when they encounter one another. When an idea is transferred, a new link is made between the transferring parties, and then any existing network links are forgotten. People occasionally develop new ideas. When they do so, they forget all of their old links. The quality of ideas is rated and tracked over time. The question posed to the simulation is: what is the quality of ideas over time?

In the sections that follow we discuss the appropriateness of using agent-based modeling to simulate the problem of interest, decompose the problem into components a using a series of design questions, and then show how the resulting implementation proceeds step-by-step. The combination of these steps produces a simple illustrative model.

When to use agent-based modeling

Agent-based modeling is a good choice compared to other modeling techniques when the problem naturally consists of agents or when the decision makers to be modeled meet the following criteria:

- Must have real individual behaviors;
- Adapt, change, or learn;
- Form dynamically changing relationships;
- Form organizations;
- Have spatial interactions;
- Have arbitrarily large populations; or
- When structural change is an output, not an input.

Our example satisfies two of the criteria: our innovation diffusion CAS is naturally composed of agents, and the agents form dynamically changing relationships.

Essential agent-based model design questions

Once agent-based modeling has been identified as a candidate methodology for solving the problem, the following questions can be asked:

- What are the modeling question(s)?
- Who are the stakeholders?
- What outputs are needed?
- What input data are available?
- What verification and validation (V&V) is needed?
- What alternative techniques where tried?
- Who are the agents?
- What are the agent behaviors?
- What is the agent environment?

Each of these questions will be briefly addressed in our next section for our example CAS.

What Are the Modeling Question(s)? The modeling question is: how does the quality of ideas change over time in the innovation network?

Who Are the Stakeholders? The stakeholders in this demonstration case are simply the modelers themselves.

What Outputs Are Needed? The output will be the history of the quality of innovation. This output will be displayed as a time series chart.

What Input Data Are Available? For this model, the input will be the probability of developing a new innovation on a given time step.

What V&V is Needed? The V&V step is key. It can be said that V&V turns toys into tools (North and Macal 2007). V&V is like a *properly prepared* court case in that modelers select a standard of evidence and then prepare arguments to meet the standard. To be properly prepared, countervailing facts must also be presented!

There are many sources of potential evidence for the arguments, including by:

- Conducting design and code walkthroughs;
- Performing unit testing;
- Conducting face validity checks;
- Docking against other models;
- Trying real-world and tricky artificial cases; and
- Conducting parameter sweeps and sensitivity analyses.

Simple face validity checks will be used for the innovation model.

What Alternative Techniques Were Tried? Many models of innovation diffusion have been developed (Meade and Islam 2006). Our model offers a simple, agent-based approach to this problem. Much more detail would be needed to translate this tutorial model into a full-scale model for studying innovation diffusion.

Who Are the Agents? The agents are individual people in the innovation network.

What Are the Agent Behaviors? The agents will randomly circulate on a continuous two-dimensional surface. When an agent with an innovation comes close enough to another agent, they will transfer their ideas. When an idea is transferred, a new link is made between the transferring parties. Any existing network links in the recipient will be deleted.

Agents develop at the rate given by the input innovation probability. When they do so, they delete all of their links.

Each idea is represented using a number. The value of the number represents the relative quality of the idea. The value also automatically maps into a display color for the agent and its links.

What Is the Agent Environment? The agent environment is a simple, continuous 2D surface with support for a network between agents.

Model implementation

Next we will illustrate how you can implement a Repast Simphony model. The model implementation follows directly from the design questions answered in the previous section.

To begin, you can create a new ReLogo project by clicking the new project wizard icon in the toolbar. This icon has a blue folder with red "RL" text in the upper left corner. The new project wizard dialog will appear with options for the project name and other advanced project options. The only required change in this dialog is the project name, which should be changed to "Innovation." You can click "Finish" to complete the new ReLogo project setup. Once the project setup process is complete, a new project named "Innovation" will appear in the left windowpane with a set of project components below. Two folders are visible — "src," which contains the model code; and "shape," which contains a set of optional icons that may be assigned to agents.

Expand the "src" folder to view the model components. The "src" folder contains the package "innovation.relogo," which itself contains the automatically generated model components such as the "UserObserver," "UserPatch," "UserTurtle," and others. The first step is to create a set of basic user controls that will appear in the model runtime window. You can double-click on the "UserGlobalsAndPanelFactory.groovy" file and modify the "addGlobalsAndPanelComponents" method as follows:

```
public void addGlobalsAndPanelComponents (){
  // Add the needed control buttons.
  addButton("setup")
  addToggleButton("go")
```

```
    // Add the model input.
    addInput("innovationProbability", 0.001)
}
```

The generated "UserGlobalsAndPanelFactory.groovy" file may contain additional example code that will not be activated given that it is commented. Commented code lines may begin with either "//" or "/*" and will be ignored by the simulation runtime. The three new lines added above create graphical buttons in the model runtime window that provide functions to setup and run the model. Changes to the "UserGlobalsAndPanelFactory.groovy" file are now complete. The "UserLink.groovy" file does not require any modifications at this time.

Next, you should open the "UserObserver.groovy" file, which will contain code that defines how to initialize the model and what happens at each simulation tick. Again, there will be example code in the generated "UserObserver" that can be ignored or deleted. The "UserObserver" requires two methods: "setup" and "go," which are listed below:

```
// Define the model builder.
def setup(){
    // Remove any existing turtles.
    clearAll()
    // Build a new set of turtles.
    createTurtles(20){
        setShape("person")
        setColor(gray())
        forward(1)
    }
    // Distribute an initial innovation.
    ask (turtles().first()) {
        innovation = yellow()
        setColor(innovation)
    }
}
// Define the model time step routine.
def go(){
    // Give each turtle a chance to act.
    ask(turtles()){ act() }
}
```

The "setup" method is called when the "setup" button in the runtime window is pressed. The button was previously created in the "UserGlobalsAndPanelFactory," and the functionality is provided in the "UserObserver." The setup routine first clears the model of any existing turtles using "clearAll()" for the case when the model had already been run. Next the "setup" method creates a set of 20 turtles, sets the shape of each turtle to the "person" icon shape, sets the initial color of the icon to gray, and moves the turtle agent one step forward. Finally, setup retrieves the first turtle and sets its "innovation" parameter and color to yellow. The "go" method asks each turtle agent to execute its "act" method once each time the simulation is stepped. The turtle's "ask" method will be defined next. The "UserPatch.groovy" file does not require any additional modification.

You should open the "UserTurtle.groovy" file to edit the turtle agent class that will contain all of the Turtle agent behaviors. The "UserTurtle" agent behavior is defined entirely within the "act" method and "innovation" parameter, which are described below:

```
// Define the innovation tracker.
int innovation = 0
// Define the activity.
def act() {
  // Move forward randomly.
  left(random(15))
  right(random(15))
  forward(1.1)
  // Spontaneously create innovations.
  if (randomFloat(1.0) > (1.0-innovationProbability)) {
    // Develop the new innovation.
    innovation = 10 * random(11) + 25
    setColor(innovation)
    // Clear out my old links.
    ask (myLinks()) { die() }
  }
  // Check for an innovation.
  if (innovation > 0) {
    // Diffuse an innovation from "myself" to
    // nearby turtles ("self").
    ask (other(turtlesHere())) {
      // Get the innovation.
      innovation = myself().innovation
      setColor(myself().getColor())
      // Clear out the old links.
      ask (myLinks()) { die() }
      // Build a new link.
      UserLink newLink =
        createLinkWith(myself())
      newLink.setColor(myself().getColor())
    }
  }
}
```

The "UserTurtle" "act" method first moves the turtle forward by 1.1 units in a direction that is randomly determined by generating a random integer between 1 and 15 and then rotating the turtle left and right and finally moving 1.1 units forward. Next, spontaneous innovations are randomly created within the turtle by generating a random integer between zero and one and checking this value against the input threshold value (1.0 – innovationProbability). If a new innovation is to be created, then the "innovation" value is set based on the random value, the turtle's icon color is set to the new innovation value, and the turtle's existing links are destroyed.

In the event that a new innovation is created, the turtle next asks for other turtle agents nearby (located in the same patch) to which the new innovation will be transferred. The innovating turtle sets its neighboring turtles' "innovation" and "color" to the new "innovation" value, destroys the neighboring turtles' existing links, and generates a new link between the turtle and its neighbor. Editing of all of the model code is now complete.

Next, you can add a "Total Innovation" chart. New charts begin with data sets. Data sets collect data from agents. As previously discussed, charts, files, or external analysis tools can use the resulting data. To begin creating the chart, you can right-click on the "Data Sets" item in the "Scenario Tree" in the lower-left panel of the Repast Simphony runtime interface. You should then type in "Total Innovation" as the "Data Set Id." You should then click "Next" and choose the "Method Data Sources" tab on the next wizard page. You can click "Add" to add a column to the data set. You should type in "totalInnovation" as the "Source Name," select "UserTurtle" as the "Agent Type," and "getInnovation" for the "Method." "Sum" should be automatically selected as the default aggregate function. The results of this step will direct Repast Simphony to add up all of the "innovation" values across all of the turtles for each time step. You should choose "Next" and then "Finish" to finalize the work. You can now proceed to add the chart itself.

To add the chart, you should right-click on the "Charts" item in the "Scenario Tree" in the lower left panel of the Repast Simphony runtime interface. You should then choose "Add Time Series Chart" from the popup menu. You should type in "Total Innovation" for the "Name" and then choose "Next." You should then click on the "totalInnovation" line on the chart properties wizard page and again choose "Next." You should type in "Total Innovation" for the chart title and "Innovation" for the Y-Axis title on the next wizard page and then click "Finish." To complete this step you, should save the chart definition by clicking on the floppy disk icon on the upper right corner of the Repast Simphony runtime interface.

To run the innovation model, the user can select the "run" icon from the toolbar. The run button has a green circular icon with a white triangle in the middle. The model runtime window will appear and can by initialized using the "Initialize Run" button in the toolbar. This button has a blue circular icon with a white circular "on" symbol in the middle. After the model has been initialized, the "User Panel" pane of the model runtime window will contain the "setup" and "go" buttons used to control the model. Clicking the "setup" button executes the "setup" method of the "UserObserver" described above, which creates an initial population of turtles in a circular arrangement in the center of the display panel on the right side of the simulation runtime window. One of the turtle icons is a different color than the others since the "UserObserver" set its "innovation" value to the nonzero value. Selecting the "go" button runs the model continuously. The turtles start to spread out in a circular pattern, and some additional turtles near the initial innovator may be colored according to the adopted "innovation." The runtime window for a model that was run continuously for several hundred time

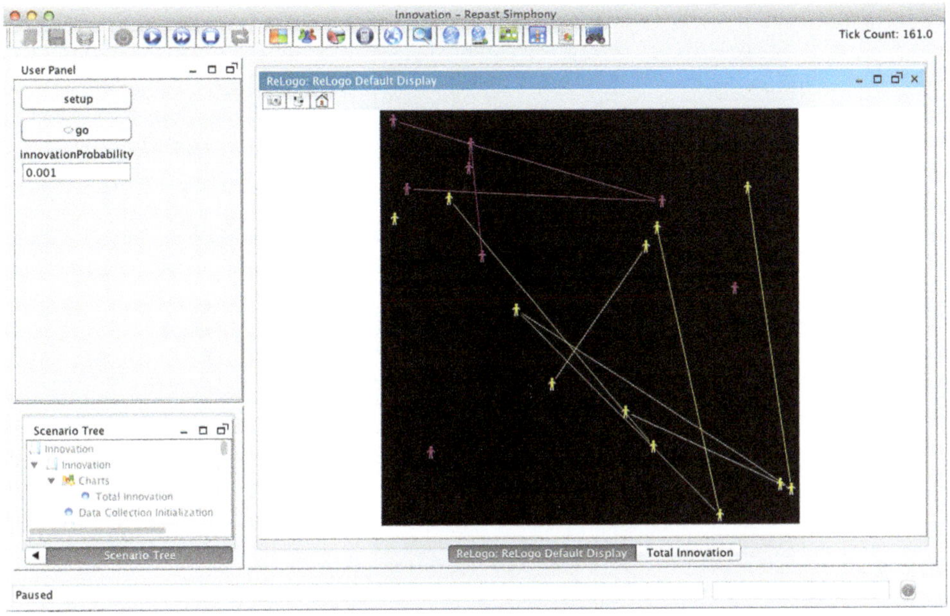

Figure 4 The example Repast Simphony innovation model.

steps is shown in Figure 4. Turtles that share the same "innovation" value are similar colors, and the links between these turtles indicate the "innovation" pedigree. Clicking on the "Total Innovation" chart tab on the lower right shows the resulting total innovation.

Conclusions

There are several directions for future work using the open source Repast Simphony agent-based environment for modeling in CASs, including to provide more scalable support for large-scale modeling; to simplify model specification; and to provide more facilities to support model verification and validation. In terms of scalability, the Repast team has recently completed a C++ Message Passing Interface implementation of the core Repast feature set that targets supercomputer execution. This new library, Repast for High Performance Computing (Collier and North 2011; Collier and North 2999), has recently been shown to have excellent weak scalability on Argonne National Laboratory's IBM Blue Gene/P.

Abbreviations

2D: Two-dimensional; 3D: Three-dimensional; API: Application programming interface; AST: Abstract syntax tree; BSD: Berkeley software distribution; CAS: Complex adaptive system; CRS: Coordinate referencing system; EndEC: Endogenous emergence of coordination; ESRI: Environmental systems research institute; GIS: Geographic information system; GPL: General public license; GRASS: Geographic resources analysis support system; JOGL: Java binding for the OpenGL; JoSQL: SQL for Java Objects; JPF: Java plug-in framework; JUNG: Java Universal Network/Graph Framework; MADCABS: Monitoring analysis, and diagnosis of distributed processes with agent-based systems; NQP: Not Quite Python; OGC: Open geospatial consortium; P&G: Procter & gamble; Repast: REcursive porous agent simulation toolkit; ROAD: Repast organization for architecture and design; SAF: Simphony application framework; SDK: Software development kit; SLD: styled layer descriptor; SQL: Structured query language; V&V: Verification and validation; WWJ: World wind Java; XML: Extensible markup language.

Competing interests

The authors declare that they have no competing interests.

Authors' contributions

MJN, NTC, JO, ERT, MB, and others developed the Repast Simphony code. MJN, NTC, JO, ERT, and MB wrote parts of the paper. MJN wrote the innovation demonstration model. All authors MJN, NTC, JO, ERT, CMM, MB, and PS read and approved the final manuscript.

Acknowledgments

The submitted manuscript has been create by UChicago Argonne, LLC, Operator of Argonne National Laboratory ("Argonne"). Argonne, a U.S. Department of Energy Office of Science laboratory, is operated under Contract No. DE-AC02-06CH11357. The U.S. Government retains for itself, and others acting on its behalf, a paid-up nonexclusive, irrevocable worldwide license in said article to reproduce, prepare derivative works, distribute copies to the public, and perform publicly and display publicly, by or on behalf of the Government.

The authors also wish to acknowledge the many developers, supporters, and contributors to the Repast family of agent-based modeling software.

References

Artel A, Teymour F, North MJ, Cinar A: **A multi-agent approach using perceptron-based learning for robust operation of distributed chemical reactor networks.** *Int Sci J Eng App Artif Intell* 2011, **24:**1035–1045.

Bonabeau E: **Agent-based modeling: Methods and techniques for simulating human systems.** In *Proceedings of the National Academy of Sciences*, Volume 99(3). Washington, D.C. USA: National Academy of Sciences Press; 2002:7280–7287.

Bozada T, Perkins T, North MJ, Simunich KL, Tatara E: **An applied approach to representing human behavior in military logistics operations.** In *Proceedings of the fall simulation interoperability workshop; September 10–15, 2006*. Edited by Weber R. Orlando, FL USA: Curran Associates; 2006:669–678.

Branting S, Wu Y, Srikrishnan R, Altaweel MR, SHULGI: **A geospatial tool for modeling human movement and interaction.** In *Proceedings of the Agent 2007 Conference on Complex Interaction and Social Emergence*. Edited by North M, Macal C, Sallach D. Argonne, IL: Argonne National Laboratory; 2007:258–273.

CASOS – *ORA: 2999,: http://www.casos.cs.cmu.edu/projects/ora/.

Chiba S, Nishizawa M: **An easy-to-use toolkit for efficient java bytecode translators.** In *Proc. of 2nd international conference on generative programming and component engineering (GPCE '03). Springer Lecture Notes in Computer Science 2830*. Edited by Pfenning F, Smaragdakis Erfurt Y. Germany: Springer-Verlag; 2003:364–376.

Collier NT, North MJ: **Repast HPC: A library for large-scale agent-based modeling.** In *Large-Scale Computing Techniques for Complex System Simulations*. Edited by Dubitzky W, Kurowski K, Schott B. Hoboken, NJ: Wiley-IEEE Computer Society Press; 2011.

Collier NT, North MJ: **Parallel agent-based programming with Repast for High Performance Computing.** *Simulation* 2999, in press.

Collier N, Howe T, North MJ: **Onward and upward: The transition to Repast 2.0.** In *Proceedings of the first annual North American Association for Computational Social and Organizational Science conference*. Edited by Carley K. Pittsburgh: Carnegie Mellon University; 2003. Electronic Proceedings.

Colt Home Page: 2999, http://acs.lbl.gov/software/colt/.

Eclipse Home Page: 2999, http://www.eclipse.org/.

ESRI – Understanding our world: 2999, http://www.esri.com.

Feurzeig W, Papert S, Bloom M, Grant R, Solomon C: **Programming-languages as a conceptual framework for teaching mathematics.** *Interface* 1970, **4:**#2.

GeoTools – The Open Source Java GIS Toolkit: 2999, http://geotools.org/.

GRASS GIS – The world's leading Free GIS software: 2999, http://grass.fbk.eu/.

Groovy: A dynamic language for the Java platform: 2999, http://groovy.codehaus.org/.

Harvey B: *Computer Science Logo Style*. Boston: MIT Press; 1997.

Holland JH: **Studying complex adaptive systems.** *J Syst Sci Complex* 2006, **19:**1–8.

Jaspersoft Community – iReport Designer: 2999, http://jasperforge.org/projects/ireport.

Jaspersoft Community – JasperReports Library: 2999, http://jasperforge.org/projects/jasperreports.

Java Plug-in Framework (JPF) Project: 2999, http://jpf.sourceforge.net/.

jogl – Java Binding for the OpenGL API: 2999, http://jogamp.org/jogl/www/.

JoSQL – What is JoSQL?: 2999, http://josql.sourceforge.net/.

JUNG – Java Universal Network/Graph Framework: 2999, http://jung.sourceforge.net/.

Ke J, Holland JH: **Language origin from an emergentist perspective.** *Applied Linguistics* 2006, **27:**691–716.

LOG4J – Logging Services: 2999. http://logging.apache.org/log4j/.

Luke S, Cioffi-Revilla C, Panait L, Sullivan K, Balan G: **MASON: A multiagent simulation environment.** *SIMULATION* 2005, **81:**517–527.

Macal CM: **Agent-based Modeling and Artificial Life.** In *Encyclopedia of Complexity and System Science*. Edited by Meyers RA.: Springer; 2009:112–131. ISBN 978-0-387-75888-6.

Macal CM, North MJ: **Tutorial on agent-based modeling and simulation.** *J Simul* 2010, **4:**151–162.

Mahalik MR, Conzelmann G, Stephan CH, Mintz MM, Veselka TD, Tolley GS, Jones DW: **Modeling the transition to hydrogen-based transportation.** In *Proceedings of the agent 2007 conference on complex interaction and social emergence*. Edited by North M, Macal C, Sallach D. Argonne, IL: Argonne National Laboratory; 2007.

Meade N, Islam T: **Modelling and forecasting the diffusion of innovation – A 25-year review.** *Int J Forecast* 2006, **22:**519–545.

Minar N, Burkhart R, Langton C, Askenazi M: *The Swarm simulation system: A toolkit for building multi-agent simulations*. Santa Fe: Santa Fe Institute; 1996. Working Paper 96-06-042.

NASA World Wind: 2999, http://worldwind.arc.nasa.gov/java/.

NetLogo User Manual – FAQ (Frequently Asked Questions): 2999, http://ccl.northwestern.edu/netlogo/faq.html.

Networks/Pajek: Program for Large Network Analysis: 2999, http://vlado.fmf.uni-lj.si/pub/networks/pajek/.

North MJ, Macal CM: *Managing business complexity: Discovering strategic solutions with agent-based modeling and simulation*. Oxford, U.K.: Oxford University Press; 2007.

North MJ, Macal CM: **Foundations of and recent advances in artificial life modeling with Repast 3 and Repast Simphony**. In *Artificial Life Models in Software*. 2nd edition. Edited by Adamatzky A, Komosinski M. Heidelberg: Springer; 2009:37–60.

North MJ, Howe TR, Collier NT, Vos RJ: **The Repast Simphony runtime system**. In *Proceedings of the agent 2005 conference on generative social processes, models, and mechanisms*. Edited by Macal C, North M, Sallach D. Argonne, IL: Argonne National Laboratory; 2005:151–158.

North MJ, Collier NT, Vos RJ: *Experiences creating three implementations of the Repast agent modeling toolkit*, ACM Transactions on Modeling and Computer Simulation. 16(1)th edition. New York: ACM; 2006:1–25.

North MJ, Howe TR, Collier NT, Vos JR: **A declarative model assembly infrastructure for verification and validation**. In *Advancing social simulation: the first world congress*. Edited by Takahashi S, Sallach DL, Rouchier J. Heidelberg: Springer; 2007:129–140.

North MJ, Macal CM, St. Aubin J, Thimmapuram P, Bragen M, Hahn J, Karr J, Brigham N, Lacy ME, Hampton D: **Multi-scale agent-based consumer market modeling.** *Complexity* 2010, **15**(5):37–47.

OGC – Making Location Count: Standards: 2999, http://www.opengeospatial.org/standards/sld.

OGC – Making location count: 2999, http://www.opengeospatial.org/.

OpenGL – The Industry's Foundation for High Performance Graphics: 2999, http://www.opengl.org/.

Ozik J, North MJ: **Modeling endogenous coordination with a dynamic language**. In, Simulating interacting agents and social phenomenon: the second world congress on social simulationEdited by Takadama K, Cioffi-Revilla C, Deffuant G. Heidelberg, FRG: Springer; 2010:265–276.

Parker MT: **Ascape: Abstracting complexity.** *Nat Res Environ Issues* 2001, **8:**21–30.

Piccolo2D – A Structured 2D Graphics Framework: 2999, http://www.piccolo2d.org/.

Repast – The Repast Suite: 2999. http://repast.sourceforge.net/.

Resnick M: **StarLogo: An environment for decentralized modeling and decentralized thinking**. In *Conference companion on human factors in computing systems: common ground*. Edited by Tauber MJ. New York: ACM; 1996:11–12.

Sallach D, Macal C: **The simulation of social agents: An introduction.** *Soc Sci Comput Rev* 2001, **19**(3):245–248.

Standish RK: **Going stupid with EcoLab.** *Simulation* 2008, **84:**611–618.

Swarm: license: 2999, http://www.swarm.org/index.php/Swarm:_license.

Tao G, Minett JW, Jinyun K, Holland JH, Wang WSY: **Coevolution of lexicon and syntax from a simulation perspective: Research articles.** *Complexity* 2005, **10:**50–62.

Tatara E, North MJ, Hood CS, Teymour F, Cinar A: **Agent-based control of spatially distributed chemical reactor networks**. In *Engineering self-organising systems: third international workshop revised selected papers*, Lecture Notes in Computer Science Series, Volume 3910. Edited by Brueckner SA, DiMarzo Serugendo G, Hales D, Zambonelli F. Heidelberg: Springer; 2006:222–231.

Tisue S, Wilensky U: **NetLogo: Design and implementation of a multi-agent modeling environment**. In *SwarmFest*. Edited by Riolo R. Ann Arbor, MI: Swarm Development Group; 2004.

VisAD – McIDAS-V: 2999, http://www.ssec.wisc.edu/~billh/visad.html.

WEKA – Machine Learning Group at the University of Waikato: 2999, http://www.cs.waikato.ac.nz/ml/weka/.

Wilensky U: *NetLogo. Evanston, IL: Center for Connected Learning and Computer-Based Modeling, Northwestern University*. 2012.

XStream Home Page: 2999, http://xstream.codehaus.org/.

Permissions

All chapters in this book were first published in CASM, by Springer; hereby published with permission under the Creative Commons Attribution License or equivalent. Every chapter published in this book has been scrutinized by our experts. Their significance has been extensively debated. The topics covered herein carry significant findings which will fuel the growth of the discipline. They may even be implemented as practical applications or may be referred to as a beginning point for another development.

The contributors of this book come from diverse backgrounds, making this book a truly international effort. This book will bring forth new frontiers with its revolutionizing research information and detailed analysis of the nascent developments around the world.

We would like to thank all the contributing authors for lending their expertise to make the book truly unique. They have played a crucial role in the development of this book. Without their invaluable contributions this book wouldn't have been possible. They have made vital efforts to compile up to date information on the varied aspects of this subject to make this book a valuable addition to the collection of many professionals and students.

This book was conceptualized with the vision of imparting up-to-date information and advanced data in this field. To ensure the same, a matchless editorial board was set up. Every individual on the board went through rigorous rounds of assessment to prove their worth. After which they invested a large part of their time researching and compiling the most relevant data for our readers.

The editorial board has been involved in producing this book since its inception. They have spent rigorous hours researching and exploring the diverse topics which have resulted in the successful publishing of this book. They have passed on their knowledge of decades through this book. To expedite this challenging task, the publisher supported the team at every step. A small team of assistant editors was also appointed to further simplify the editing procedure and attain best results for the readers.

Apart from the editorial board, the designing team has also invested a significant amount of their time in understanding the subject and creating the most relevant covers. They scrutinized every image to scout for the most suitable representation of the subject and create an appropriate cover for the book.

The publishing team has been an ardent support to the editorial, designing and production team. Their endless efforts to recruit the best for this project, has resulted in the accomplishment of this book. They are a veteran in the field of academics and their pool of knowledge is as vast as their experience in printing. Their expertise and guidance has proved useful at every step. Their uncompromising quality standards have made this book an exceptional effort. Their encouragement from time to time has been an inspiration for everyone.

The publisher and the editorial board hope that this book will prove to be a valuable piece of knowledge for researchers, students, practitioners and scholars across the globe.

List of Contributors

Muaz A Niazi
Bahria University, Shangrila Road, E-8, Islamabad, Pakistan

Aemen Lodhi
School of Computer Science, Georgia Institute of Technology, Atlanta, USA

Amogh Dhamdhere
The Cooperative Association for Internet Data Analysis, University of California, San Diego, USA

Constantine Dovrolis
School of Computer Science, Georgia Institute of Technology, Atlanta, USA

Omid Askari-Sichani and Mahdi Jalili
Department of Computer Engineering, Sharif University of Technology, Tehran, Iran

Enrique Hernández-Lemus
Computational Genomics Department, National Institute of Genomic Medicine, Mexico City, México
Complexity in Systems Biology, Center for Complexity Sciences, National Autonomous University of México, Mexico City, Mexico

Jesús M Siqueiros-García
Ethics, Legal and Social Studies Department, National Institute Genomic Medicine, Mexico City, México

Umme Habiba, Rahat Masood and Muhammad Awais Shibli
Department of Computing, School of Electrical Engineering & Computer Science, National University of Sciences &
Technology (NUST), Sector H-12, Islamabad, Pakistan

Muaz A Niazi
COMSATS Institute of IT, Islamabad, Pakistan

Evangelos Pournaras
Faculty of Electrical Engineering, Mathematics and Computer Science, Delft University of Technology, Mekelweg 4, 2628 CD, Delft, Netherlands

Martijn Warnier and Frances MT Brazier
Faculty of Technology, Policy and Management, Delft University of technology, Jaffalaan 5, 2628 BX Delft, Netherlands

Pedro A Ortega
GRASP Laboratory, Electrical and Systems Engineering Department, University of Pennsylvania, Philadelphia, PA 19104, USA

Daniel A Braun
Max Planck Institute for Biological Cybernetics and Max Planck Institute for Intelligent Systems, Speemanstrasse 38, Tübingen 72076, Germany

Giuliano Armano and Marco Alberto Javarone
DIEE - Department of Electrical and Electronic Engineering University of Cagliari Piazza d'Armi, Cagliari - 09123, Italy

Giuseppe Vizzari, Lorenza Manenti and Luca Crociani
Complex Systems and Artificial Intelligence research center, Universita degli Studi di Milano–Bicocca, Milano, Italy

Matteo Convertino
School of Public Health, Environmental Health Sciences Division and Public Health Informatics Program, HumNat Lab, University of Minnesota, Minneapolis, MN, USA
Institute on the Environment, University of Minnesota, Minneapolis, MN, USA
Institute for Engineering in Medicine, University of Minnesota, Minneapolis, MN, USA
Academic Health Center, Mayo Memorial Building, Suite 1232 12th fl., MMC807, 420 Delaware St. SE, 55455 Minneapolis, MN, USA

Yang Liu and Haejin Hwang
School of Public Health, Environmental Health Sciences Division and Public Health Informatics Program, HumNat Lab, University of Minnesota, Minneapolis, MN, USA

Michael J North, Jonathan Ozik and Charles M Macal
Decision and Information Sciences Division, Argonne National Laboratory, 9700 S. Cass Avenue, Argonne, IL 60439-4867, USA
University of Chicago, Chicago, IL, USA

Nicholson T Collier and Eric R Tatara
Decision and Information Sciences Division, Argonne National Laboratory, 9700 S. Cass Avenue, Argonne, IL 60439-4867, USA

Mark Bragen and Pam Sydelko
Decision and Information Sciences Division, Argonne National Laboratory, 9700 S. Cass Avenue, Argonne, IL 60439-4867, USA